MONOGRAPHS ON CRYOGENICS

in collaboration with

the British Cryogenics Council

Editor: R. G. Scurlock

University of Southampton

MONOGRAPHS ON CRYOGENICS

1. R. G. Rhodes and B. E. Mulhall: *Magnetic levitation for rail transport*
2. Martin N. Wilson: *Superconducting magnets*
3. Randall F. Barron: *Cryogenic systems, second edition*
4. René A. Haefer: *Cryopumping: theory and practice*
5. W. H. Isalski: *Separation of gases*

SEPARATION OF GASES

W. H. ISALSKI

Process Engineering Manager
Costain Engineering Ltd

CLARENDON PRESS · OXFORD
1989

M
22/1/90
JM2

Oxford University Press, Walton Street, Oxford OX2 6DP
Oxford New York Toronto
Delhi Bombay Calcutta Madras Karachi
Petaling Jaya Singapore Hong Kong Tokyo
Nairobi Dar es Salaam Cape Town
Melbourne Auckland
and associated companies in
Berlin Ibadan

Oxford is a trade mark of Oxford University Press

Published in the United States
by Oxford University Press, New York

British Library Cataloguing in Publication Data
Isalski, W. H.
Separation of gases.
1. Gases. Separation
I. Title II. Series
660.2′842
ISBN 0–19–854811–7

Library of Congress Cataloging in Publication Data
Isalski, W. H.
Separation of gases/W. H. Isalski.
p. cm.—(Monographs on cryogenics; 5)
Bibliography: p.
Includes index.
1. Gases—Separation. I. Title. II. Series.
TP242.I83 1989 660.2′842—dc19 88–13922
ISBN 0–19–854811–7

Typeset by Cotswold Typesetting Ltd, Cheltenham
Printed in Great Britain
at The University Printing House, Oxford
by David Stanford
Printer to the University

PREFACE

Since the early 1900s, the community has benefited from the products of gas-separation plants. The majority of these products have been generated by plants which operate at very low temperatures. This has led to the development of 'cryogenic technology' which is a fascinating field for a relatively small proportion of scientists and engineers. To date, many discoveries have been made concerning the effect of very low temperatures on the behaviour of fluids and metals. Such discoveries revolve around superconductivity in some metals and superfluidity of certain liquids. Some of these phenomena have already been applied commercially. For example, large, highly efficient, superconducting magnets are widely used for magnetic resonance imaging in medicine. However, a larger proportion of the milestone discoveries in cryogenics have been used and studied only in the laboratory, and will probably remain there for several years to come until appropriate funding is available to develop these ideas into commercially useful ventures.

There are many texts written by eminent authors including M. Ruhemann and R. B. Scott, who describe cryogenic technology in its widest sense. This monograph will confine itself to the commercial-scale separation of industrial gases, but will not limit itself to cryogenics only. Recent years have seen newer technologies, such as adsorption and permeation, entering the commercial field of gas separation. These will be considered in the text as both stand-alone technologies and processes which may be synergistic with low-temperature processes.

The monograph consists of ten chapters, with three basic themes. First gas-separation technology is introduced and the sources and uses of industrial gases are described. This gives the reader some idea of where he may find gas-separation plants and their likely size. The second part includes a description of the industries which use gas separation and an analysis of the gas-separation processes themselves. The text will include examples in SI units to provide greater awareness of the technology. Some very basic thermodynamics is covered in this part to give the reader an insight into design thinking. The last part describes the plant hardware and some of the constraints placed upon the designer in their selection.

The monograph is aimed at a wide variety of readers. The undergraduate who has an interest in gas separation or has a related design project may find this text useful in gaining an initial overview with a source of further references. The technologist who works in a variety of industries can use the text to identify how gas separation may be of assistance in solving some of his problems. The contents also provide a simple description of cryogenic and non-cryogenic gas processes which will hopefully eliminate some of the fears in using such processes in new applications in refineries, petrochemical works, fertilizer plants, and allied industries.

The unit operations considered here are classical chemical engineering and range from absorption, adsorption, compression, distillation, dephlegmation, expansion, and heat exchange through to permeation, pumping, and reaction. It is assumed that the reader will be familiar with these principles and their terminology, though not perhaps with the hardware used to achieve a given processing objective. It is also anticipated that the reader will be familiar with the concepts of phase equilibrium, enthalpy, entropy, and the basics of thermodynamics. The non-specialist reader may not be familiar with the manner in which these principles are put together to achieve a result. It is hoped that this text will enlighten.

The author was introduced to cryogenics at the beginning of his career at Petrocarbon Developments Ltd—now called Costain Engineering Ltd. Continuing interest was inspired by Dr M. Ruhemann who at that time was Consultant to Petrocarbon Developments Limited and was one of its founders. The author extends most sincere thanks for all the help with ideas for the text to many of the colleagues at Costain Engineering Limited, in particular Dr M. Ruhemann, Dr R. Banks, Dr A. Harmens, M. Duckett, and R. D. Openshaw. Thanks are also due to Costain Engineering Ltd for permission to publish this monograph.

Word processing and the production of diagrams always takes longer than anticipated because authors inevitably change the content at the last possible moment. The author's thanks are extended to Margaret Ellis, Andrea McGill, Elizabeth Wood, Michelle Mulcair, and Pam Howard in the Word-processing Department, and Louise Lynes for compilation, all at Costain Engineering Ltd, Manchester; and to Judy Griffiths and Jane Foster of the Engineering Department, also at Costain Engineering Ltd, Manchester, for the painstaking care with which the drawings were completed.

Sincere thanks are also due to Professor R. G. Scurlock, Director of the Institute of Cryogenics at Southampton University, for encouragement and guidance leading to the completion of the monograph.

Finally, the author wishes to commend his wife Barbara and family's great patience throughout the time of research and editing of this monograph. The preface would not have been the same without them!

Manchester W. H. I

July 1988

CONTENTS

1
INTRODUCTION

The roots of gas separation lie in the early cryogenic technology. It is, therefore, appropriate to begin with the history when developing the subject of gas separation. In the past, the word cryogenics has generally implied the attainment of temperatures below about −150°C (123 K), although other definitions suggest it to be technology which processes gases which cannot exist as liquid at ambient temperature. Many processes for the purification of nitrogen, oxygen, argon, krypton, neon, xenon, hydrogen, methane, and helium reach temperatures below −150°C and are therefore clearly defined as cryogenic. There are also several other processes where operating temperatures are warmer than −150°C, but the general principles, equipment, and materials used are similar to those adopted by processes attaining much lower temperatures. Therefore these also merit inclusion in the cryogenic sector and many natural gas and refinery gas processes utilize this technology at temperatures ranging from −60°C to −110°C.

Cryogenic technology requires the use of metals that do not become brittle at low temperatures. Carbon steels cannot therefore be used and so the designers resort to metals that are more expensive but are without embrittlement problems, such as copper, stainless steel, nickel steels, and aluminium. This use of materials is perhaps the most practical way of defining the technology, and gives it an upper value of around −50°C. Hence any equipment that is required for processes operating at below −50°C should be considered as cryogenic equipment.

Gas separation is a mature technology and very few recent 'discoveries' in gas processing can be called 'outstanding' or 'milestones' in its development. However, with rising energy costs, severe competition, and profitability becoming harder and harder to sustain, gradual development of cryogenic processes has occurred over more recent years. This development has been largely aimed at improving the efficiency of the processes and this has led to more sophisticated hardware which is often more expensive initially. It has also led to much greater process integration and the use of loss analysis techniques and heat exchanger network analysis. Even though exergy analysis was postulated by M. Ruhemann in the 1940s it is only the 1970s and 1980s that have seen the development of this technique to a significant extent in the process industries. Cryogenic process specialists have often used exergy analysis, sometimes unknowingly, to improve their processes.

Currently, these techniques are much more widely applied throughout all industries. Their use will be taken largely for granted in this monograph, although they will be mentioned and used from time to time in the text. All of this has produced much improved, more integrated and more efficient cryogenic technology which has an increasing place in the process industries of the future.

A necessary part of this technology is the compression of gases and their pretreatment in order to make their conditions suitable for the cryogenic process to operate continuously, safely, and efficiently. Both of these operations take place at ambient or above-ambient temperatures, and will also be discussed in some detail because of their importance in the feed preparation. Adsorption and permeation are processes that often operate at ambient or above-ambient temperatures and merit serious mention. Adsorption is a fundamental part of cryogenic process pretreatment methods, and, along with permeation, offers synergistic processes with cryogenic technology.

Typical cryogenic processes consist of both cryogenic and non-cryogenic parts. Often the cryogenic parts are relatively compact and sometimes less expensive than the non-cryogenic parts needed to prepare the feed for low temperature processing. Hence, it is inevitable that due consideration is given to both the areas of gas processing. A typical air separation plant may be drawn up as a series of process areas as shown in Fig. 1.1. A real plant is shown in Fig. 1.2.

The first three areas shown in Fig. 1.1 hold no great mysteries for a good chemical engineer. The same is true for storage and vaporization of products. However, the cryogenic unit is sometimes shrouded in a little mystery, perhaps because it is usually housed in a plate and frame structure filled with insulation (a cold box—shown in Fig. 1.3) which renders it rather inaccessible to the curious engineer. Historically, cryogenic processes have been a physicist's territory.

Even today, some chemical engineers are reluctant to wrestle with the thermodynamic fundamentals to unravel the workings of a cryogenic process. As a result there is a tendency for cryogenic technology to be produced and operated by specialist companies who then often have a

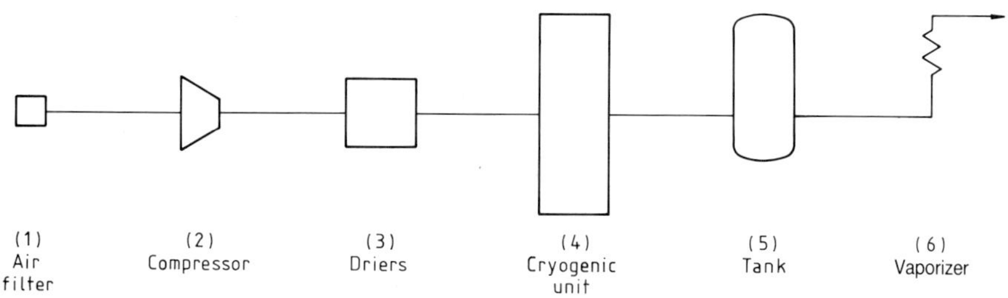

FIG. 1.1. Block flow diagram of an air-separation plant.

FIG. 1.2. An air-separation plant (courtesy of Costain Engineering Ltd).

difficult task in explaining the principles to unfamiliar ears. The cryogenic process, therefore, remains in the mysterious 'cold box' only to be looked at during scheduled shutdowns and maintenance work. This monograph hopes to unravel some of the mysteries, in order to make the processes more understandable and more readily acceptable to industries normally confined to above-ambient operating temperatures. It is further hoped that by bringing a greater understanding of cryogenic processes and their high efficiencies, a better insight into process efficiency can be gained. This would be of benefit to many 'hot' processes when looking for ways of reducing energy consumption. Many of the techniques used today for energy conservation analysis (Pinch Technology, etc.) are derived from analyses routinely carried out by cryogenic process designers.

It should be noted that permeation and adsorption processes also compete effectively with cryogenic processes in certain areas. Adsorption is also becoming a mature technology for gas separation; it has, for example, found a very effective market in the recovery of pure hydrogen from low molecular weight gas mixtures. Permeation is perhaps the most exciting technology at the time of writing. The early 1970s saw the advent of permeation as a means of recovering hydrogen from ammonia purge gas with the installation of a recovery unit by Du Pont. This was the only unit of its kind until about 1980 when Monsanto launched its

FIG. 1.3. Manifolded plate-fin exchangers in the frame during cold box assembly (courtesy of ALTEC International, Inc.).

FIG. 1.4. The Space Shuttle at take-off (courtesy of NASA).

version of membrane process. Currently there are a number of companies offering membrane technology all of whom are continuing research into better membranes. This technology is currently economic for bulk separation rather than for pure product. However the future will see great strides in membrane technology which will encroach on the territories of cryogenic technology. Both adsorption and permeation are considered in more detail in later chapters.

It should not be forgotten that rocketry and space exploration have depended on cryogenic technology for a number of decades from World War II onwards. The NASA Space Shuttle program of the 1980s still utilized the basic products of cryogenic processes, namely hydrogen and oxygen. Both of these products are required in liquid form for economic volume utilization in the propulsion units, which necessitate cryogenic processing. Figure 1.4 shows one such shuttle flight taking off, confirming the use of cryogenics at the frontiers of technology.

2

SOURCES AND USERS OF INDUSTRIAL GASES

2.1 Introduction

There are many low molecular weight gases which have a wide application in industrial, health-care, and even domestic areas. These gases are considered in this chapter in alphabetical order. Three of these are grouped together under the title of 'rare gases': they are krypton, neon, and xenon. Another group, LPGs (including propane, butanes, and heavier hydrocarbons), which have relatively high molecular weights, are often recovered from gas mixtures by cryogenic technology and are thus covered here. The term 'industrial gas' is therefore loosely applied to any fluid which exists as a gas at normal temperatures and pressures and is used in industry for a myriad of applications.

Most of the gases described here are marketed by industrial gas companies who either produce the gases in their own plants, or buy the gases and sell to users. For example, nitrogen and oxygen are produced by separation from air using cryogenic technology. The air-separation plants are often owned by the gas distribution company. The distribution company then produces a wide variety of purities or specifications for each of the gases. Each of the gases produced may have a variety of uses and to meet these uses, different gas qualities are often required. Therefore it is often advantageous to be in control of the manufacture, storage, sendout, distribution, and marketing of industrial gases. In this way a company can ensure that it is aware of the changes in the market-place and be able to accommodate new or reduced demands and stay as a profitable venture.

In some cases, a single user of an industrial gas may be large enough to justify the construction of a plant to meet his needs. This can be done in a variety of ways which broadly group into two categories. The first is where a well known industrial gas company constructs a cryogenic plant to meet the single user's needs. In this case, the user must enter into an arrangement where he pays for the gas he uses. This means that if the user no longer requires the gas, the gas company or the user suffers financially. The second category is where the gas user builds his own plant to recover or generate his gas. In this case the user employs a reputable contractor to design and construct his cryogenic plant and subsequently train the operators to run the unit. This situation carries risk only for the gas user who should be in a better position to assess the risk of such a venture.

The numerous factors which govern the route adopted by the industrial gas user to get his supply of gas and the interrelationship between the user and the supplier include the following:

(1) location of source with respect to user;
(2) reliability of source;
(3) amount of gas required;
(4) purity or purities of gas required;
(5) variation in demand;
(6) lack of cheaper alternatives;
(7) size of user company;
(8) size of supplier;
(9) interrelationship of supplier with third party users;
(10) availability of contractor to build gas supply plant.

Each of those points, and many more, is carefully considered before a venture to make, sell, or buy industrial gas is undertaken. The following text in this Chapter gives technical data (courtesy of The BOC Group), some historic data, sources, uses, and quantities typically required. Each of the gases is considered separately because of the diversity of sources, applications, and quantities involved.

2.2 Argon

2.2.1 *Technical data*

Atomic weight	39.95
Boiling point	-185.9°C at 101.3 kPa
Specific volume	0.604 $m^3\,kg^{-1}$ at 20°C, 101.3 kPa
Critical temperature	-122.4°C
Critical pressure	4860 kPa
Heat capacity	521 $J\,kg^{-1}\,K^{-1}$ at 25°C, 101.3 kPa
Flammable limits	not flammable
TLV	considered non-toxic

Argon is a highly inert substance over a very wide range of conditions, both at cryogenic and very high temperatures. It is a colourless and odourless gas at ambient conditions and is a relatively dense clear liquid at cryogenic temperatures. Although it is itself non toxic, argon is an asphyxiant if no oxygen is present in the inhaled gas.

2.2.2 *Sources of argon*

All argon is ultimately obtained from air which contains 0.932 mole per cent argon and is readily available in unlimited quantities. The steel industry requires large quantities of oxygen which is purified by cryogenic technology from air.

Therefore, to satisfy argon and oxygen requirements, the large air-separation units that produced oxygen for steelmaking also recovered argon to satisfy the argon market. In the 1960s and 1970s the air separation units satisfied the world demand for argon. However, the late 1970s and 1980s saw a decline in the basic steel industry. This meant that the demand for oxygen declined significantly. More importantly, however, the steel industry was undergoing a change with more emphasis on higher quality steels including special and stainless steels. In many of these cases argon is required as an inerting medium and so rather than declining, the argon market continued to grow. The result was that some large air separation units continued to operate inefficiently to maintain argon production even though all of the potential oxygen could not be utilized by the steel plants. In these circumstances it is very costly to produce argon from air.

A second source of argon which satisfies the above disparity is the fertilizer industry. This industry relies on the fixation of nitrogen from air, and vast amounts of air are compressed into the ammonia process so that nitrogen can be reacted with hydrogen to make ammonia. The air that is introduced contains argon which acts as an inert in the ammonia synthesis loop.

The hydrogen production stages of the ammonia process also introduce methane into the synthesis loop, which does not react within it. Therefore many ammonia synthesis processes purge off the inerts from the loop to prevent a significant buildup of them. A typical ammonia purge gas contains 2.5–6.5 mole per cent argon. This gas can be purified by cryogenic technology to recover argon at very high purities and recoveries. A typical 1000 metric ton per day ammonia plant can be the source of up to 12 tonnes per day of argon. This process is described in more detail in Chapter 4.

2.2.3 *Uses of argon*

Typical applications for argon are in the steel-making, light bulbs, electronics industries, welding, and gas chromatography. A significant part of argon is used in the welding industry as a shielding gas in both TIG welding and MIG welding (semi-automatic and automatic). In these cases the argon is passed over the area being welded to prevent oxidation at the welded joint. For high quality welding, for example, in aluminium joints, argon is necessary because the alternative, nitrogen, can react to form nitrides and result in a poorer joint quality.

The light bulb industry uses argon to fill light bulbs. This gives long life to the filament because the argon does not react, even at high temperatures. The electronics industry also uses argon as an inert medium or carrier gas in the manufacture of semiconductors and other components. Gas chromatography is a relatively small application of very pure argon where it is used as a reference gas.

The steel industry is probably the widest user of argon. Because of its inert properties it is used to inertize moulds in pressure die-casting and to protect molten metal streams in continuous casting. Moulds are inertized by pouring in a small amount of argon which evaporates and displaces the air from within the mould. During evaporation, the mould cools and the surface of the resultant casting from it is greatly improved due to the initial cooling of the mould. In die-casting the argon is used as a pressure gas to meter given quantities of metal into the die. In continuous casting, argon is used to shield the molten metal stream between the tundish and the ingot mould from the atmosphere. Lances which blow argon and other gases into the molten metal are often used in steel-making. The reasons for this are many, for example—to agitate the melt and bring undesirable solids to the surface, to mix the melt and distribute desulphurising agents, or to reduce carbon by injecting argon with oxygen.

A steel-making process that has contributed to the rise in the demand for argon in the mid 1970s and later is the AOD (argon-oxygen-decarburization) process developed by Union Carbide. About 90 per cent of stainless-steel is made in the USA by this process, and it is estimated that about 15–25 per cent of the argon made in the USA is routed to it. About 460 000 tonnes of argon were produced in the USA in 1984, and the AOD process was probably the largest user, taking some 100 000 tonnes of liquid. The AOD process uses an injection of oxygen and argon into molten metal to reduce carbon impurities. This also reduces the loss of chromium through oxidation which leads to greater productivity of the electric furnace. It also provides energy savings by using some of the heat generated during refining to melt part of the cold charge metal.

2.2.4 *Argon purities and forms*

Argon is sold in three forms—gaseous cylinders, liquid cylinders, and bulk liquid via road or rail tankers. Gaseous cylinders are available in a range of capacities in both solid drawn steel and for higher purity argon in aluminium. Typical solid drawn-steel cylinders for argon range in size from 0.4 to 50 litres water capacity, the largest of which can hold some

10.38 m^3 of argon (measured at 15°C and 101.325 kPa). Aluminium cylinders from 50 to 2000 litres of argon capacity are also available.

The above is a typical range of sizes offered by one of the leading UK gas suppliers and distributors, BOC Limited. The cylinders are pressurized with the gaseous argon to very high pressures—some 20 000 kPa for steel cylinders and some 6000 kPa for aluminium ones. In the USA, the situation is very similar, with the most common argon cylinders containing 300 cubic feet (8.5 m^3) of argon.

Liquid cylinders (often called dewars) are made in stainless steel, and are available in 25 litre to 2000 litre capacity. When choosing to use liquid cylinders, usage must be considerable to justify the evaporation losses such cylinders incur. Typically a 50 litre dewar can lose 2 per cent of liquid per day and a 2000 litre cylinder can lose 0.9 per cent per day. Thcsc dewars are double walled vessels with super-insulation in between. A more detailed description of these and other storage facilities is given in Chapter 9.

The third category is bulk liquid argon, where shipments of 3000 Nm^3 to 16 000 Nm^3 are common via road transport and 4 times this quantity are possible via railcars. A bulk liquid customer may bc a single large user or a utility company which supplies gas to a number of local users. In the USA 80–90 per cent of argon was shipped in bulk liquid form in the late seventies and early 1980's.

Typical argon specifications are given in Table 2.1. The data is taken from the BOC Special Gases Specification (UK) and the CGA Compressed Gas Association (USA) data.

In general the higher quality specifications are used in the electronics industry, special steel applications, or where a very high quality of aluminium welding is required. The poorer quality argon is used for less stringent inerting applications or the AOD process.

TABLE 2.1. Argon specifications.

	BOC grade			CGA specification		
Impurity (v.p.m.) (maximum levels)	N4.8	N5.5	N6.0	C	E	H
H_2	1	1	—	3	1	1
O_2	2	1	0.1	5	5	1
N_2	6	2	0.4	20	20	5
CO_2	1	1	0.05	—	—	0.5
Total hydrocarbon	1	1	0.05	5	3	0.5
H_2O	2	1	0.05	16	10.5	1.5
Balance argon (vol.% min.)	99.998	99.9997	99.9999	99.995	99.997	99.999

2.3 Carbon dioxide

2.3.1 *Technical data*

Molecular weight	44.01
Sublimation point	-78.5°C at 101.3 kPa
Specific volume	0.547 $m^3\ kg^{-1}$ at 20°C, 101.3 kPa
Critical temperature	31°C
Critical pressure	7382 kPa
Heat capacity	850 $J\ kg^{-1}\ K^{-1}$ at 25°C, 100 kPa
Flammable limits	not flammable
TLV	5000 p.p.m. by volume
Triple point	-56.6°C at 518 kPa

Carbon dioxide is mildly toxic and does not support combustion. It is a colourless and slightly pungent gas at ambient conditions. As a solid (dry ice) it is translucent white in colour and has a 'biting' taste. Below temperatures of -56.6°C, CO_2 exists only as vapour and solid phases.

2.3.2 *Sources of carbon dioxide*

A substantial amount of carbon dioxide is derived from processes which combust hydrocarbons or reform hydrocarbons with the aid of oxygen (or air) and steam. A good example of this is a fertilizer plant where natural gas or naphtha is reformed to produce hydrogen and carbon oxides. The equations governing this process of primary reforming, secondary reforming, high and low temperature shift conversion are given below.

$$C_nH_m + nH_2O \rightarrow nCO + (n + 0.5m)H_2 \tag{1}$$

$$CO + H_2O \rightleftharpoons CO_2 + H_2 \tag{2}$$

$$CO + 3H_2 \rightleftharpoons CH_4 + H_2O \tag{3}$$

The carbon dioxide formed in this process is then removed using an amine wash solution or alternative which can then be regenerated to free the carbon dioxide (CO_2). This CO_2 can then be vented, recovered, or used in another process to make urea. A typical 1000 tonnes/day ammonia plant produces approximately 1100 tonnes/day of CO_2 assuming methane is the feedstock to the ammonia plant. However, if the fertilizer complex aims at urea manufacture all of this CO_2 can be used up, because each tonne of NH_3 uses 1.29 tonnes of CO_2 to make urea according to the equation:

$$2NH_3 + CO_2 \rightleftharpoons NH_2CONH_2 + H_2O \tag{4}$$

Therefore the majority of CO_2 produced in fertilizer plants is used within the complex. However, there are some cases where urea is not the prime product and large amounts of CO_2 are available for purification and other uses.

Another significant source of CO_2 is in the brewing industry, where it is produced during the manufacture of ethanol-containing beverages. Another area which ferments grains, sugarbeet, and sugar cane is the gasohol manufacturing industry. In certain parts of the world (e.g. Brazil) this represents an abundant supply of CO_2.

Many oil refineries have processes which convert hydrocarbons to hydrogen and carbon oxides which ultimately are oxidized further into CO_2. These are also potential sources of CO_2 which can be upgraded and sold as a higher value product because there are few refinery processes that consume CO_2. There are some petrochemical processes which produce CO_2 in quantities large enough to recover economically. For example, the direct oxidation of ethylene to produce ethylene oxide results in the formation of appreciable quantities of CO_2 by the following reactions:

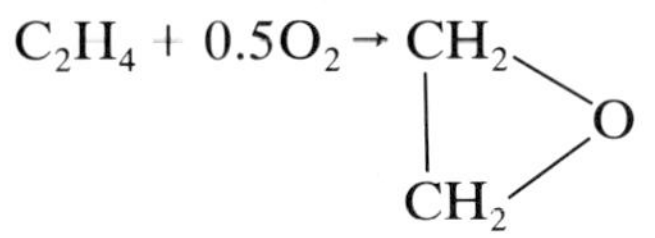

(5)

$$C_2H_4 + 3O_2 \rightarrow 2CO_2 + 2H_2O \qquad (6)$$

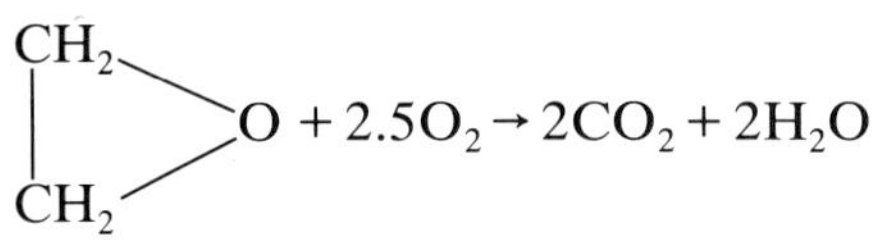

(7)

This CO_2 can be recovered and purified into a food grade product using a cryogenic process as described in Chapter 6. However, petrochemical processes are only minor sources of CO_2.

There are many natural gas wells which contain a high CO_2 concentration. Removal of the CO_2 is a necessary part of processing of the natural gas in order to bring its calorific value to acceptable values for injection into a natural gas grid. This CO_2 can be re-injected into the gas well to keep pressure and hence capacity up, or be purified for alternative uses.

CO_2 is sometimes generated in very large quantities for injection into depleted oil wells as a means of enhanced oil recovery (EOR). This CO_2 is often produced by a combustion process aimed at making a moderately pure product of 98 per cent by volume pure CO_2 for EOR applications.

2.3.3 *Uses of carbon dioxide*

Carbon dioxide (CO_2) is used for refrigeration, beverage carbonation, EOR, chemicals manufacture, metals fabrication, concentration of wine, and inerting. Other uses are preparation of carbonates, neutralization, paints, and aerosols, and to a very minor extent in the medical profession.

The food industry is a significant user with the extensive development of 'fast foods', which use CO_2 for freezing food, particularly meats, in order to preserve it. Solid CO_2 has been used widely in refrigeration of foods in coolers where no mechanical refrigeration was installed. The advantage of CO_2 over water ice blocks was the fact that it sublimed and did not form any liquids in the base of the cooler. The 1970s and 1980s saw the development of 'fast food' stores to such an extent that the demand for CO_2 was dominated by this market. In 1985, 40 per cent of the CO_2 was routed as liquid into refrigeration of foods. By contrast, in 1978, 52 per cent of CO_2 was used in refrigeration, 37 per cent in the food industry, and 15 per cent for industrial refrigeration. Currently, in the USA, some 80 per cent of the CO_2 is produced in liquid form which is the preferred phase for food freezing applications.

Beverage carbonation is another major use of CO_2, with some 25 per cent of CO_2 being diverted to this industry in the USA in 1978 and 20 per cent used for this purpose in 1985. Gaseous CO_2 is required for this application to pressurize fruit juices, giving them an effervescent 'texture' for the consumer. This is considered to be a much more appealing beverage than still fruit juices.

The late 1970s saw the advent of enhanced oil recovery (EOR) when it was considered that oil resources were running down. CO_2 was considered a suitable, high molecular weight gas which could be injected into a depleting well in order to keep the pressure up and hence maintain oil production capacity. Many tests were carried out in the late 1970s and 1980s to find out how different types of wells behave under CO_2 injection.

Large amounts of CO_2 were needed for this application. Much of the CO_2 was trucked in as liquid to the test sites, but the larger test projects used on-site methods for generating CO_2. This was often by the combustion of hydrocarbons to form CO_2 and water and then purified CO_2 was injected into the well. In some cases the remote locations of the test sites made it uneconomic to ship even relatively small amounts of CO_2 and on-site generation prevailed in such circumstances.

Chemicals manufacture is also a user of CO_2, but only the fertilizer industry consumes appreciable quantities. This could all be used to make urea (NH_2CONH_2). Nitrogen fixation is one of the fundamental problems in the fertilizer industry, but once the nitrogen is fixed into ammonia, it needs to be administered effectively onto the fields to reap maximum benefit. Direct injection of ammonia solution has been used for some time in small quantities in various countries. However, losses of ammonia become significant due to its high volatility, and therefore manufacturers look for methods of combining the NH_3 with other compounds so that the soil gets maximum benefit from nitrogen fixation, as well as from other elements vital to the sustained growth of plants. Urea is one such compound and is made from the main product of the ammonia plant (the NH_3) and a by-product, the CO_2, from the CO_2 stripping section, which would otherwise be lost. The extent of urea manufacture varies from country to country due to the varied soil and cultivation needs of the farming industry.

In the nuclear industry CO_2 is used in gas cooled reactors where several hundred tonnes are typically required as a charge per reactor.

In the metals industry CO_2 is often used as a blanket gas for welding steel. Depending on the quality of materials (steel) and the type of arc welding work required, CO_2 or mixtures of CO_2 and nitrogen can be used. Generally, for higher quality work and aluminium welding CO_2 cannot be used and super-inert gases like argon are needed.

The use of CO_2 in general inerting, wine conservation, preparation of carbonates, and neutralization is small and highly varied. Generally the quality of the CO_2 will be high or 'food grade' quality. CO_2 is also used as the carrier/dispersant gas in spray paints and aerosols although in this area halogen-hydrocarbons compete strongly, despite the environmental lobby concerning the breakdown of the ozone layer due to release of halocarbons into the atmosphere.

2.3.4 *Carbon dioxide purities and forms*

Commercial CO_2 is sold in solid, liquid, and, to a lesser extent, in gaseous form. Solid CO_2 or 'dry ice' forms only 10–30 per cent of the market depending on the country. The majority of solid CO_2 goes to the food and beverage industry as a coolant for cold rooms/boxes which are not mechanically refrigerated. Solid CO_2 is made by flashing liquid CO_2 to a low pressure—the resultant isenthalpic flash enters the solid–vapour region having evaporated some 50 per cent of the liquid. The mixture of CO_2 snow and vapour is of low density and not suitable for any cooling application. The vapour is recompressed into the liquefaction process whilst the solid snow is compressed by a hydraulic ram piston (dry-ice press) into blocks of uniform density which can then be used in cooling applications.

Liquid is the usual commercial form of CO_2 which is distributed to the user by the industrial gas companies. Even the consumers who require

CO_2 in gaseous form are likely to take delivery in bulk liquid form into a vacuum insulated storage tank and then draw from this tank, vaporize, and use the CO_2 for their application. Gaseous CO_2 can also be supplied by these companies in cylinders of 20 to 2000 litre capacities pressurized to 3200 kPa with pure CO_2. The purity of CO_2 sold in this way, whether solid, liquid or gas, is generally very high and of 'food grade' quality, as shown in Table 2.2.

TABLE 2.2. Carbon dioxide specifications.

Impurity (v.p.m.) (maximum values)	BOC grade N4.5	BOC grade N5.0 (research)	USA food grade
H_2	1	1	—
N_2	10	2	—
O_2	2	1	—
CO	1	1	10
Total hydrocarbons	1	1	40
H_2O	4	1	—
Balance CO_2 (vol.% min.)	99.998 (>99.995)	99.9993 (>99.999)	

The BOC N4.5 grade is somewhat better than the minimum for food applications, but this purity will then have many other potential applications. Much lower purity CO_2 is produced by the ammonia plant CO_2 removal system and can be used directly for urea manufacture. Similarly CO_2 made for EOR applications only needs to be 98–99 per cent pure because it is merely re-injected or injected into an oil well where there are no stringent consumer-imposed limitations on purity.

2.4 Carbon monoxide

2.4.1 *Technical data*

Molecular weight	28.01
Boiling point	−191.5°C at 101.3 kPa
Specific volume	0.861 $m^3\,kg^{-1}$ at 20°C, 101.3 kPa
Critical temperature	−140.24°C
Critical pressure	3500 kPa
Specific heat	1042 $J\,kg^{-1}\,K^{-1}$ at 25°C, 101.3 kPa
Flammable limits (in air)	12.5–74 per cent by volume
TLV	50 v.p.m.

Carbon monoxide (CO) is a toxic and flammable gas which is colourless and odourless and can be fatal to subjects who inhale CO-rich air. Inhaled CO combines with haemoglobin to form carboxy-haemoglobin (COHb), thereby preventing oxygen being passed to body tissues via the blood. Long term low dosages of CO, for example from smoking, also cause poor circulation of oxygen to body tissues because of the significantly increased levels of COHb formed by CO inhalation.

2.4.2 *Sources of carbon monoxide*

Virtually all of the commercially produced carbon monoxide is obtained from reforming or partial oxidation of hydrocarbons. Many of these processes are designed to produce hydrogen and carbon monoxide simultaneously. Five basic proccsscs can be used:

(1) reforming of natural gas;
(2) reforming of naphtha;
(3) partial oxidation of fuel oils;
(4) gasification of coal;
(5) blast furnace off-gas recovery.

Reforming is a starting point for many processes, including ammonia manufacture. Here CO is a catalyst poison, and therefore shift and methanation reactions are used to reduce the CO to very low values. For CO production, however, we must maximize the first of these two reforming reactions:

$$CH_4 + H_2O \rightleftharpoons CO + 3H_2 \tag{1}$$

$$CO + H_2O \rightleftharpoons CO_2 + H_2 \tag{2}$$

The second reaction must, of course, be minimized. With fuel oils, heavier hydrocarbons, and coal, the amount of CO formed is potentially much greater because of the higher carbon to hydrogen ratio. For this reason, these materials are often chosen as feedstocks for CO producing plant in areas where there is low demand for hydrogen. Whatever the raw material, the resultant gas mixture still contains large amounts of hydrogen and small quantities of nitrogen, argon, and methane. A typical mixture produced from the gasification of bitumen, quench cooling, rectisol wash, and drying is as follows (mole per cent): H_2–48.2; N_2–0.1; CO–51.3; Ar–0.1; CH_4–0.3. Thus, cryogenic purification is required to produce a CO product of 98.0 mole per cent purity or better in order that it can be used in CO consuming processes.

A blast furnace can be regarded as a partial oxidation process as it produces an off-gas of similar composition to that given above. Although this source of CO-rich gas is at essentially atmospheric pressure it can still be worthwhile to recover CO because it would otherwise be lost to the vent stack in the steelworks.

This source of CO aroused considerable interest in the early 1980s because the oxidation side of the process was already a profitable steel-making venture. A moderate sized steelworks can emit 15 to 25 tonnes per hour of CO in its vent gas which could satisfy the largest of commercial chemicals plant for any CO consuming process such as acetic or formic acids. Its recovery does present some practical problems with the variety of impurities in this kind of feedstock.

2.4.3 *Uses of carbon monoxide*

The majority of CO is used in the manufacture of a significant number of intermediate chemicals. The CO is generally produced on site by reforming or otherwise, often followed by cryogenic purification making it of adequate purity for use in chemicals manufacture. Commerical processes using CO include the manufacture of:

- acetic acid;
- acetic anhydride;
- tolualdehyde/terephthalic acid;
- phosgene;
- dimethyl formamide;
- nickel refining;
- formic acid;
- synthetic tertiary carboxylic acids;
- oxo alcohols and acids.

Other processes developed or piloted in the 1970s and 1980s included the manufacture of methyl methacrylate, ethylene, glycol, vinyl acetate, styrene, ethanol, ethyl acetate, adipic acid, and gasoline; many others are also consumers of CO.

Taking the commercial processes listed above, many are the intermediate phase of polymer manufacture which results in a consumer product. Taking acetic acid as an example, the worldwide production in 1980 was 2.8 million tonnes with some 30 per cent made in Western Europe. A small but significant part of this (10–20 per cent) was made by the naphtha oxidation route described above. Vinyl acetate is used to make polyvinyl acetate acid derivatives, and their uses are:

• polyvinyl alcohol	textile sizing, supension agent;
• Vinyl chloride/vinyl acetate co-polymers	speciality PVC applications;
• Ethylene/vinyl acetate co-polymers	speciality polythene applications, elastomers, paints, adhesives;

- ethylene–vinyl alcohol co-polymers barrier resins;
- polyvinyl butaryl resins auto windscreen laminates.

If we take one of the other CO derived intermediates, we can find a similar situation with several consumer related products. These range from formic acid for fermentation of silage in the farming industry through to formation of nickel carbonyls in the nickel refining industry. This clearly shows the importance of CO as a fundamental building block in the chemical industry.

2.4.4 *Carbon monoxide purities and forms*

The majority of CO is generated on site as an intermediate product for chemicals manufacture. This is usually in the gaseous form as in synthesis gas ($H_2 + CO$) for methanol production and oxo-alcohols. It can also be separated by cryogenic means from this synthesis gas as relatively pure CO as an intermediate/feedstock to make a variety of chemicals, as described in section 2.4.3. For many of these processes it is required at only moderate purities of 97 per cent or higher, usually 98 per cent by volume. The impurities in this CO will most likely be H_2, argon, and CH_4, although the gas will be bone-dry and free from freezable compounds such as CO_2 because it will have been processed in a cryogenic unit. High purity CO is also available for other uses such as laboratories and catalyst regeneration. In this instance, however the two purity levels given in Table 2.3 are typical of requirements.

TABLE 2.3. Purities of carbon monoxide.

	BOC grade	
Impurity (v.p.m.) (maximum values)	N2.5 (technical grade)	N3.5 (research grade)
Ar	5	20
H_2	1000	5
N_2	100	80
O_2	25	10
CO_2	200	1
CH_4	750	2
H_2O	25	1
Balance CO (vol.% min.)	99.8	99.95

2.5 Ethane

2.5.1 *Technical data*

Molecular weight	30.07
Boiling point	−88.68°C at 101.3 kPa
Specific volume	0.810 $m^3\ kg^{-1}$ at 20°C, 101.3 kPa
Critical temperature	32.27°C
Critical pressure	4884 kPa
Heat capacity	1767 $J\ kg^{-1}\ K^{-1}$ at 25°C, 101.3 kPa
Flammable limits in air	3–12.4 per cent
TLV	not established (considered low toxicity)

Although of relatively low toxicity, ethane is an asphyxiant. It is believed to act as a weak anaesthetic. As a gas or liquid it is non-corrosive and colourless with practically no odour.

2.5.2 *Sources of ethane*

The source of nearly all ethane is natural gas and oil wells. Ethane is nearly always present in hydrocarbon-rich gas wells and is also usually present in the associated gas in oil wells. The concentration of ethane varies to a large extent in natural and associated gases, typically, from practically zero to several per cent and sometimes very high quantities in excess of 10 per cent. If the concentration of ethane is relatively high in a natural gas, the gas calorific value is often too large to allow its use as a fuel for consumer gas burners and in many cases ethane and heavier components must be taken out. This aspect is described more fully in Chapter 7.

2.5.3 *Uses of ethane*

Ethane is primarily used as a feedstock in petrochemical complexes. Ethylene plants can use naphtha, fuel oils, and ethane as feedstocks to make ethylene and other unsaturated hydrocarbons, but the main objective is to make the most important building block for petrochemicals—ethylene. Many ethylene plants use ethane as the main feedstock. A good example of this is the EssoChem Olefin Fife Ethylene Plant in the UK, where up to 700 000 tonnes of ethane are piped ‘over the fence’ from a gas fractionation plant which treats natural gas liquids (NGL).

Wellhead gas is first treated to extract NGLs which include ethane, propane, butanes, and heavier, saturated hydrocarbons. The lighter gas then passes to a fuel grid. The NGLs pass into a gas fractionation plant which separates them into their individual components. The propane and butane are used as a fuel in remote areas and by the camping industry, while the ethane is used to produce ethylene. A similar situation exists in the Middle East where in Saudi Arabia three ethylene complexes which produce a total of 1.6 millon tonnes per year of ethylene all use ethane as feedstock. This is probably one of the largest users of ethane in any country.

To a much smaller extent ethane is also used as a refrigerant for refrigeration units which can provide cold at about −95°C. Ethane is also used in the manufacture of halocarbons or halogenated ethanes.

2.5.4 *Ethane purities and forms*

The majority of ethane is generated on site near the end user, which may be an ethane cracker to make olefins. For example, catalytic cracking of crude oil heavy ends can yield ethylene and ethane. The ethane can be futher cracked to produce more ethylene, which is a major building-block in the petrochemical industry. The purity therefore varies considerably and the form is usually gaseous as most of these processes are high temperature ones. Smaller users of pure ethane exist in the form of halogenated ethanes (halocarbons) and to a much smaller extent low temperature refrigeration systems. These areas require high purity ethane of 99 per cent minimum by volume. Typical impurities are nearly always hydrocarbons and are listed in Table 2.4.

2.6 Ethylene

2.6.1 *Technical data*

Molecular weight	28.05
Boiling point	−103.7°C at 101.3 kPa
Specific volume	0.862 $m^3\ kg^{-1}$ at 20°C, 101.3 kPa
Critical temperature	9.9°C
Critical pressure	5076 kPa
Heat capacity	1540 $J\ kg^{-1}\ K^{-1}$ at 25°C, 101.3 kPa
Flammable limits in air	3.1–32 per cent
TLV	not established (asphyxiant)

Ethylene is a colourless practically odourless gas at ambient conditions and is a liquid at atmospheric pressure and temperatures below −103.7°C. It is flammable and slightly poisonous (generally olefins are a little more narcotic than saturated hydrocarbons).

TABLE 2.4. Purity of ethane. This form of ethane would be available in cylinders pressurized up to 3750 kPa at 15°C.

Impurity	Level (mol.%)
C_3H_6	0.25
C_3H_8	0.35
C_2H_4	0.05
CH_4	0.005

2.6.2 *Sources of ethylene*

Natural sources of hydrocarbons are nearly always saturated hydrocarbons and therefore ethylene is produced from another compound. Ethylene is the key building block in the petrochemical industry and it is therefore not surprising that it aroused the interest of chemists as early as 1795. At that time, Dutch chemists described the preparation of ethylene by heating ethanol with sulphuric acid, thereby resulting in a mixture of ether, sulphur dioxide and ethylene. In subsequent years, other ethylene preparation methods were devised, including hydrogenation of acetylene, catalytic conversion of halogenated hydrocarbons, and thermal decomposition of hydrocarbons.

During World War I two methods of industrial scale ethylene production were developed. The first was dehydration of ethanol vapour over alumina or phosphoric acid catalyst at elevated temperatures. The second was recovery of ethylene from coke oven gases by cryogenic means. Coke oven gas was primarily made for its hydrogen to make ammonia, but as a result of purification of the hydrogen an ethylene by-product of moderate purity (30–40 per cent) could be attained. This was further purified cryogenically to 90 per cent or even 99 per cent for use in the manufacture of petrochemicals, as well as for war purposes, i.e. mustard gas. During World War II, two further industrial methods were developed. The first was hydrogenation of acetylene over palladium on keiselghur catalyst at elevated temperatures. The second was based on the cracking of ethane and higher hydrocarbons. Although the commercial units operating during World War II were only of 4000–5000 tonnes per annum capacity, they were the fore-runners of today's processes for ethylene manufacture, which are several orders of magnitude larger. They are described more fully in Chapter 6.

After World War II some ethylene was made by dehydration of ethanol and other methods up to the mid-1950s. However, the cracking of hydrocarbons increasingly dominated the scene to the extent that in the 1980s nearly all the ethylene was made in this way.

2.6.3 *Uses of ethylene*

The production of ethylene around 1970 was approximately 10 million tonnes in the free world. In 1985, the total ethylene capacity in Europe, inclusive of Eastern Europe, was considerably larger at 17 million tonnes per annum. Often both ethylene and propylene are utilized on the same site because the typical ethylene plant also makes significant quantities of propylene. In Europe (Eastern and Western), over 9 million tonnes per annum of propylene were produced as a by-product. Therefore, ethylene chemistry is often linked to propylene chemistry by location and processes using both compounds (e.g. ethylene–propylene co-polymers).

Ethylene is a fundamental building block of any developed petrochemical industry and is the raw material for a variety of organic chemical intermediates and polymer products. Figure 2.1 shows the variety of products and intermediates that are made from ethylene.

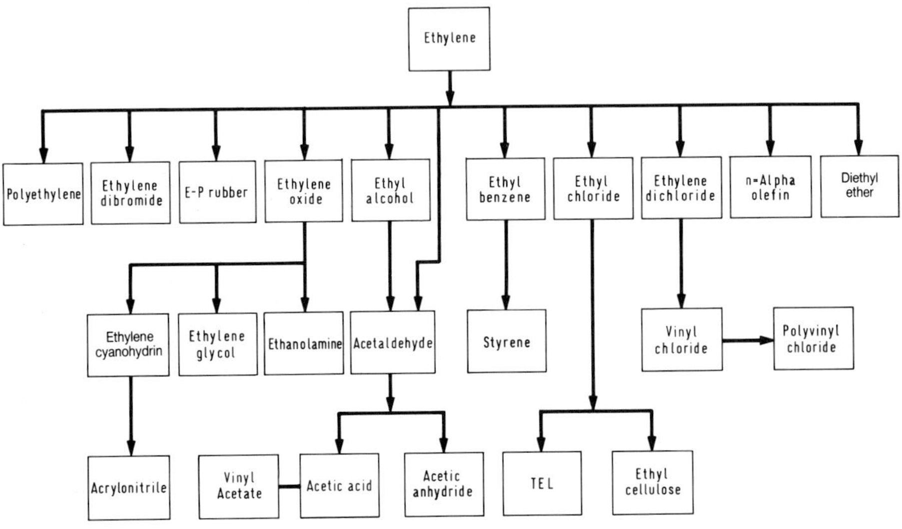

FIG. 2.1. Derivatives produced commercially from ethylene.

A typical petrochemical complex may have a large ethylene plant at the 'front end' of the processing train. Taking a West European example of a site with 940 000 tonnes per annum (tpa) ethylene and 450 000 tpa propylene capacity we can see that most of the products use ethylene as one of the feedstocks. This petrochemical complex produces the following range of petrochemicals:

- low density polyethylene;
- linear low density polyethylene;
- styrene;

- ethylene oxide;
- ethylene glycol;
- cumene;
- ethylene amines;
- ethylbenzene.

Not all ethylene plants provide product to adjacent or 'over the fence' petrochemical processes. The 500 000 tpa ethylene plant at Mossmorran in Scotland routes much of its product several hundred miles to ICI, Wilton and Teeside as well as Shell, Stanlow, and Carrington in north-west England. At Carrington, this ethylene is converted to low density polyethylene.

There are other minor uses for ethylene in the food and refrigeration industries. Ethylene is sometimes used to accelerate the ripening of fruit. For these applications it is required at relatively high purities. Another area for the use of high purity ethylene is in low temperature refrigeration where cooling down to about -100°C is required.

2.6.4 *Ethylene purities and forms*

Ethylene is supplied to petrochemical processes in gaseous form, although many plants have a buffer storage tank of liquid ethylene. This is to cater for shortfall in supply or peaks in demand by the process. In such cases the ethylene is liquefied by refrigeration cycle incorporating a compressor, condenser, and expansion valve to provide the necessary cold. The ethylene is then evaporated using a steam or ambient air varporizer for use in the process. The refrigeration industry will also use ethylene in gaseous form to charge the cycle and liquid is only formed when the refrigeration compressor is started up. The fruit industry uses ethylene as a low pressure vapour to promote ripening.

The gas distribution companies sell ethylene in a very pure condition, typically 99.9 per cent ethylene minimum. This grade of purity can contain no more than the contaminants in p.p.m. by volume given in Table 2.5.

The vast majority of ethylene is consumed by petrochemical processes which do not always require such high purities. However, to meet the requirements of most of the processes, the usual purity expected is 99.9 per cent by volume with essentially the same levels of impurities as above with relaxations in some areas. It should be noted that this purity requirement developed due to market needs of the 1950s. Prior to that time, purities of 97.5 per cent by volume were adequate for most of the processes in operation at that time. Those processes which needed a higher purity carried out their own upgrading.

TABLE 2.5. Contaminants in commercial grade ethylene. For research grades the required levels of these impurities are often an order of magnitude lower.

Contaminant	Level (v.p.m.)
Methane and ethane	700
Acetylene	50
Hydrogen	5
Oxygen	3
Carbon monoxide	3
Water	10
Carbon dioxide	20
Methanol	10

2.7 Helium

2.7.1 *Technical data*

Atomic weight	4.00
Boiling point	-268.93°C at 101.3 kPa
Specific volume	6.03 $m^3\ kg^{-1}$ at 20°C, 101.3 kPa
Critical temperature	-267.95°C
Critical pressure	228 kPa
Heat capacity	5238 $J\ kg^{-1}\ K^{-1}$ at 25°C, 101.3 kPa
Flammable limits in air	non-flammable
TLV	not established (considered non-toxic)

Both gaseous and liquid helium are colourless and odourless, and there is no taste upon inhalation of helium gas. The dangers of helium inhalation are therefore those of suffocation, since its presence may go unnoticed.

2.7.2 *Sources of helium*

Although air contains 5.2 p.p.m. volume of helium, it is not economic to recover because the concentration is so low. Today helium is obtained from natural gases. Many natural gases contain small concentrations of helium which are in some cases high enough to recover. Table 2.6 gives some examples of natural gases which contain helium.

Natural gases containing substantial amounts of nitrogen often also contain helium in appreciable quantities which warrant its recovery. Taking the Polish natural gas with 0.4 per cent helium, we can see that

TABLE 2.6. Typical natural gases containing helium.

Source of gas	He (mol.%)	CH_4 (mol.%)	N_2 (mol.%)	C_2H_6 and heavier
Poland	0.40	56.01	42.75	balance
Kansas, USA	2.00	—	23.0	balance
Holland	0.045	81.30	14.35	balance
France	0.001	97.1	0.3	balance
Texas, USA (Panhandle)	0.70	73.2	14.3	balance
Wustrow, Germany	0.04	42.5	56.5	balance
North Sea, Leman Bank	0.03	94.7	1.3	balance

for a gas field with a resource of 125 000 $Nm^3\,h^{-1}$ up to 500 $Nm^3\,h^{-1}$ of helium can be recovered, which represents a substantial proportion of West European consumption.

The USA is the largest source of helium and has produced it in appreciable quantities since 1918. In the first twenty years of helium production its main use was for the inflation of balloons and airships used mainly in military operations. Only small amounts were used in diving making the total helium consumption very small, around 250 000–500 000 Nm^3 per annum. American involvement in World War II raised its helium consumption to over 3 000 000 Nm^3 per annum. After the War, helium usage halved for several years until it began a steady rise from 1950, reaching 35 000 Nm^3 per day in 1960. The onset of the American space programme and the development of ballistic missiles made a large impression on helium consumption. In 1966, the USA helium usage breakdown was as follows:

National Aeronautical and Space Adminstration	60 per cent
Army, Navy and Airforce	20 per cent
Atomic Energy Commission	9 per cent
Weather Bureau	2 per cent
Others and non-federal	9 per cent

The 1980s USA helium production capacity is approximately 70 million Nm^3 per annum and represents around 90 per cent of the world's production capacity.

The limited resources in natural gas and the rapidly rising demand for helium gave rise to a helium conservation programme initiated by the federal government in the USA. This programme was initiated in September 1960 and it persuaded many companies to store helium underground if production exceeded demand. The reason for this was that in many cases the prime objective of a cryogenic separation unit was to produce methane-rich gas of the correct calorific value at a given quantity. If the raw gas was rich in helium, the plant would produce a helium-rich off-gas which could be further processed to purify helium. If all of this off-gas were treated to produce helium, it would exceed local demand. Therefore only the required amount was processed, and the rest of the helium-rich off-gas was either compressed back into the well or stored elsewhere underground. Europe has been an importer of USA helium for many years and its own resources are not likely to exceed demand in the near future.

2.7.3 *Uses of helium*

Helium gas is highly inert, very light, and penetrating, and is therefore used in a variety of applications where these properties are exploited. For example, gaseous helium is used in welding, diving, ballooning, medicine, leak detection, and gas chromatography.

The armed forces have been users of helium since as early as the first quarter of this century and became the reason for its commercial production. The helium was used mainly to fill surveillance balloons, blimps, and dirigibles. Smaller quantities were used for diving and it was only in the 1930s that helium was used in welding applications. In the USA, helium was used almost exclusively (99 per cent) by federal establishments up to 1946. Thereafter as the quality from producers improved from 98.5 to 99 per cent and subsequently 99.5 per cent more users became interested in helium. In the 1950s ballistic missile and spaceflight programmes in the USA increased the demand for helium dramatically. At this time the purity approached 100 per cent thereby widening its uses.

The diving industry uses mixtures of oxygen and helium in the breathing equipment bottles. If compressed air is used instead, significant amounts of nitrogen are also absorbed by the blood when diving at moderate depths. When the diver surfaces, the desorption of the nitrogen from the blood causes bubbles to form, which are exceedingly painful and often fatal to the diver. Helium is much less soluble in blood, and so divers can descend to greater depths using helium–oxygen mixtures. Breathing helium–oxygen mixtures leads to a harmless phenomenon—the diver's voice changes to a much higher pitch. This is related to the low molecular weight and the frequency vocal chords adopt in a helium-rich medium.

The welding industry uses helium shielding gas for specialist welding applications for aluminium and stainless steel. Helium is used in similar instances to argon for gas shield of non-consumable electrode welding.

Also, many elements and alloys become superconducting at well-defined low temperatures. The disappearance of electrical resistance and the appearance of perfect diamagnetism has a lot of appeal to electrical engineers in the design of superconducting motors and frictionless bearings. Diamagnetism or perfect magnetic insulation, known as the Meissner effect, occurs when the magnetic field is expelled from within the material when it becomes superconducting. Helium is often used as a cooling medium for superconducting motors and magnets. These are often used in research establishments in atomic physics research. Consideration is also being given to the use of superconducting magnets in the power generation industry. This is clearly a very specialized application of helium which requires it in liquid form. A major growth area for helium usage in the 1980s is magnetic resonance imaging (MRI) in diagnostic 'body scanners'. Here liquid helium is used to cool superconducting magnets. Recent discoveries of superconducting ceramics which do not require liquid helium temperature may moderate this use of helium.

Helium is also used in medicine for the diagnosis of lung disorders. Helium–oxygen mixtures are used to check for blockages in lung airways and to check the distribution of air in lungs. Although related in mixture quality to the diving industry, the quantity used in medicine is much smaller.

Because helium gas is very penetrating, i.e. it diffuses through very small clearances or cracks, it is often used to detect leaks in chemical plant equipment. Halocarbons are also used for this purpose, but helium is considered to be better for this application. Helium is also used in laboratories as a reference gas in chromatography. For this application, very pure helium is required.

Manned hot-air and hydrogen balloons first took to the air in 1783 with passenger transatlantic flights becoming a regular event in the 1910s and 1920s. In 1930 the British R101 airship crashed and in 1937 the large passenger airship, Zeppelin Hindenburg, caught fire in an electric storm in New Jersey and exploded. The heavy loss of life marked the end of regular service airship flights. The Hindenburg had been filled with lower cost hydrogen which kept it buoyant with its very low molecular weight of two, and its explosion marked the end of hydrogen-filled balloons for manned flights. Since then, many safer hot-air balloons of a much reduced payload have served enthusiasts. The weather forecasting specialists continued to use helium-filled balloons to deliver electronic equipment to high altitudes but it is only in the 1970s and 1980s that interest has grown in airships of more substantial size.

Modern airships all contain helium and are mainly used for surveillance by military establishments and for advertising. The advertising side is growing in popularity because the balloons are very large with only a small payload. The advertising slogans are painted onto the large area of the helium-filled buoyancy aid. The balloons are charged with relatively pure helium and after some weeks of service, air permeates through the balloon skin reducing its buoyancy. The helium can then be replaced, or, more economically, it can be re-purified by a semi-permeable membrane process thus restoring the balloons buoyancy without buying more helium. This operation was started by Airship Industries in England in 1984 and continues to be an economic method of helium recovery.

2.7.4 *Helium purities and forms*

Nearly all helium is recovered from natural gas and purified by cryogenic technique. Most ultimate users of helium require it in the gaseous form. Helium is often trucked to a distribution point as liquid in double- or triple-walled road or rail tanks. These tanks, therefore, have two insulation cavities; the outer contains a vacuum and the inner liquid nitrogen to minimize helium boil-off. The distributor then sells some to laboratories or research establishments as liquid for use in super-conductivity applications. The remaining helium (usually the majority) is vaporized and put into cylinders for use by diving, medical, metal fabrication, and other industries.

When helium was first produced it was only 98–99 per cent pure, which is very crude by today's standards. It was only in the late 1940s that the demand for better helium purity gradually raised product purities through 99.5 per cent to today's high values, typically those shown in Table 2.7.

TABLE 2.7. Typical helium purities currently sold by gas distributors (maximum values in v.p.m.). Gaseous cylinders of various capacities are available, ranging from 50 litres to 135.9 m^3 of gas. The cylinders are pressurized from 1000 kPa up to 20 000 kPa with helium prior to despatch to a user.

	BOC Grade			
Impurity	Grade A	Grade CP	Research	N6.0
Neon	5	2	1	—
N_2	10	2	1	0.4
O_2	2	1	1	0.1
Ar	1	1	—	—
H_2	1	1	1	—
CO_2	1	1	—	0.5
Total H/C	1	1	1	0.5
H_2O	2	1	1	0.5
He (balance) (%)	99.995 (min.)	99.999 (min.)	99.9995 (min.)	99.9999 (min.)

2.8 Hydrogen

2.8.1 *Technical data*

Molecular weight	2.02
Boiling point	−252.77°C at 101.3 kPa
Specific volume	11.967 $m^3\ kg^{-1}$ at 20°C, 101.3 kPa
Critical temperature	−239.77°C
Critical pressure	1298 kPa
Heat capacity	1432 $J\ kg^{-1}\ K^{-1}$ at 25°C, 101.3 kPa
Flammable limits in air	4–74.5 per cent
TLV	not established (considered non toxic)

2.8.2 *Sources of hydrogen*

Gaseous hydrogen can be made using several commercially feasible processes. These include:

- electrolysis of water;
- thermal–catalytic dissociation of ammonia;
- thermal–catalytic dissociation of methanol;
- steam–iron conversion of reducing gas;
- partial oxidation of hydrocarbons;
- catalytic steam-reforming of hydrocarbons;
- the water-gas process.

There are a number of additional processes which produce hydrogen as a by-product. These include caustic-chlorine manufacture, coke-oven gas, and a variety of refinery and petrochemical processes. Some of these processes are described in more detail in Chapters 5 and 6. Each of these produces a hydrogen-rich gas which needs to be further purified before it can be liquefied. The choice of hydrogen production method depends on many factors. The cost of raw material or a particular utility is often a major factor. For example, natural gas or naphtha are the major cost in the reforming method of hydrogen production. Nevertheless, the electrolysis route generally produces more costly hydrogen because of the high electricity consumption. Typical costs for some of the more common hydrogen production methods are given in Table 2.8.

Clearly, at these prices of raw materials the steam reforming route is by far the most economic. However, when a source of electricity is very cheap, for example, near hydro-electric power stations, the electrolysis route may be attractive. Also the electrolysis route provides pure hydrogen with only a small amount of water vapour. Therefore, the pretreatment of hydrogen prior to liquefaction will be a lot cheaper than for the steam reforming or partial oxidation routes. However, energy or

TABLE 2.8. Production costs for various methods for a typical capacity of 240 tonnes of hydrogen per day (basis = 1980).

	Process		
	Steam reforming	Partial oxidation	Electrolysis
Raw material cost (£/GJ)	Natural gas 1.25	Residual oil 1.6	Power 5.75
Plant capital cost	£50 × 10^6	£100 × 10^6	£180 × 10^6
Overall thermal efficiency (%)	74	82.7	75.7
Cost of hydrogen manufacture (£/tonne)	439	733	1800

raw material cost is generally the main reason for selecting a given process for manufacturing hydrogen feedstock for a liquefier.

In processes where hydrogen is consumed *in situ* only moderate purities are often required, but usually in large quantities. Moreover, the hydrogen is generally required in gaseous form. This situation lends itself to partial oxidation and steam reforming routes for hydrogen manufacture and these are by far the predominant source of hydrogen.

2.8.3 *Uses of hydrogen*

A significant amount of gaseous hydrogen is produced by steam reforming of natural gas or naphtha and consumed in a downstream process. One notable example is in the fertilizer industry, where ammonia synthesis gas is produced by steam reforming of hydrocarbons followed by air addition and secondary reforming to produce a mixture which contains hydrogen and nitrogen in a ratio approximately 3:1. This process is described more fully in Chapter 6 and is a major producer and user of hydrogen. A second important area of the process industry which produces and uses hydrogen is oil refining. Many refineries have reformers which produce relatively pure hydrogen for use on site by hydrogen-consuming processes. These processes include hydrotreating, hydrodealkylation, hydrodesulphurization, hydrogenation, and hydrocracking of heavy residues.

These processes require hydrogen of 90–99 per cent purity by volume and all take place at ambient and higher temperatures, and so do not often use cryogenic technology. Typical operations involved may be reforming, CO_2 removal by absorption and final upgrading by methanation, and sometimes adsorption. The refinery hydrogen balance is described in more detail in Chapter 5.

The bulk of the hydrogen is made at moderate purities but in vast quantities and is consumed *in situ* or in downstream processing units. The hydrogen balance is one of the fundamental parts of the make-up of a sophisticated refinery and many cryogenic processes have been effectively utilized to recover medium purity hydrogen to improve the hydrogen balance when refineries turn to more severe processing of their crudes.

The remaining users of hydrogen are highly varied, but have two things in common: first, they consume much smaller amounts of hydrogen; secondly they require much higher purities.

Also many of the small consumers require liquid hydrogen and hence require cryogenic technology for its production. These areas include:

- liquid hydrogen engines;
- liquid hydrogen rockets;
- metallurigical applications.

Liquid hydrogen engines were researched extensively in the 1950s in the USA with encouraging results. The test program began with military intentions but the results were also applied by the NASA space program. The program began with conversions of aircraft jet engines to be fired on hydrogen and resulted in rocket engines of the 1970s and 1980s which ultimately carried the famous Shuttles into space. Hydrogen was advantageous as a fuel due to its high performance with minimum weight for fuel storage in the booster rockets. The use of hydrogen and oxygen propellants is not without its hazards, as highlighted in the ill-fated Challenger flight in January 1986 which resulted in the deaths of seven astronauts.

Although hydrogen/oxygen combustion is a common means of propulsion in rocketry, another more effective method has also been seriously investigated. Liquid hydrogen can also be used as a coolant and propellant in a nuclear reactor and this energy provides the thrust for a rocket. An extensive test programme to investigate this was undertaken in the USA in the late 1950s. Liquid hydrogen is also used in a variety of test programmes in atomic physics areas, but the quantities involved are much smaller than the military and space programmes. In all cases, liquid hydrogen is of a very high purity, as necessary in order to allow liquefaction without freezing problems.

High purity gaseous hydrogen is also used in some chemical and metal industries, e.g. hydrogenation of fats and edible oils and in nickel refining.

2.8.4 *Hydrogen purities and forms*

Hydrogen is available in purities which broadly fall into two groups, the first having purities up to 99.9 per cent and the second greater than 99.9

per cent by volume. The majority of hydrogen is produced at purities up to 99 per cent and is usually consumed as soon as it is produced in a downstream process, e.g. ammonia synthesis. This hydrogen is invariably made only in gaseous form. Typical uses and purities are given in Table 2.9.

The second category of hydrogen product is generally for special uses such as welding, hydrogenation of edible oils and fats, reduction of metal oxides, research, and as a rocket fuel. BOC Limited produces four typical grades of high-purity gases and these are listed in Table 2.10. These grades are usually supplied in gas bottles, although research and space exploration establishments use liquid hydrogen. In order to liquefy this hydrogen, a comparable purity must be generated.

TABLE 2.9. Uses and purities of hydrogen.

Hydrogen user	Typical purity (%)	Typical impurities (%)
Ammonia synthesis	74–75*	24% nitrogen—used in ammonia synthesis Approx. 1–1.5% argon + methane
Hydrocracking	97–99	Methane
Hydrodealkylation	90–98	Methane Traces of ethane
Hydrotreating	90–98	Methane and other light hydrocarbons
Methanol synthesis	65–75	Carbon oxides used in reaction to form methanol

*85–90 per cent for hydrogen recycled from ammonia purge gas recovery.

TABLE 2.10. Commercial grade hydrogen purities (maximum impurity level in v.p.m.).

	Grade			
Impurity	N4.0	N5.0	N5.5	N6.0
Oxygen	2	1	1	0.1
Nitrogen	30	5	1	0.4
Hydrocarbons	32	1	1	0.05
Carbon dioxide	3	0.05	0.05	0.05
Carbon monoxide	1	1	1	0.05
Water	2	2	1	0.5
Minimum hydrogen content (mol.%)	99.99	99.999	99.9995	99.9999

2.9 Krypton, neon, and xenon

2.9.1 *Technical data*

	krypton	neon	xenon
Atomic weight	83.80	20.18	131.3
Boiling point (°C at 101.3 kPa)	−153.35	−246.05	−108.1
Specific volume ($m^3\ kg^{-1}$ at 20°C, 101.3 kPa)	0.288	1.196	0.180
Critical temperature (°C)	−63.75	−228.75	16.583
Critical pressure (kPa)	5502	2765	5840
Heat capacity ($J\ kg^{-1}\ K^{-1}$ at 20°C, 101.3 kPa)	247	1029.6	159

All three of these gases are non-flammable and this property is exploited to the full. They are all colourless, odourless, and tasteless, and hence their hazard is asphyxiation.

2.9.2 *Sources of krypton, neon, and xenon*

These rare gases are usually obtained from air and thus are often recovered in air-separation units. However, some krypton and xenon is recovered from ammonia plant purge gases. Ammonia purge gas contains rare gases because of the air addition stage in the production of synthesis gas. The process routes for krypton, neon, and xenon recovery from both sources are complex in order to extract the very low concentrations that are present in air. Air has the composition given in Table 3.1—it contains 1.1 v.p.m. krypton, 18.0 v.p.m. neon, and 0.08 v.p.m. xenon.

Neon boils at temperatures much lower than nitrogen and is therefore present as an incondensible in the lower column of a double column air separation plant. Traces of helium are found in this mixture and could theoretically be recovered, although economics clearly dictate that helium should be sourced from natural gas and only neon should be recovered from here. The helium–neon–nitrogen mixture can be treated by adsorption and fractionation to remove nitrogen and separate helium from the neon. It is interesting that in this case helium is produced as a by-product in very small quantities.

Krypton and xenon boil at temperatures much warmer than oxygen and therefore can accumulate in the sump of the upper column of a double column air separation unit. A liquid drawn from this point can yield a mixture which contains the krypton and xenon. This is then further processed to remove trace hydrocarbons by catalytic oxidation and distilled to separate krypton and xenon. If the source of inert gases is an ammonia purge gas there is no place for recovering neon, but the krypton and xenon accumulate in the methane product and can be purged off in a liquid from the sump of the methane column in the same way. The subsequent process to recover krypton and xenon is essentially very similar to that for air separation.

2.9.3 *Uses of krypton, neon, and xenon*

Because these gases are highly inert (sometimes termed 'super-inert') they are all used in different areas as part of the electric lighting industry. Each can be used as part or all of the fluorescent medium which provides lighting in a large part of our lives.

Krypton has replaced argon in a large amount of the lighting industry, because it has a higher molecular weight. Incandescent light bulbs have a thin filament which emits heat and light and in turn 'boils' itself giving off particles that then deposit on the inside wall of the bulb. The rate of degradation of the filament determines the life of the bulb. Filling the bulbs with krypton greatly reduces this rate of degradation compared with argon, and currently around 90 per cent of krypton is used in this area. The remaining krypton is used in high energy physics, lasers, and 'sputtering'. Sputtering is a process by which unwanted particles are removed from etched areas of electronics components. Krypton-filled fluorescent lamps can be designed to stimulate plant growth by giving off light of the required wavelength. Krypton and argon mixtures are often used in fluorescent lighting of this kind.

The first use of neon was as the gas medium in fluorescent lighting with the advent of the range of coloured 'strip-lights' in the 1920s. More recently, neon has found a use in high-energy physics. In the quest for sub-atomic particles physicists use bubble chambers filled with liquid hydrogen, in the centre of which resides a target. Large accelerators are used to project atomic particles into the target area in the hope that reactions of interest occur upon collision with the target. Liquid hydrogen is used because of its characteristic of creating 'tracks' when such an event takes place. During some of these experiments, it was found that some of these 'events' produced particles that passed through the chamber containing the liquid hydrogen. At this point it was decided to put a denser medium outside the hydrogen chamber. This led to the selection of liquid neon as the outer layer which contained many of these 'straying' particles. In this application the neon is a 'one charge' situation with boil-off being recovered, liquefied, and recycled. Gaseous neon is also used in the physicists' laboratories in combination with helium inside spark chambers in an application similar to the liquid hydrogen bubble chambers. In military applications liquid neon is used in infrared detection equipment to enhance its sensitivity to objects being monitored by reducing molecular motion within the detection device itself.

Xenon is the rarest element recovered from air and has the highest boiling point and molecular weight of the rare gas group. When electrically charged it emits a light pattern similar to sunlight making it highly suitable for electronic flash units in photography. This means that a photographer can dispense with filters when using an electronic flash with daylight film used indoors. Due to its high-intensity light characteristics, xenon-filled flashing lights are used to guide pilots to runways. The field of medicine has also benefited from xenon where X-ray photography can be used with shorter patient exposure times. Xenon flash lamps are used in laser technology to 'energize' or start laser lights.

2.9.4 *Krypton, neon, and xenon purities and forms*

Generally purities of each of the rare gases is high because of the delicate nature of their applications. For example, impurities such as oxygen or hydrocarbon would quickly reduce the life of both incandescent and flourescent bulbs. Typical purities available in the market are given in the Table 2.11.

TABLE 2.11. Commercial grades of rare gases (maximum impurity levels in v.p.m.)

Impurity	Krypton (BOC grade N4.0)	Neon (BOC grade N5.0)	Xenon (BOC grade N4.3)
N_2	10	1	5
O_2	1	1	1
CO_2	1	0.1	1
Hydrocarbons	1	1	1
H_2	1	1	1
H_2O	1	1	1
Krypton	99.99 mol.% (min.)	—	40
Neon	—	99.999 mol.% (min.)	—
Xenon	20	—	99.993 mol.% (min.)
Argon	—	—	1
Helium	—	2	—

2.10 Liquefied petroleum gases

2.10.1 *Technical data*

Liquefied petroleum gases (LPGs) are low molecular weight hydrocarbons including propylene ($C_3^=$), propane (C_3), butylenes ($C_4^=$), n-butane (n-C_4), i-butane (i-C_4), and pentanes (C_5), to a lesser extent. LPGs can be pure C_3 or C_4 products, mixtures of C_3 and C_4, or mixtures of all of the above with C_5 and heavier in relatively small concentrations. LPGs sold commercially are required to conform with vapour pressure specifications rather than actual purities and are rarely used in pure single component form. Table 2.12 shows the technical data for each of the components likely to constitute LPG.

Propylene is a colourless, narcotic gas and burns in air with a yellow, soot-forming flame. In high concentrations, it has a slight peculiar odour. Propane is also colourless and non-poisonous but is odourless and can act as an anaesthetic when inhaled in high concentrations.

The hydrocarbon α-butylene is also colourless and non-poisonous but has a narcotic effect, whereas γ-butylene or $CH_2=C(CH_3)_2$ is similar but anaesthetic. Both n-butane and i-butane are colourless, practicaly odourless, non-poisonous, and anaesthetic.

2.10.2 *Sources of LPGs*

LPGs are present in varying small quantities in crude oil wells and to a greater extent in gas wells. Thus when stabilizing crude prior to refining some LPGs are made. Natural gases often have fairly significant concentrations of LPGs typically ranging from 1 to 15 per cent C_3 and heavier. LPGs can be recovered from natural gases by low temperature processing, as described in more detail in Chapter 7. They can also be recovered by oil wash processes with lower recovery efficiencies, but in both cases yielding C_3 and heavier product which can be further treated to produce commercial propane and butanes. Similarly condensate from a gas well can also be treated to recover LPGs using distillation at ambient and above ambient temperatures, nearly always yielding saturated hydrocarbons as products.

Crude refining and cracking processes maximize the output of hydrocarbon compounds compatible with gasoline or automotive fuel, i.e. C_5, C_6, C_7, and C_8. This means that C_3 and C_4 hydrocarbons are formed in the cracking processes. These processes often yield unsaturated as well as saturated hydrocarbons. Therefore refineries and steam crackers are the major sources of propylene, butylenes and other double-bonded hydrocarbons. In 1985, refineries in Europe produced some 70 per cent of LPGs with 20 per cent originating from North Sea gas and 10 per cent from sources outside Europe. By contrast, Scandinavian refineries produced only some 10 per cent of LPGs with the North Sea and outside sources providing the remainder. The split of sources varies considerably from country to country very widely because it depends signficantly on the extent of refining and downstream production.

TABLE 2.12. Technical data for LPGs.

	Propylene $C_3^=$	Propane C_3	Butylenes *$C_4^=$	n-Butane n-C_4	Isobutane i-C_4	n-Pentane n-C_5
Molecular weight	42.01	44.09	56.11	58.12	58.12	75.15
Boiling point (°C at 101.3 kPa)	−47.7	−42.1	−6.25	−0.5	−11.7	36.07
Specific volume ($m^3\ kg^{-1}$ at 20°C and 101.3 kPa)	0.565	0.531	0.423	0.40	0.413	0.385
Critical temp. (°C)	91.8	96.8	146.4	152.0	135.0	196.5
Critical pressure (kPa)	4600	4200	3925	3796	3810	3370
Heat capacity ($J\ kg^{-1}\ K^{-1}$ at 20°C and 101.3 kPa)	2690	1678.0	1556	1727	1729	1734
Flammable limits in air (%)	2.0–11.1	2.2–9.5	1.6–9.3	1.9–8.5	1.9–8.5	—
TLV	—	—	—	600 p.p.m.	—	600 p.p.m.

*$C_4^=$ assumed to be α-butylene or $CH_2=C_3H_6$

2.10.3 *Uses of LPGs*

Uses of LPGs can be broken down into four main areas:

(1) domestic and commercial fuel;
(2) petrochemicals;
(3) various industrial and downstream refinery uses;
(4) transportation.

Domestic, commercial, and recreational fuels make up the largest group of users at about 40 per cent, and consist mainly of propane and butanes. In countries which have areas of low population density, e.g. Australia, Canada, and some parts of the USA, there is no National Grid or gas pipeline to supply homes and industrial consumers with fuel. As a result, LPG storage facilities are purchased by consumers so that they may benefit from propane or butane as a fuel. This can vary from a family home, through to farms, restaurants, and larger industrial concerns. In these cases both propane and butane find applications. In the recreation area, cooking facilities for campers use small butane bottles. These may either be portable or fitted onto caravans or mobile homes.

Petrochemicals is another large user of LPGs weighing in at just over 30 per cent. Uses include in ethylene plants, which crack propane and butane, and a variety of processes which use propane, propylene, and butanes. For example, commercial grade propylene (generally better than 99.7 per cent purity) is used as feedstock to make polypropylene, acrylonitrile, cumene, isopropyl alcohol, oxo alcohols, propylene oxide, synthetic rubber, and several others. The major consumption of propylene in the petrochemical field is in polymerization. Propylene and propane are also used as solvents and refrigerants. Butanes are used as aerosol propellants and in the manufacture of several organic chemicals.

With the advent of lead reduction programmes in the late 1970s and early 1980s, refineries have looked for many alternatives. One such alternative is methyl tertiary butyl ether (MTBE) which allows significant reduction of lead in automotive fuel without reducing its anti-knock properties or octane number. Unsaturated C_4 is needed for MTBE manufacture. LPGs can therefore be used in the synthesis of methanol and isobutylene to form MTBE. Isobutylene is the product of fluid catalytic cracking, but can also be made by dehydrogenating isobutane.

LPG is widely used as an automotive fuel in countries such as Australia where in certain areas population density is very low. In such applications, mixtures of propane and butane are used varying in concentration from around 40 to 60 per cent propane, depending on the season. Europe has also found a limited application for LPGs as automotive fuel, where mixtures of propane and butane are generally used with compositions varying from 20 to 100 per cent propane, depending on country and application, which is generally limited to taxis and commercial vehicles. In total only about 8–10 per cent of LPGs are used for this purpose.

2.10.4 *LPG purities and forms*

In large scale industrial applications LPGs are often produced on site and consumed in gaseous form in downstream processes. Purities for these applications vary widely depending on processes used. For example polypropylene manufacture requires 'polymer grade' propylene of greater than 99.7 per cent purity. If LPGs are to be transported any distance, they are produced as liquids, transported to user site, vaporized and used as required. The purities again vary widely depending on application. Table 2.13 shows typical purities of each of the LPGs.

TABLE 2.13. Typical LPG component purities (mol.%).

Grade	C_3H_6	C_3H_8	n-C_4H_{10}	i-C_4H_{10}
Technical	—	96	—	—
'Pure'	99.5	99.35	99.4	99.4
Instrument	—	99.6	99.5	99.5
Research	99.99	99.98	99.95	99.95

Another way of expressing LPG quality is in terms of vapour pressure specification, volatile residues, and residual matter as determined by a series of ASTM test methods. These are commonly used in determining LPGs for automotive fuel applications and are shown in Table 2.14.

TABLE 2.14. Gas Processor Association (GPA) LPG specifications (USA).

	Commercial Propane	Commercial Butane	Commercial P–B mix
Vapour pressure at 37.8°C (kPa gauge)	1434	483	1434
Temp. at max 95% evaporation °C max	−38.3	2.2	2.2
Butane and heavier liq. vol.% max	2.5	—	—
Pentane and heavier liq. vol.% max	—	2.0	2.0
Residual matter:			
residue on evaporation of 100 ml max.	0.5 ml	—	—

In the UK, commercial propane and butane specifications are according to BS5250:1975 and are as follows.

Commercial propane

(1) Total content of C_2 hydrocarbons less than 5 mole per cent;
(2) ethylene content less than 1 mole per cent;
(3) total content of C_4 and heavier hydrocarbons less than 10 mole per cent;
(4) C_5 content less than 2 mole per cent;
(5) vapour pressure at 45°C to be not more than 1760 kPa.

Commercial butane

(1) 95 per cent by volume shall evaporate at 2.2°C or lower at 760 mm Hg;
(2) vapour pressure at 45°C shall be no greater than 687 kPa determined by BS3342; or 586 kPa for portable containers;
(3) total dienes to be not more than 10 per cent (molar);
(4) total acetylenes to be not more than 2 per cent (molar).

In addition, both UK and USA specifications quote sulphur compound, mercaptan, and H_2S limits.

2.11 Methane

2.11.1 *Technical data*

Molecular weight	16.04
Specific volume	1.479 $m^3\ kg^{-1}$ at 20°C, 101.3 kPa
Critical temperature	−82.62°C
Critical pressure	4596 kPa
Heat capacity	2224 $J\ kg^{-1}\ K^{-1}$ at 22°C, 101.3 kPa
Flammability limits	5–15 per cent in air
TLV	not established, considered low toxicity

2.11.2 *Sources of methane*

The principle of 'digestion' by anaerobic bacteria has been taking place underground over four billion years, making methane the world's most pervasive hydrocarbon. To illustrate this, everyday examples include swamps which bubble methane as a result of decaying plant matter (will-o'-the-wisp or marsh gas), decaying sewage produces methane, silage on farms produces methane, landfill decays underground producing methane and CO_2 (landfill gas).

Petroleum and associated gas was formed in organically rich sediments over millions of years on margins of long-vanished oceans. Partially decayed by bacteria, organic matter was converted into oil and gas by heat, pressure, and reaction as layer upon layer pressed down on the decayed matter. Therefore, searches for oil and gas go deep underground with a high risk of not finding any deposits. Often oil wells have a significant amount of associated gas and sometimes wells are predominantly gaseous.

The vast majority of methane is obtained from natural gas wells or gases associated with oil from oil wells. Invariably natural gases contain saturated compounds which include methane, ethane, propane, and heavier hydrocarbons in ever-decreasing quantities, plus carbon dioxide and nitrogen and sometimes helium. Table 2.6 shows helium-bearing natural gases with typical methane content. Many natural gases do not have helium but will usually contain most of the other components listed above to varying degrees.

2.11.3 *Uses of methane*

The largest users of methane or methane-rich gases are steam reformers which produce hydrogen for many uses, including

- ammonia synthesis;
- methanol synthesis;
- hydrogen manufacture for use in a refinery for a variety of hydrogen consuming processes;
- metallurgical applications;
- liquid hydrogen fuel;
- gaseous pure hydrogen for hydrogenation of oils and fats for domestic consumption.

As can be seen above, the uses of methane are very much tied to the uses of hydrogen because methane is a primary feedstock for hydrogen manufacture.

A second, subtantial use for methane is directly as a fuel in both domestic and industrial environments. In both cases it is used in gaseous form, although it is often liquefied for buffer storage or transportation.

Other minor uses of methane are in the testing of gas appliances, engine testing, and as a coolant in reactors.

2.11.4 *Methane purities and forms*

The major use of methane, i.e. steam reformers in ammonia plants, methanol plants, and refineries, can accept a crude methane feedstock. This is usually natural gas which has only been desulphurized and fed directly into the steam reformers. It may have substantial amounts of ethane and heavier components but those are reformed along with the methane. Typical natural gases will often have a methane content in excess of 90 per cent, but sometimes they will contain substantial amounts of ethane, propane, butanes, and heavier as well as varying amounts of nitrogen, carbon dioxide, helium, and traces of hydrogen sulphide. On some occasions the natural gases have large amounts of hydrogen sulphide and/or carbon dioxide and in such cases it is uneconomic to process them. The gases usually emanate from the wellhead at relatively high pressures and often have to be let down to 2500–7000 kPa to be admitted to the user process. The gases are usually saturated with water.

In cases where the natural gas has been extracted from wells where there are no downstream processes, the gas is transported by long gas pipeline to the user or is liquefied and shipped by sea. Liquefied natural gas (LNG) is of a fairly consistent purity because it has been cooled to −163°C and nothing must freeze in it at those temperatures. Consequently, it must have a very low propane content and generally not more than 0.5 per cent ethane. It must also have a calorific value set by the users. Upon arrival at its destination the LNG is vaporized before being utilized for heating or reforming.

Industrial gas companies also offer a number of grades of methane but for smaller users. This methane is available in gas bottles ranging in contents from 20 litres to 120 m^3 and pressures varying from 1400 kPa to 13 700 kPa. Typical purities of methane marketed in this form are given in Table 2.15.

TABLE 2.15. Bottled methane purities.

Component	Technical grade	Technical grade II	Research grade
CH_4	96.0%	99.2%	99.995%
O_2	< 0.5%	< 1 v.p.m.	< 1 v.p.m.
N_2 + Ar	< 0.8%	< 0.8%	< 40 v.p.m.
CO_2	< 2.5%	< 1 v.p.m.	< 1 v.p.m.
H_2O	< 50 v.p.m.	< 2 v.p.m.	< 1 v.p.m.
H_2	—	< 2 v.p.m.	< 1 v.p.m.

2.12 Nitrogen

2.12.1 *Technical data*

Molecular weight	28.01
Boiling point	−195.8°C at 101.3 kPa
Specific volume	0.855 m^3 kg^{-1} at 20°C, 101.3 kPa

Critical temperature	-146.95°C
Critical pressure	3400 kPa
Heat capacity	1040 J kg^{-1} K^{-1} at 25°C, 101.3 kpa

Nitrogen is a colourless, odourless, tasteless, and inert gas. At first reading one may wrongly assume that nitrogen is harmless; however it is because of this that it presents a particular danger of asphyxiation. Because of its extensive use as an inerting and purging medium it is sometimes mishandled, resulting in asphyxiation of operators.

2.12.2 *Sources of nitrogen*

Air contains 78.084 per cent nitrogen and it is therefore the obvious source of nitrogen. An extensive cryogenic industry has been established for the separation of components of air. The late 1970s and early 1980s saw the advent of adsorption and membrane technology for air separation, but to date this has been on a small scale.

Inert gas generation by combustion of hydrocarbons in air to yield N_2 and CO_2 is an accepted form for producing nitrogen, but cryogenics offers the greatest flexibility in this area giving the purest nitrogen which is increasingly required in today's industries. Therefore, air separation using cryogenic technology remains the predominant source of nitrogen for the widest applications.

The fertilizer industry fixes large amounts of nitrogen by steam reforming of natural gas to make hydrogen, secondary reforming with air addition, and synthesis of H_2 and N_2 (from the air) to make ammonia. This is a 'hot' process which does not make nitrogen as a product and will therefore not be discussed at length here.

2.12.3 *Uses of nitrogen*

In the early years of air separation, oxygen was the most sought-after component for use in the steel industry. This situation began to change only in the 1950s when nitrogen was no longer considered a waste product and its inert properties could be exploited as a blanketing gas. Therefore, nitrogen began its useful life outside the fertilizer industry as a purging medium in the metals industry and in hydrocarbon processing. Today, nitrogen is used in a host of applications, including:

- blanketing of tanks with flammable or toxic materials;
- blanketing metal reduction processes;
- purging process equipment;
- food freezing;
- finishing of rubber goods;
- cooling in helium or hydrogen plants;

- oil well fracturing to permit drilling;
- freeze grinding;
- medical applications;
- soil freezing prior to excavation;
- fragmentation of scrap metal;
- electronics industry;
- various minor users of gas and liquid nitrogen.

The above applications cover both gaseous and liquid nitrogen, but the majority of users take gaseous nitrogen even though they may have it shipped in as a liquid. Purging and blanketing are probably the most popular uses of nitrogen gas. Tanks filled with hydrocarbon liquids have a vapour space in the top of the tank which collects hydrocarbon vapours. In atmospheric tanks, air can be drawn in when emptying the tank and it would be easy to ignite this by static electricity or friction. Purging the vapour space with nitrogen prevents this. Corrosive liquids such as sulphuric acid are also purged, but in this case to prevent corrosion taking effect upon air ingress. The metals industry uses nitrogen as a blanket gas to prevent metal oxidation during smelting.

Foundries also use nitrogen, but in liquid form, to cool casting moulds before the metal is poured into them. The liquid nitrogen purges the mould of any oxygen and makes it more resilient to collapse while the metal is being poured in.

The electronics industry is a rapidly growing new area for nitrogen applications. Here nitrogen is used at very high purities to ensure perfectly inert (dust-free) conditions for the production of chips and other complex miniature components.

Carbon dioxide has for a long time been used in food freezing, but the food industry's desire to improve quality has made nitrogen a competitor. Although thermodynamically inefficient, liquid nitrogen has found good use in the flash freezing of strawberries and fish, because it does not affect their taste. Liquid nitrogen is sometimes sprayed onto the payload of a truck which keeps the cargo of food frozen until it reaches its destination. Nitrogen is also 'sparged' into food to drive out oxygen from the vapour spaces, thus preventing spoilage.

Liquid nitrogen is used in the medical industry to freeze blood, organs, embryos, and semen for artificial insemination. Although significant, this application of liquid nitrogen is much smaller than the food industry.

Heavy industry uses liquid nitrogen to freeze rubber tyres so that they can be pulverized into small pieces before recycling the rubber. Scrap metal from compressed cars can be embrittled by cooling with liquid nitrogen making it easier to crush. The building industry occasionally uses liquid nitrogen to cool concrete mixtures on very hot days to ensure setting within given times.

The oil exploration industry uses liquid nitrogen for well fracturing. A technique that pumps nitrogen together with a foam and sand mixture is used. The mixture then props open the material down in the well allowing further drilling to take place. Enhanced oil recovery techniques (EOR) use nitrogen in a number of ways. For example, nitrogen and hydrochloric acid are mixed and pumped into a well which has limestone formations. The acid combines with the limestone and results in improved oil flow characteristics. In other oil field applications, nitrogen is injected to clear debris from a well being drilled so that it can be assessed for further drilling. Gaseous nitrogen has been used in EOR applications to raise the pressure of depleted wells. Although EOR consumed a quarter of the nitrogen produced in the USA in the mid 1980s, carbon dioxide and other methods remained dominant in the EOR industry.

To give an idea of the distribution of nitrogen consumers, this list gives typical values for the USA in 1985:

Blanketing/purging	25 per cent
Enhanced oil recovery	25 per cent
Electronics	15 per cent
Metals industry	15 per cent
Aerospace	5 per cent
Food freezing	5 per cent
Other varied users	Rest

Although the use of nitrogen for EOR in the UK and Europe is much lower, the breakdown of other users will still follow the above pattern.

2.12.4 *Nitrogen purities and forms*

Nitrogen is sold extensively in both liquid and gas forms, although some of the liquid sold to users is vaporized on site before being consumed. Liquid nitrogen is more convenient to transport when reasonable distances are involved. Moderate to large users of gaseous nitrogen are usually only a short pipeline away from the cryogenic unit producing it. Small users of gas can be supplied with bottles of compressed nitrogen. The form in which the nitrogen is supplied depends on many factors including amount used, peak demand values, distance from plant, and nature of application.

Generally, the purities in Table 2.16 are expected for food, medical, and especially electronic applications. Nitrogen purging and blanketing on a chemical plant can usually be satisfied with less pure nitrogen, often containing several hundred v.p.m. of oxygen and other impurities.

The purities of both forms follow the same pattern, and typical purities are given in Table 2.16.

TABLE 2.16. Typical nitrogen purities. The grades indicated are BOC designations. The data are component content in v.p.m. and are 'not to be exceeded' values.

	Grade		
Component	Commercial	Research	N6.0
Oxygen	3	0.5	0.2
Carbon dioxide	1	0.5	0.05
Total hydrocarbons	1	0.5	0.5
Hydrogen	1	0.5	—
Nitrogen oxides	0.1	0.1	—
Carbon monoxide	1	0.5	0.2
Water	2	0.5	0.5

2.13 Oxygen

2.13.1 *Technical data*

Molecular weight	32.00
Boiling point	− 182.97°C
Specific volume	0.755 m^3 kg^{-1} at 20°C, 101.3 kPa
Critical temperature	− 118.6°C
Critical pressure	5043 kPa
Heat capacity	918 J kg^{-1} K^{-1} at 25°C, 101.3 kPa
Flammable limits in air	non-flammable
TLV	not established

Gaseous oxygen is colourless, odourless, and tasteless, but unlike nitrogen, it is very active. Oxygen does not burn itself, but readily combines with other elements by oxidation. This is often a very rapid process of burning which leads to explosion. Liquid oxygen has a transparent blue colour and has magnetic properties.

2.13.2 *Sources of oxygen*

Oxygen is the second most abundant element in the atmosphere and makes up around half of the total matter on the Earth's surface. This makes it by far the most abundant element in the world. Water contains about 90 per cent oxygen by weight and the human body is about two-thirds oxygen.

Theoretically, oxygen can be obtained from both the air and water. Because water electrolysis is expensive, virtually all oxygen is derived from the air. Adsorption and cryogenics can be used to purify oxygen from air, but the predominant method is cryogenics. This is because cryogenic technology is able to produce oxygen of very high purity more economically. Another, perhaps more important, factor is that cryogenic technology can produce liquid oxygen. This is much more useful to merchants who distribute small amounts of oxygen to a large number of users.

Adsorption technology has found a small market niche in the sewage treatment area. Only medium purity, gaseous oxygen is required for this application in relatively small quantities.

2.13.3 *Uses of oxygen*

Oxygen was the first of the atmospheric gases sought by the pioneers of cryogenic technology. Oxygen was first liquefied by Cailletet and Pictet in 1877. In 1983, the Kraków University in Poland celebrated the Centenary of the liquefaction of nitrogen and oxygen by Wróblewski and Olszewski. Therefore, oxygen began its useful life in industry very early in the twentieth century.

Oxygen was used in the steel industry in the early twentieth century, but today it has a variety of uses. Table 2.17 shows the uses of oxygen along with the approximate proportion that is consumed in each case.

One of the first uses of oxygen was in the fabrication and cutting of metals using the oxy-acetylene torch. This technique is still used today in small businesses and many field applications. Another early user of oxygen was the medical profession as an aid to breathing for patients with lung disorders.

The consumption of oxygen rose dramatically after World War II due to the rapid expansion of the steel industry in developed countries. Pure oxygen was injected into the furnaces to improve combustion efficiency. This use made the steel industry the largest single consumer of oxygen, making it economic to build large air separation plants on the steel mill site solely to supply oxygen to the furnaces. As a result, oxygen plants

TABLE 2.17. Uses of oxygen. The data is for the USA in 1985, although it is probably a good yardstick for most developed countries.

User description	Proportion (%)
Primary metals manufacturing	55
Chemicals manufacturing	20
Partial oxidation processes	10
Metals fabrication	5
Others, including pollution control, medicine, fabricated metal products, welding, fuel cells, glass manufacture, etc.	10

typically producing several hundred tonnes of pure oxygen per day were built to satisfy steel mills.

The late 1970s and early 1980s saw a decline of the steel industry in developed countries, and many of these plants could not operate economically. Other products such as argon were therefore sought from the same plants. New plants built in the 1980s are typically multi-product units to provide a more varied product slate in order to survive the variances of industry.

Coal gasification is another large single user of gaseous oxygen. Some nations experienced shortages of oil and gas supplies early in their development and turned to coal as a base feedstock for their industry. Technologies were developed to convert coal into transport fuels, chemicals, electric power, fuel gas, substitute natural gas (SNG), and synthesis gas. Many of these processes consume large amounts of oxygen. The most notable of these is perhaps the SASOL II complex in South Africa which uses about 13 650 tonnes per day of oxygen, all produced by cryogenic plants.

Six cryogenic units, each capable of supplying 2275 tonnes per day of pure gaseous oxygen, supply the coal gasification/liquefaction plants on site. This represents about 30 per cent of the oxygen production in South Africa, the largest single-site oxygen facility, and the largest single train in the world.

Coal is gasified by adding oxygen and steam to pulverized coal in a gasifier which may be of the fluidized bed type. The heat generated in the gasifier produces a gas mixture which consists mainly of hydrogen and carbon monoxide. The exit gas is first cooled to recover heat, scrubbed to remove sulphur compounds, and treated by various unit operations to produce either SNG or industrial fuel gas. The hydrogen formed in the coal gasification steps can then be used to 'liquefy' coal into liquid hydrocarbon fuels. This is the way in which oxygen, water, and coal form the starting point for automotive fuel manufacture.

Chemicals manufacture is another large consumer of oxygen, but it is divided into many areas. Probably the largest chemical process consumer is the manufacture of ethylene oxide, which takes over 40 per cent of the oxygen devoted to chemicals. The manufacture of acetylene takes almost 20 per cent of this oxygen, while titanium dioxide, propylene oxide, and vinyl acetate take 10–15 per cent each, with other minor users taking up the balance.

Oxygen is used in metals manufacture mainly in oxy-acetylene welding. This will generally be commercial grade gaseous oxygen in bottles. The same grade of oxygen is sold to hospitals to aid breathing in patients with respiratory problems.

In glass manufacture, oxygen is added to enrich the combustion air in glass-melting furnaces. This results in more efficient use of fuel because enrichment with oxygen means combustion at higher temperatures, better heat transfer, and hence lower losses. Many of the old glass-melting furnaces can be converted, without any major modifications, to use oxygen-rich combustion air. Gaseous oxygen is also used in the treatment of liquid municipal wastes and sludges. Oxygen is needed to promote growth of the bacteria that decay the waste. This purity can readily be attained by PSA adsorption techniques and therefore this application is contested by both PSA and cryogenic technologies.

Jet aircraft used for high altitude missions are equipped with oxygen systems for breathing atmospheres. The charge of oxygen is in liquid form, but it is obviously vaporized prior to use.

2.13.4 *Oxygen purities and forms*

All of the above users require gaseous oxygen, the larger ones having on-site plants to supply it. The very small users will buy gaseous oxygen from a distributor who will himself buy liquid, vaporize it, bottle it, and sell it to the local market. The application that can be most accurately described as a use of liquid oxygen is in a rocket. Liquid oxygen is put on board a rocket such as the space shuttle, as a fuel oxidizer and life support. The American space programme at its peak consumed about 400 000 tonnes annually, with a typical Apollo flight to the moon consuming 2000 tonnes of oxygen.

The small users use gas bottles of sizes typically varying from 20 litres to over 10 cubic metres. Liquid oxygen is supplied in double-walled vacuum insulated vessels as described in more detail in Chapter 9. Required purities of oxygen vary a great deal depending on the application. Sewage treatment requires only 90–95 per cent pure oxygen, while metal, chemical, and medical applications generally need at least 99.5 per cent oxygen content. Table 2.18 shows typical grades of oxygen available on the open market in the UK.

TABLE 2.18. Purities of oxygen. Trace impurities are maximum values.

	Grade		
Component	Argon free	Zero	Research
O_2 (mol.%)	99.5	99.7	99.985
H_2 (mol.%)	0.35	1 v.p.m.	1
N_2+Ar (v.p.m)	190	0.3%	150
CO_2 (v.p.m.)	1	1	1
CO (v.p.m.)	0.5	0.5	1
THC* (v.p.m.)	1	20	1
H_2O (v.p.m.)	5	2	1

*THC = Total hydrocarbons.

Bibliography

ABBOTT, P. (1973). *Airship.* Adams and Dart, Jupiter Books, England.

BARRON, R. F. (1985). *Cryogenic systems* (2nd edn). Oxford University Press.

BOC special gases catalogue, BOC Limited, England.

BOND, P. (1987). *Heroes in space.* Basil Blackwell, Oxford.

BRITISH CRYOGENIC COUNCIL (1982). *Cryogenics safety manual, a guide to good practice.* 2nd Ed. Mechanical Engineering Publications Ltd., London.

CHEMICAL AND ENGINEERING NEWS (1978). Argon review. 21 Aug.

COMPRESSED AIR MAGAZINE (1979). 'Air separation review,' Sept. and Oct.

COMPRESSED AIR MAGAZINE (1979). 'There's more to it than cold air,' Aug.

COMPRESSED GAS ASSOCIATION. *Argon specification,* Spec. No. G11.1, p. 3.

GARDNER, J. B. (1979). 'Worldwide cryogenics—UK, cryogenics in BOC'. *Cryogenics,* March, p. 131.

GLASS, D. S. (1983). 'The propylene feedstock story.' *Process Economics Intnl.,* **4** (3), 38.

GREEK, B. F. (1985). 'Industrial gases: producers face mixed demands.' *C & EN,* July 29, p. 9.

GUMZ, J. P. (1983). 'Gases used in medicine.' *Linde Reports Sci. Tech.,* **36**, 22.

HANDS, B. A. (1986). *Cryogenic engineering.* Academic Press, London.

HOBSON, G. D. (1986). *Modern petroleum technology* (5th edn), Part 2, p. 838. John Wiley & Sons, New York.

HODGSON, B. (1978). 'Natural gas: the search goes on.' *National Geographic,* **154** (5), 632.

HOULDCROFT, P. T. (1967). *Welding processes.* Cambridge University Press.

Industrial gases data book (5th edn). Airco. USA.

INFORMATIONS CHIMIE (1974). *Hydrogène: quoi de nouveau?,* **134**, June, 77.

KREPSCHOT, R. H., BIRMINGHAM, B. W., and MANN, D. B. (1968). *Technology of liquid helium.* National Bureau of Standards Monograph 111. US Dept. of Commerce.

LATIMER, R. E. (1967). Distillation of air. *Chem. Eng. Process,* **63** (2), 35.

MAROE, K. M. (1986). 'The Scandinavian LP gas market.' *G.P.A. Proc. 65th Conv.,* p. 32, GPA, USA.

NOBLES, J. E. and STANCIK, J. W. (1983). Purifying food grade, naturally occurring CO_2. *Oil & Gas Journal,* Dec 26, p. 156.

NOVIS SMITH, W. and SANTANGELO, J. G. (1980). *Hydrogen: production and marketing.* ACS Symposium Series 116.

SCOTT, R. B., DENTON, W. H., and NICHOLLS, C. M. (1964). *Technology and uses of liquid hydrogen.* Pergamon Press.

SPRINGMAN, H. (1978). 'The liquefaction of oxygen, nitrogen and argon'. *Linde Reports Sci. Tech.* **28**, 23.

URBAN, R. and BOSCAINO, V. (1986). 'LPG in the future automotive scenario'. *G.P.A. Proc. 65th Conv.,* p. 104.

VERANNEMAN, G. (1980). 'The use of industrial gases in foundries'. *Linde Reports Sci. Tech.* **31**, 32.

3

AIR-SEPARATION PROCESSES

3.1 The development of the air-separation industry

Liquefaction of air was first achieved in 1877 when Cailletet and Pictet simultaneously but independently liquefied oxygen, which appeared as a fine mist for a very short time. Six years later Olszewski and Wroblewski produced stable liquid oxygen within a tube shielded in liquid ethylene which in turn was insulated by ethylene vapour passing through the annular space surrounding the ethylene container. To emphasize the importance of Wróblewski and Olszewski's work it is interesting to note that some low temperature storage techniques still use this principle. Moreover, it illustrates the basics of cascade liquefaction as used today.

The first large-scale liquefaction of air was carried out by Linde in 1895 using Joule–Thomson expansion. Subsequently, Claude used a reciprocating engine to achieve the same end.

The early part of the 1900s saw many developments in cryogenics with the first milestone occurring in the field of air separation as early as 1902 when the first single-column air separation plant was operated by Linde. Shortly afterwards, in 1910, the first double-column air separation plant was also commissioned by the Linde Company. Fränkl introduced the use of cold regenerations, thus allowing the construction of large air separation units from around 1925. Subsequently the air separation industry grew with the aim of producing large quantities of oxygen and, later, rare gases using both 'single-column' and 'double-column' processes.

'Single-column' means the use of a single distillation tower with a feed, reboiler or condenser or both, operating at a single pressure producing two products. 'Double-column' means two distillation towers operating at different pressures. The condenser of the column operating at higher pressure acts as the reboiler for the column operating at lower pressure. This means that the columns are usually physically placed one over the other with the lower column operating at the higher pressure. This system is linked by two further features: firstly, the lower column bottoms product is fed to the upper column as feed, and secondly reflux to the upper column is provided by liquid top product from the lower column. Chapter 9 gives a fuller description of this unusual unit operation.

A further milestone in air separation technology was the use by Linde in 1936 of a turbo-expander to provide refrigeration, followed by improved designs suggested by Kapitza in 1939. This was a timely addition to cryogenic technology because the demand for oxygen for the steel industry was growing rapidly, far beyond the earlier medical requirements.

In the first half of this century air separation was a means of producing oxygen in the main, nitrogen being of lesser importance. However, during the late 1920s, the ammonia industry became the first to use appreciable quantities of nitrogen generated by air-separation units. At that time hydrogen for ammonia synthesis was derived from coke oven gas, and the nitrogen performed three duties:

(1) provided refrigeration for coke oven gas separation;
(2) removed CO, Ar, and H_2 from coke oven gas using liquid nitrogen wash;
(3) raised the level of nitrogen to a stoichiometric H:N ratio.

With the introduction of steam reforming and the use of natural gas as fuel and raw material, air separation and thus in particular nitrogen production ceased to feature in the ammonia industry, though recent developments suggest its possible return.

The use of nitrogen from air separation as an inert gas for purging equipment and for conducting operations on materials requiring an oxygen-free atmosphere is a development of the last 35–40 years.

The 1950s saw the advent of small, single column nitrogen plants used for inert gas generation. These became competitors to the combustion process for inert gas generation. This was another milestone in air-separation technology as it embodied the use of a single-column process operating at a higher pressure than usual and air compression to the much lower than usual pressure of 800 kPa. During the 1960s a range of these plants became available for use in a number of industries, including oil and chemicals, semi-conductors, LNG storage, and breweries. Today users of up to about 20 tons per day have the option of trucking in liquid or producing nitrogen on site, the economics depending on local transport facilities and cost of power. Larger users operate their own plants, or if suitably situated, may avail themselves of pipelines from large distribution centres.

Pipelines to distribute nitrogen and oxygen are well established methods of gas distribution in industrial areas of the USA, and there are some systems like those in Europe, the most famous being that in Northern France and Belgium, which has many oxygen plants supplying gas into the pipeline. The largest of these plants delivers 2275 tonnes per day of oxygen in Dunkerque. However, the largest air-separation plant constructed in single trains are at SASOL in South Africa. These units have the capability of simultaneous production of 2275 tonnes per day oxygen and 1250 tonnes per day of nitrogen.

Today's air separation industry is a highly developed technology—some would call it 'mature'. However, the benefits of long experience are evident in the high efficiency and degree of sophistication in the units designed today. It has become a precise technology where accurate equilibrium data and feed gas conditions are essential for effective design. Even trace components require careful definition as they could have significant consequences for the cost of the separation unit. The location of the plant can have a major effect on the design of the unit. For example altitude may affect the suction pressure and water content. Carbon dioxide and trace impurity levels can also be different depending on location. Table 3.1 gives normal and maximum impurities in air along with its normal composition. Today's designers use this information as the starting point for design of the air-separation units.

TABLE 3.1. The composition of air.

	Normal	With maximum impurities
Nitrogen	78.084 mol.%	78.028 mol.%
Oxygen	20.946 mol.%	20.931 mol.%
Argon	0.934 mol.%	0.933 mol.%
Carbon dioxide	350 v.p.m.	1000 v.p.m.
Water	variable*	saturated*
Neon	18 v.p.m.	18 v.p.m.
Krypton	1.1 v.p.m.	1.1 v.p.m.
Xenon	0.08 v.p.m.	0.08 v.p.m.
Helium	5.3 v.p.m.	5.3 v.p.m.
Hydrogen	0.5 v.p.m.	0.5 v.p.m.
Acetylene	0.1 v.p.m.	1.0 v.p.m.
Ethylene	0.01 v.p.m.	2.0 v.p.m.
Propylene	nil	0.2 v.p.m.
Methane	2.5 v.p.m.	10 v.p.m.
Ethane	0.02 v.p.m.	0.1 v.p.m.
Propane	nil	0.1 v.p.m.
Butane	nil	0.1 v.p.m.
Carbon monoxide	nil	35 v.p.m.
Nitrogen oxides	nil	0.5 v.p.m.
Sulphur compounds	nil	0.1 v.p.m.

*Air composition data are on dry basis.

3.2 Small- to medium-capacity nitrogen generators

Nitrogen is used widely as a purging, blanketing, or inerting medium in many branches of industry. In the 1950s and 1960s this nitrogen was frequently produced by the combustion technique, which yielded a relatively pure nitrogen stream, although it was contaminated with water and carbon dioxide. In later decades, air separation by cryogenic means became the main source of pure nitrogen. The main advantages of this process were that clean, dry nitrogen was produced at a higher purity than from the combustion process. This meant that the uses of nitrogen became more varied and numerous among the refinery, pharmaceutical, gas processing, and fertilizer industries. Many cryogenic design companies saw the growing need for nitrogen at a variety of flowrates and qualities and developed a range of small to medium-sized plants utilizing highly efficient and proven cryogenic and adsorption technology. The size range of these plants varies from 3 to over 100 tonnes per day of gaseous nitrogen.

Many users consume nitrogen on an intermittent basis or with variable requirements. For example, pharmaceuticals often involve batch processes and require very small amounts of liquid or gaseous nitrogen. For them it is economic to buy liquid nitrogen from a large distributor. Similarly, a small fertilizer complex only requires nitrogen during its annual shut-down for purging purposes for about 20 to 30 days per year. Table 3.2 shows typical requirements for quantity, phase, and purity for nitrogen users in various parts of industry.

TABLE 3.2. Typical users, quantities, and purities of nitrogen.

Application	Nitrogen requirement (tonnes/day)	Purity (p.p.m. O_2)	Phase (gas/liquid)
Refineries, hydrogen storage blanketing	0–6	5000	gas
Pharmaceuticals, foods, and drinks	0–50	1–200	gas and liquid
Electronics	0–10	0.5–100	gas and liquid
Well fracturing	50–100	1–10	liquid

A typical process that has been developed for this application by a number of cryogenic unit designers is described below, and the cryogenic section for this process is shown in Fig. 3.1. The plant requires filtered air, compressed to about 900 kPa in C1. The air is then chilled using a packaged refrigeration unit, R1, dried in molecular sieves, V1, and passed into the cryogenic unit.

The dry, oil-free feed air then passes into the cold box, which houses the cryogenic equipment, and downwards through the main heat exchanger, E1, where it is cooled against the returning colder nitrogen and waste gas streams. The feed air then enters the distillation column at its dewpoint (−167°C). In this distillation column the air is separated

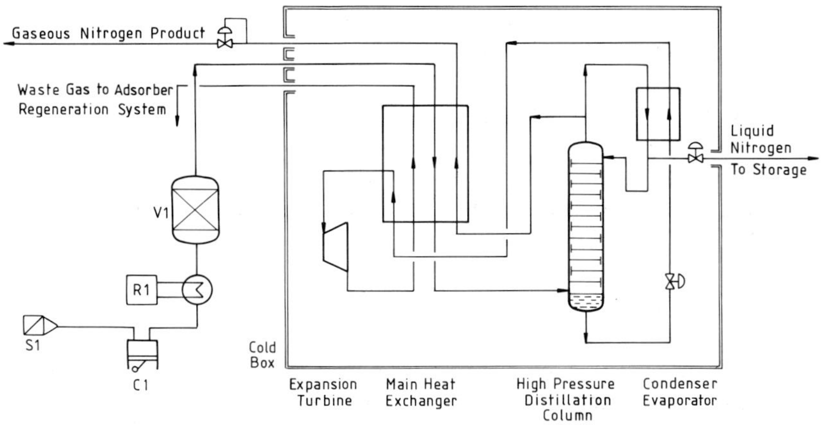

FIG. 3.1. A low recovery, single-column nitrogen plant.

into a pure nitrogen overheads product and a 35 per cent oxygen bottoms product. The liquefied bottoms product is expanded to an intermediate pressure and evaporated against a stream of condensing nitrogen in a condenser/evaporator. The condensed nitrogen at the higher pressure is fed back to the column as reflux, and a small quantity may also be withdrawn from the cold box as product and stored in a vacuum-insulated storage tank.

The 35 per cent oxygen (waste gas) stream leaving the condenser/evaporator is reheated slightly in the main exchanger before being passed through an expansion turbine, where it expands to a pressure slightly above atmospheric. The turbine uses gas bearings such that the shaft and rotors 'float' on a cushion of gas, and thus there is no requirement for a lube oil system. In the turbine, the cold waste gas stream is made to do work by compressing a warm waste gas stream at the opposite end of the shaft. This results in a fall in the temperature of the cold waste gas stream leaving the expansion wheel, thereby producing the refrigeration necessary to support the process.

Gaseous nitrogen product leaving the top of the distillation column at the higher pressure is similarly re-heated up to ambient temperature in the main heat exchanger as it leaves the cold box. The product is then available at the battery limits at a pressure of up to 800 kPa. The same process produces a product of 1–200 v.p.m. oxygen, depending on user requirement.

These plants are designed to be versatile, simple, and to have low maintenance cost requirements. Therefore, their power consumption is not very low at about 0.35 kW Nm^{-3} of nitrogen product. Bearing in mind their relatively low capacity (up to 100 tonnes per day nitrogen) and that they produce 5–8 per cent of their product as liquid they are, taking everything into account, of optimum design.

3.3 Large nitrogen generators

3.3.1 *Low-temperature process section*

The plants described in section 3.2 above are based on a single-column cycle and produce a limited nitrogen recovery because of the equilibrium conditions at the column feed point. When operating at about 700 kPa and feeding the air in at its dewpoint, only about 40 per cent of the treated air is recovered as pure nitrogen product with the ensuing waste gas containing some 35 per cent oxygen. For small plants, this low recovery is not very important, but as we go to larger units of several hundred tonnes per day, feed compression becomes a large consumer of energy and reducing air feed rate for a given nitrogen output is important. This can be done in two ways:

(1) using a double-column process;
(2) using lower operating pressures in single-column process.

In both cases the recovery of nitrogen increases to different degrees, the double-column process (described later in this chapter) giving the highest potential recovery, which at first sight suggests it is the more attractive solution. In practice, however, it depends on the application, of which there are three basic categories:

(1) the fertilizer industry;
(2) enhanced oil recovery (EOR);
(3) merchant plants supplying a variety of products in both gas and liquid form.

Typically, the merchant plants which require a variety of products are double-column units because they produce multiple pure products, and this becomes the dominant reason for their double-column system. Both the fertilizer industry and enhanced oil recovery applications require nitrogen at pressures higher than the air separation process needs. Therefore, N_2 product compression is often a significant production cost. One might deduce from this that the economics are favoured by carrying out the nitrogen separation part of the process at the highest pressure possible. However, the exact opposite is true. Producing nitrogen at an elevated pressure requires the compression of air to at least that pressure. Since the flow of feed air is necessarily greater than the flow of product, it is cheaper to compress the nitrogen than the air. It is thus economically sound to reduce the pressure to which the air is compressed, although this involves increasing the pressure ratio of the nitrogen compressor.

In a single-column plant the nitrogen product is available at a pressure only marginally lower than the delivery pressure of the air compressor, whereas in a double-column plant the pressure in the upper column has necessarily to be below that of the lower column, and hence well below the delivery pressure of the air compressor. Between the two columns the pressure has to be let down. In simple terms the product has to be compressed twice. As a result the power consumption of a large single-column nitrogen plant can be less than that of an equivalent double-column unit.

However, other factors have contributed to this result. In particular an increase in the amount of pure nitrogen recovered from a given stream of air. As stated previously, with a saturated vapour feed the minimum nitrogen content of the bottoms product is given by its concentration in the liquid in equilibrium with atmospheric composition air. This cannot be achieved in practice since it corresponds to minimum reflux, requiring an infinite number of trays in the column. However, a low column pressure favours nitrogen recovery, as it increases the relative volatility of nitrogen with respect to oxygen and argon so that the liquid in equilibrium with atmospheric air contains less nitrogen as the pressure decreases. But since it is the evaporation of this liquid that condenses the reflux, a decrease in its nitrogen content, which means a reduction in rich liquid flow, also reduces the amount of reflux available. A way has therefore to be found to maintain the reflux ratio in spite of the decrease in the flow of bottoms product. This has been achieved by the following modification to the conventional process. Figure 3.1 shows the standard arrangement of the cold box equipment in a single-column nitrogen plant. The oxygen-rich liquid issuing from the base of the column is expanded to an intermediate pressure, evaporated in the reflux condenser, superheated in the main exchanger, and finally expanded to near atmospheric pressure in the turbine, thereby producing the refrigeration required to maintain the process.

As drawn, the cold box is seen to consist of two sections: an exchanger section and a column section. The expansion turbine is located in the exchanger section. A heat balance around the individual sections shows that, under these conditions, the feed to the column is slightly wet, i.e. it contains a few percent of liquid. Moreover the expansion of the oxygen-rich liquid results in a considerable flash in the expansion valve, so that only about 90 per cent of this liquid is available for the production of nitrogen reflux in the condenser.

In the modified arrangement illustrated in Fig. 3.2, the turbine, C2, has been moved to the column section and is positioned immediately downstream of the reflux condenser, E4. The turbine is thus fed with saturated vapour and some 5 per cent of the liquid is contained in the turbine exhaust, E5, subcooling the oxygen-rich liquid, so that no vapour is formed in the expansion valve, JT1. As a result the whole of the latent heat in the oxygen-rich liquid is available for the production of reflux in

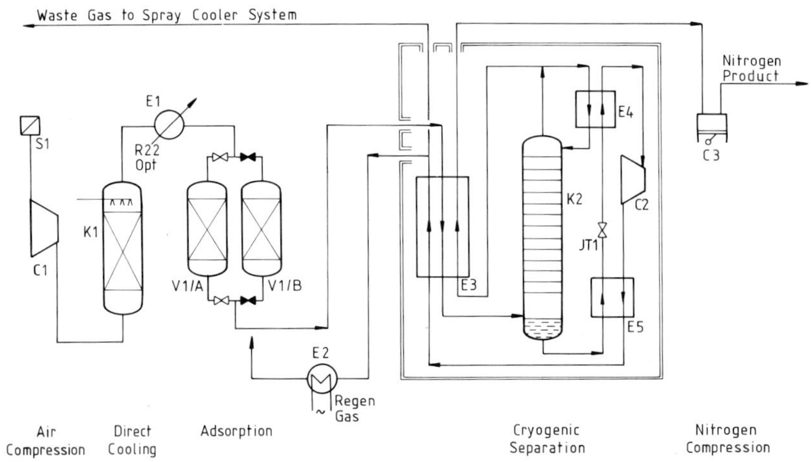

FIG. 3.2. A flow-sheet for gaseous nitrogen production. The process is for medium and large capacities.

E4, so that the same amount of reflux as before is now produced with a smaller flow or rich liquid.

This arrangement is now possible owing to the development of turbines capable of handling two-phase fluids without loss of efficiency.

A heat balance around the two sections of the cold box now shows that the feed to the column is no longer wet but, on the contrary, slightly superheated. The superheat is removed on the bottom tray, which now acts as a reboiler, further reducing the minimum nitrogen content of the oxygen-rich liquid.

Double-column processes can show even greater recoveries than the low pressure single column, but should only be used if nitrogen at low pressure is required.

3.3.2 *Pretreatment of air prior to cryogenic processing*

Two basic methods are available for air pre-purification prior to distillation:

(1) chilling with Freon followed by molecular sieve adsorption to remove residual water and carbon dioxide;
(2) admission of wet air with carbon dioxide into the cold box through reversing exchangers which alternately freeze and sublime the impurities.

Traditionally nearly all small air-separation plants utilize route (1), whilst most of the large air-separation units (500 tonnes and larger) utilize route (2). However, a few large air-separation plants use the molecular sieve route for pretreatment, showing that it is feasible for large units. The advantages of reversing exchangers are in reduced pressure drops through the plant and hence lower feed compression power.

However, nitrogen plants designed for maximum recovery may not lend themselves readily to the use of reversing exchangers for two main reasons:

(1) the low operating pressure may not give sufficient reversing pressure difference to assure efficient removal of impurities;
(2) there may not be enough waste gas to achieve proper regeneration of the reversing exchangers, thus requiring more flash air and hence yielding a poorer process efficiency.

The currrent status is that for large units reversing exchanges are still widely used, although molecular sieve plants are beginning to be used more and more on the medium-sized units. Most modern small plants, up to 100 tonnes per day of nitrogen, utilize molecular sieves. Chapter 8 covers pretreatment aspects in more detail.

3.3.3 *Comparison of energy consumption using various process routes for nitrogen generation*

Table 3.3 gives power consumption figures for air and nitrogen compression in three plant variants when providing nitrogen at 20 000 kPa pressure with 10 p.p.m. oxygen content in quantities above 10 000 Nm^3 h^{-1}. The units in Table 3.3 are kWh/100 Nm^3 of nitrogen production at battery limit. This nitrogen is aimed at enhanced oil recovery (EOR) and therefore illustrates the benefits of a single column plant operating at the lowest pressure possible.

It should be pointed out that a small amount of additional power is needed in all these variants to perform certain ancillary duties in the pretreatment section. This is mainly associated with pressure losses in

TABLE 3.3. Power consumption for N_2 processes (in kW/100 Nm^3).

	Process		
	Medium pressure, single column (Fig. 3.1)	Low pressure, single column (Fig. 3.2)	Low pressure, double column (Fig 3.6*)
Air compression	30.8	16.1	16.3
N_2 compression	14.9	17.6	19.8
Total compression	45.7	33.7	36.1

*The process as shown in Fig. 3.6 but adapted for nitrogen production and using molecular sieves as pretreatment.

the system outside the cold box and in the pretreatment of the gas prior to cold box entry. The losses are assumed to be similar in each of the options listed in Tables 3.3 and 3.4 and do not affect the comparative positions of the air separation process adopted. The tables assume the following:

(1) the adsorption route is used as pretreatment;
(2) the pressure losses in the system are similar;
(3) the design basis for the pretreatment is similar;
(4) the design basis for the cryogenic unit is similar.

The cryogenic section of each process option was studied by the use of exergy analysis to calculate the process efficiency. Exergy analysis used the second law of thermodynamics and if applied in detail can show where the largest irreversibilities occur in a given process. Table 3.4 shows the efficiencies of each of the process options and the net effect on the overall power consumption to produce 20 000 kPa nitrogen as the ultimate product.

TABLE 3.4. Efficiency of the cryogenic sections.

	Fig. 3.1	Fig. 3.2	Fig. 3.6*
Minimum work of separation (kWh/100 Nm^3 N_2)	0.93	1.08	1.42
Losses (kWh/100 Nm^3 N_2)	11.1	3.86	4.74
Efficiency (%)	7.7	21.9	23.0
Nitrogen recovery (%)	40	45	65

*The process as shown in Fig. 3.6 but adapted for nitrogen production and using molecular sieves as pretreatment.

It is clear from Tables 3.3 and 3.4 that the efficency of the cryogenic section alone does not determine that of the process as a whole, since the power consumption of process 3.2 as shown in Table 3.3 is lower than that of process 3.6. This is not surprising, since the actual work of separation, which is given by the sum of the minimum work and the losses, is only a small fraction of the total work performed. The major portion of work is consumed in compressing the air and the nitrogen, and Table 3.3 makes no allowance for this.

Table 3.4 shows that, although the minimum work of separation is higher in process 3.6 than in 3.2, owing to the higher nitrogen recovery, the losses are also greater, owing, primarily, to the expansion from lower to upper column pressure. Hence in process 3.2 even though the losses are smaller than in 3.6, the process efficiency is lower than process 3.6 because of the lower recovery of N_2 and hence lower minimum work of separation.

Two things may be concluded from this analysis. The first is that single-column processes appear to have some merit in producing nitrogen for use at high pressures. Secondly, exergy analysis should not be confined to the cryogenic process when assessing the process route to be used. The complete picture should be analysed.

3.4 Oxygen plants

3.4.1 *Single-column oxygen plants*

Oxygen was first liquefied by Cailletet and Pictet in 1877, and, in 1902, Linde put into operation the first commercial, single-column process to produce purified liquid oxygen from air. The basic process covered air compression, drying, and carbon dioxide removal, followed by the cryogenic unit. The cryogenic unit consisted of a three-stream wound coil-type exchanger and a distillation column. The cryogenic section depicted in Fig. 3.3 shows high pressure air after compression and drying entering the cold box in the main heat exchanger, E1. The cooled gas is then partially condensed in reboiler E2, and flashed to low pressure across JT1. This produces a drop in temperature across the valve which provides the necessary refrigeration for the process. The two-phase mixture resulting is administered to the top of column K1 where the liquid acts as a reflux to drive the distillation process. Waste air is drawn from the top of K1 and reheated in E1 to be vented at ambient conditions to the atmosphere and to regenerate driers in the air pretreatment section. Gaseous oxygen is drawn off from the column K1 just below the lowest tray, reheated in E1, and supplied to the consumer. Depending on the pressure the feed gas has been compressed to, a small amount of liquid oxygen may be withdrawn from the sump of K1.

Generally, if gaseous oxygen is required, the air is only compressed to

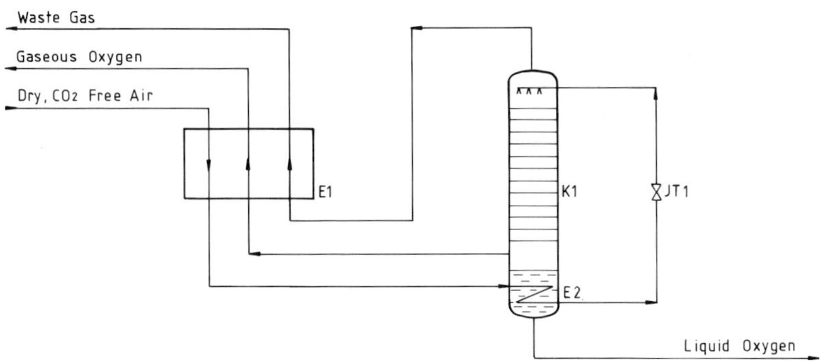

FIG. 3.3. A Linde single-column oxygen plant cold box.

3000–6000 kPa, but if a substantial amount of refrigeration is required across JT1 to produce liquid oxygen, pressures of 20 000 kPa are required. In each case the column K1 operates at pressure in the region of 150 kPa.

Following the commercialization of the Linde double-column process, most air separation was carried out in this way. It was only in the mid 1950s that the single column process was 'reinvented', but for a specific application—to generate pure nitrogen. This process was commercialized successfully from Dr M. Ruhemann's concept to drawing board by Petrocarbon Developments, now Costain Engineering Ltd, and subsequently by many other plant designers. Meanwhile, oxygen was made using double-column processes.

There is one notable exception in the United Kingdom where liquid oxygen is produced in a single high-pressure column process. BOC Limited designed and constructed a single-column plant at Widnes, UK to produce liquid oxygen and nitrogen. This probably unique design was effected in the 1970s and is capable of producing 550 tonnes per day of liquid oxygen and 600 tonnes per day of liquid nitrogen. The process, shown in Fig. 3.4, is based on a reversing heat exchanger design for the pretreatment. Air is filtered and compressed to about 800 kPa in C1 before entering the reversing exchangers in E1 in the cold box, and is then further cooled in E2 before entering the distillation column, K1, in which the nitrogen and oxygen are completely separated. Some waste gas is withdrawn as a vapour side stream from column K1, reheated a little in

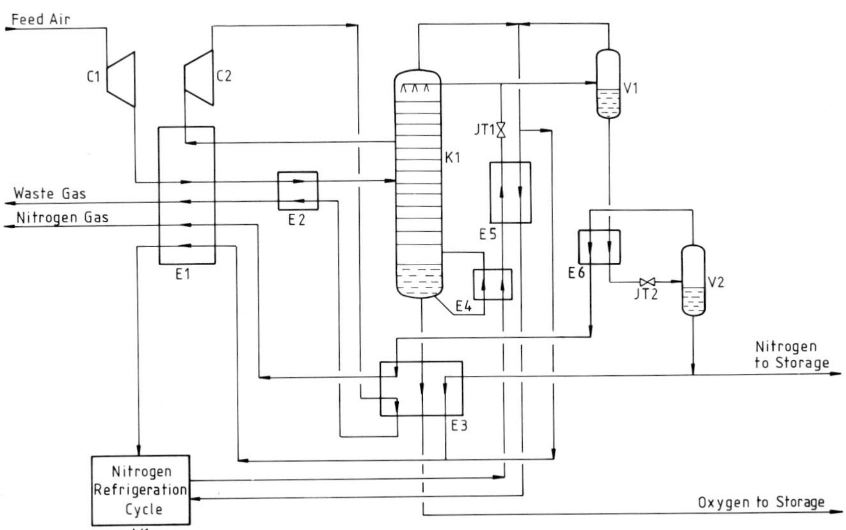

FIG. 3.4. The BOC air-separation plant at Widnes, England (courtesy of BOC Limited England).

E1, and expanded in C2 to provide some of the refrigeration. The expander outlet waste gas is then used to aid subcooling of oxygen product in E3, and is then reheated in E2 and E1 before being discharged to atmosphere at low pressure. The nitrogen product from the top of K1 is combined with vapour from V1 at about 700 kPa and reheated in cycle subcooler E5 and main exchanger E1. Both these medium pressure sources of nitrogen return to the nitrogen cycle heat exchanger–compressor–expander system, W1. High pressure nitrogen is produced by compression and cooling within cycle system W1, condensed in reboiler E4, and subcooled in E5 before being flashed across JT1 to provide the necessary reflux for K1 and sufficient liquid for the liquefier set downstream of V1, i.e. E6, JT2, and V2. The liquid nitrogen from V1 is subcooled in E6, flashed across JT2, and separated into liquid N_2 product and refrigerant in V2, whilst the vapour is reheated in E6, E3, and E1 to form the nitrogen gas product. Liquid nitrogen refrigerant from V2 is used to aid subcooling of oxygen product in E3.

This process lends itself to the use of one column to separate nitrogen and oxygen completely because it is largely a liquefier. Liquefiers are best served with high-pressure cycle gases (~3000–7000 kPa) to provide the best specific power. In the earlier parts of this century Heylandt-type plants used pressures as high as 14 000 kPa as a means of generating refrigeration. These pressures make it relatively easy to provide reboil in K1 at pure oxygen temperatures at the same time liquefying significant amounts of cycle gas.

3.4.2 *Double-column oxygen plants*

The single column is many chemical engineers' answer to the separation of two or more components. In normal practice, such separation is carried out in a distillation column in which the pressure is such that cooling water can be used in the condenser and steam in the reboiler. However, in low-temperature processes, especially for this application, the condenser and reboiler must operate under considerably below ambient conditions. This means that considerable energy must be expended to bring the condensing and reboil media to the column at the right temperature in order to make it function. Most of the single column processes described above are limited in the respect that each of them has what appears to be only half a column, i.e. the stripping section or a rectification section but not both at once. The exception to this rule is the liquefier described in Fig. 3.4, which recovers high proportions of both nitrogen and oxygen. This process is economic primarily because it produces mainly liquid products and the liquefaction energy consumption masks any inefficiencies in the gas separation part. The nitrogen and oxygen production single column processes each recover rather low proportions of products. Typical values are given in Table 3.5.

In the two cases tabulated in 3.5, only one pure product is produced. Although this is often acceptable for small plants which are designed to supply one customer, when a large plant is contemplated for the supply of nitrogen or oxygen, efficiency becomes much more important. This means that it becomes desirable to maximize product recovery in order to reduce the feed compression. This argument holds true for both O_2 and N_2 in separate plants (see section 3.2).

TABLE 3.5. Typical product recoveries of the single column process.

	Product	
	Nitrogen	Oxygen
Process route	Fig. 3.1	Fig. 3.3
Column pressure (kPa)	750	110
Product recovery based on feed rate (%)	40	12.2
Product recovery based on component level in feed (%)	51	58
Waste gas O_2 content (%)	35	10

This problem was solved with the invention of the double column in 1910 by Carl von Linde. The classical system is shown in the cryogenic section of the Heylandt Oxygen Production process shown in Fig. 3.5. For small to medium-sized oxygen plants, of up to 100 tonnes per day, the Heylandt process had been used for many decades. The process is based on compression of air to very high pressure (typically 15 000 kPa) after filtering the air in S1 and scrubbing it with caustic soda solution in K1. The air is then cooled in E1 and the water which condenses out is separated in V1. Alumina driers V2/3 are used to remove any remaining water before the dry air is admitted to the cryogenic unit where liquid oxygen is to be recovered from the air.

The dry air enters E2 and E3 where it is cooled against returning nitrogen product in wound-coil two-stream heat exchangers. Because the oxygen is required as liquid for easy shipment, a significant amount of cold must be generated within the process. To this end, a reciprocating expansion engine C2 is used taking some of the air leaving E2, expanding it to about −170°C and approximately 600 kPa before it joins the remaining air to enter the lower column K2. The air that cools against return nitrogen in E3 is expanded across Joule–Thomson valve JT1 before joining C2 outlet air and entering lower column K2.

K2 is the lower column of a typical Linde double column system in which it acts as the stripping column which effects the production of

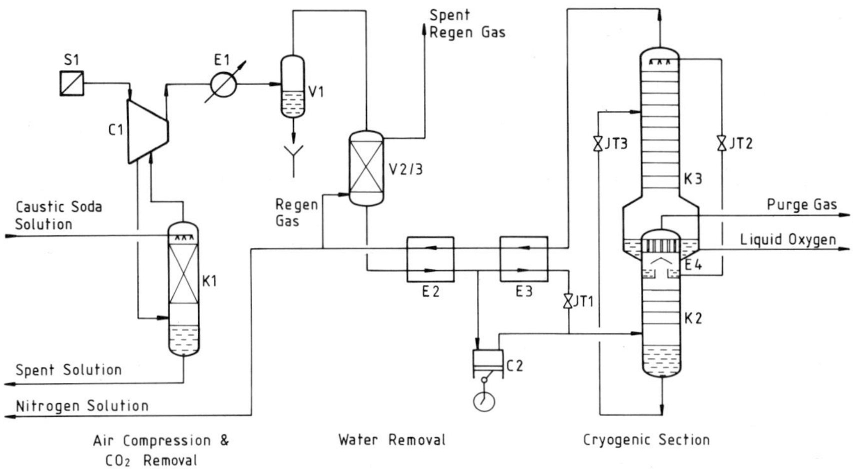

FIG. 3.5. The Heylandt cycle for liquid oxygen production.

pure nitrogen at a pressure of about 600 kPa. One can see the similarity of this part of the process to that described in Fig. 3.1 for small single column nitrogen plants. In the nitrogen plant case, the liquid from the base of the column was flashed and used as a means of providing reflux in a condenser in the top of the column. In the double column system, the liquid from the base of K2 is flashed across Joule–Thomson valve JT3 and admitted to the upper column, K3, which operates at about 150 kPa. Reboil to K3 is provided from condensing nitrogen in the top of K2 in reboiler/condenser, E4. Nitrogen condenses in the top of K2 (i.e. in E4) at 600 kPa and −175°C, whereas pure oxygen boils in the base of K3 (i.e. also in E4) at 150 kPa and −178°C. This gives the necessary driving force for the boiling of pure oxygen and condensation of pure nitrogen. Therefore, two duties are carried out on product streams using one heat exchanger, thereby making the process more efficient. A small purge is drawn from the top of K2 to prevent helium build-up over a long time and reduction of the efficiency of E4. The nitrogen condensed in E4 returns down column K2, with part of it being used as reflux in K2 and part drawn off as a sidestream, flashed in JT2 and passed as reflux to column K3. In this way two pure products can be drawn off from a system without any external heat pump cycle to drive a single-column condenser and reboiler which would otherwise be much less efficient. To complete the process, gaseous nitrogen product is drawn off from the top of column K3, and reheated through exchangers E3 and E2 respectively. These plants are suitable for small to medium-sized liquid oxygen generators of up to 100 tonnes per day oxygen. This oxygen is often supplied to several consumers as a liquid and vaporized on site by the users.

When much larger capacities of oxygen are required, on-site generation plants are used of up to 2200 tonnes per day gaseous oxygen output in a single train plant. Because the capacity is so large, process efficiency is paramount in the design of such systems in order to be as competitive as possible and to ensure availability of large machinery to carry out the air compression and expansion. As a result designers have selected a low pressure route which requires air compression of up to only 650 kPa. A typical low pressure process is shown in Fig. 3.6, with a somewhat different pretreatment.

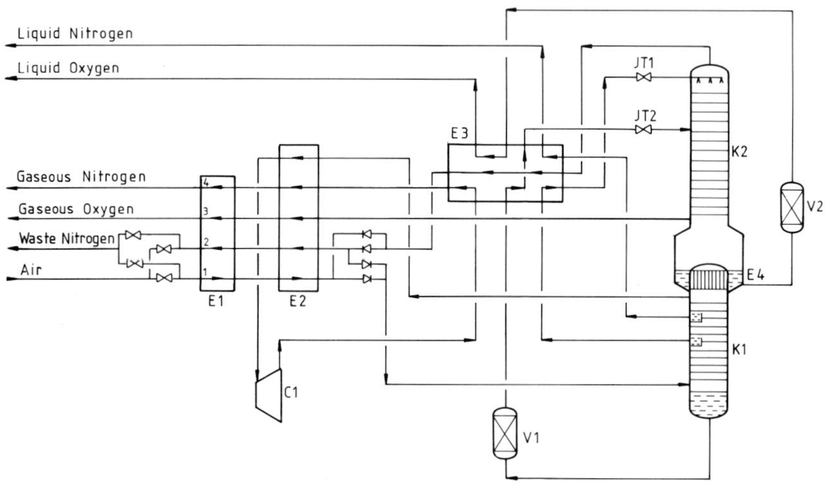

FIG. 3.6. A large capacity low-pressure oxygen process.

In this process air is compressed and cooled in order to condense out some water vapour. The air is still wet and at a pressure of about 650 kPa as it enters the cryogenic unit. The remaining water and all of the CO_2 are removed in the cold box in reversing heat exchangers. These are described in more detail in Chapter 8. The basic principle relies on cooling of the feed gas in an upward direction so that water and CO_2 freeze and deposit on the walls of the heat exchanger (E1 in this case). Dry waste nitrogen is used to remove the water and CO_2 at a low pressure of 120 kPa. The air feed stream 1 and waste nitrogen stream 2 are interconnected and valved such that their duties can be reversed—when stream 1 has been 'on line' for a few minutes, it is then automatically switched over to reverse blow with dry waste nitrogen. This cycle is repeated every 4–10 minutes, depending on design. The duties of streams 3 and 4 remain unchanged. The feed air continues to be cooled in E2 and enters the lower column K1 near the base. Rich-liquid is drawn from the base of K1, passed through guard adsorbers, subcooled in E3, flashed across a Joule–Thomson valve JT2 and fed into the upper column K2. Nitrogen product is produced from the top of K1 from condenser/reboiler E4, withdrawn from the column as liquid, reheated in E2, expanded in C1 to provide refrigeration for the process, and introduced back into E3, E2, and E1 before being supplied to battery limit as low pressure gaseous nitrogen product.

A sidedraw from column K1 is also taken off as liquid, subcooled in E3, flashed in JT1, and supplied as reflux to column K2. Gaseous overheads from K2 provide the subcooling duty in E3, and are reheated in E2 and E1 before being routed as nitrogen-rich waste gas. This waste nitrogen is passed through either of the two feed passages as reverse flow to regenerate the reversing heat exchangers. Some liquid nitrogen can also be withdrawn from K1 just below the condenser/reboiler E4, subcooled in E3, and passed to product storage. Liquid oxygen product is passed through hydrocarbon adsorbers, V2, subcooled in E3, and routed to storage. Gaseous oxygen product is withdrawn from the vapour space just below the bottom tray in K2, and reheated in E2 and E1 before being routed to a product. This unit is very similar to the six 2275 tonnes per day oxygen plants operating in the SASOL II complex in South Africa. This is the largest battery of plants producing oxygen in the world, as well as being the largest single plant capacity. A typical oxygen plant is shown in Fig. 3.7.

FIG. 3.7. A large oxygen plant (courtesy of BOC Cryoplants Limited).

3.5 Complex air-separation units

From the foregoing discussion it becomes apparent that most of the larger plants produce more than one product, i.e. both oxygen and nitrogen in both liquid and gaseous form. Much of this can be achieved relatively simply by the addition of more trays, a larger expander, feed compression to a higher pressure, and/or modification of the process scheme in some way to achieve the objectives. The cryogenic designer skilled in the process synthesis normally has a good starting point on which to build up his process route. Depending on the application of the products, the process used to recover them becomes complex. An example of this is a unit which produces low pressure gaseous nitrogen *and* 20 000 kPa gaseous oxygen product. The process has been designed to eliminate an arduous oxygen gas compression duty and uses cryogenic pumping of liquid oxygen up to 20 000 kPa. This process is shown in Fig. 3.8, and is based on a 'split air' cycle where separate fractions of air are compressed to different pressures. Approximately 65 per cent of the air is compressed to about 650 kPa and processed in via reversing exchangers E3, with the main purification being done in the conventional double column K1 and K2, and associated subcooling, adsorption, and reboiling/condensing in E6, E7, V3, and E5.

The interesting feature of this process lies in the supply of refrigeration to the process and the means by which oxygen is produced at 20 000 kPa. First the remaining 35 per cent of the air feed is compressed up to 20 000 kPa in C2 and cooled in E2 and the warm end of E4 before being dried in molecular sieve adsorbers V1/2. The dry air is then further cooled in E4 and supplementary refrigeration such as a closed loop from Freon cycle before being cooled further, expanded to about 620 kPa in JT1, and mixed with the main feed to K1. Liquid oxygen is drawn from the base of K2, pumped in P1 up to supercritical conditions at 20 000 kPa, and reheated in E4 before being supplied at very high pressure to the product user.

Although this process consumes some 5–10 per cent more power than a normal oxygen plant which then requires oxygen compression up to 20 000 kPa, it is less machinery intensive and finds favour in some areas. There are many cases where specific requirements of a consumer mean that the optimum route for him may be a non-standard process.

3.6 Liquefiers and liquefaction principles

The plants described in the preceding sections of this chapter are often able to produce some liquid products. However, they are limited to

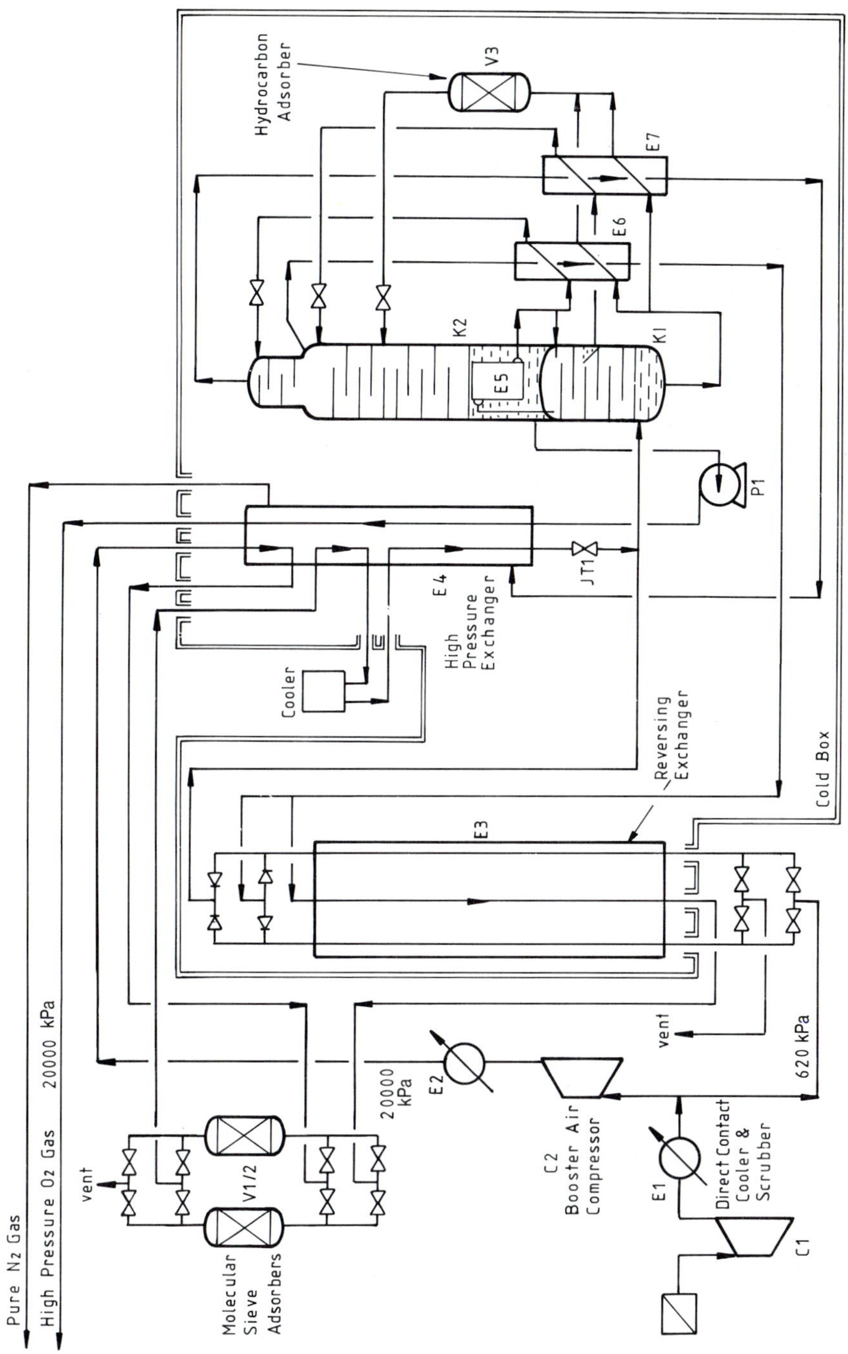

FIG. 3.8. A pumped liquid oxygen process for generating 20 000 kPa oxygen.

producing only small quantities economically (0–10 per cent of products), as they are primarily designed to produce gaseous products. Many of these plants have an expansion turbine which produces the necessary refrigeration to satisfy the cold balance around the process. Some of the high pressure processes only rely on the Joule–Thomson effect for refrigeration and require very high pressures to allow them to operate. There are several ways in which refrigeration can be supplied to a process to cool and/or condense a gas or mixture of gases. These can be grouped into the three fundamental principles used in commercial applications:

(1) the Joule–Thomson effect;
(2) compression/condensation and expansion of a refrigerant;
(3) expansion turbines or engines.

Most processes in cryogenic technology use one or more of the above principles to provide refrigeration in the same process. To prevent heat ingress into the colder parts, these processes are usually housed in a plate and frame structure called a 'cold box', which is filled with powdered or fibrous insulation. Nevertheless, even gas producing processes will require at least one of the above methods to maintain the low temperatures required to achieve and maintain the process conditions. This is because, no matter how well the cold box is insulated, some heat gets in. And since heat is exchanged between incoming and outgoing streams, the products are a few degrees colder than the incoming gas. Both of these effects need sustenance from a refrigeration source.

To gain an understanding of the three methods of generating 'cold', let us suppose we have a need to provide 1000 kcal h^{-1} of refrigeration at a level of −40°C and we have at our disposal the above three methods, any of the gases listed in Chapter 2, electricity, and cooling water at 25°C. Which is the best method of doing it? To assist us in gaining the answer the three methods should be studied.

3.6.1 *The Joule–Thomson effect*

Assuming we can select nitrogen as the working fluid we can design a cycle that would provide some refrigeration to an exchanger at a temperature not warmer than −40°C. Let us imagine a simple compression, heat exchange, and expansion system as shown in Fig. 3.9(a) and represented on a pressure–enthalpy (P–H) chart as in Fig. 3.9(b). To gain a reasonable Joule–Thomson effect across JT1 we will assume compression to 3000 kPa and expanding across JT1 from 2900 kPa to 180 kPa. The steps we now take are similar to those of a

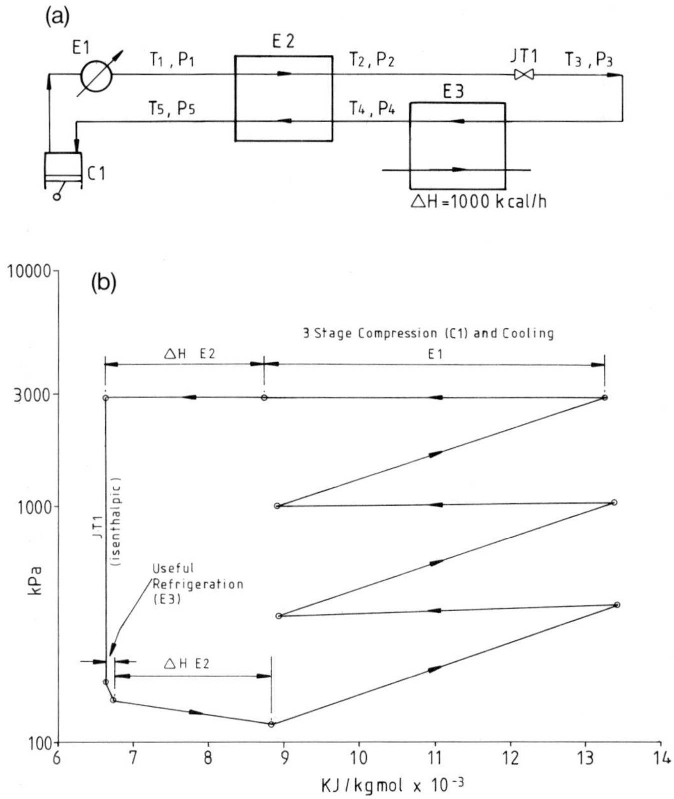

FIG. 3.9. (a) The process scheme and (b) the *P-H* diagram for a closed Joule–Thomson cycle.

cryogenic unit designer who aims to find the power consumption for a given cycle. For this design we can fix the following:

$T_1 = 35°C$ after watercooling; $P_1 = 2950$ kPa
$T_5 = 32°C$ with 3°C warm end difference; $P_5 = 120$ kPa
$T_4 = -40°C$, our desired design condition; $P_4 = 150$ kPa
$P_2 = 2900$ kPa; $P_3 = 180$ kPa
$\Delta H_{E3} = 1000$ kcal h^{-1}; C1 efficiency = 75 per cent

The unknowns in the cycle are the temperatures T_2 and T_3 and hence the cycle flow and power. Using a pressure–enthalpy (P–H) chart, or more accurately a computer with thermodynamic prediction techniques programmed into it, we can determine the unknowns and hence the cycle power consumption. First the heat balance around E1 is done to find T_2, and then the isenthalpic expansion across JT1. Once T_3 is known the enthalpy difference across E3 for the nitrogen can be found. Finally the flow through the cycle and the power of compression can be calculated.

The results for this calculation were that $T_2 = -32.9°C$, $T_3 = -43.5°C$, flow = 900 $Nm^3\ h^{-1}$ and power consumption by C1 was 150 kW. This means that to produce 1000 kcals h^{-1} at below $-40°C$, 150 kW of electricity is consumed. Obviously by repeating this calculation for a different pressure P_1 we can investigate the effect. By dropping P_1 to 800 kPa, we actually produce a negative result. This is because the amount of Joule–Thomson refrigeration produced in JT1 is less than the warm end loss in E2 (or $T_1 - T_5$). Therefore, for such levels of refrigeration there are alternatives which may be considerably more efficient, and it is the job of the cryogenic designer to find these.

3.6.2 *Compression, condensation, and expansion of refrigerant*

In order to compare this refrigeration cycle for the same level of $-40°C$ and 1000 kcal h^{-1} duty we cannot use nitrogen because it does not condense at $-40°C$. We could of course provide a cycle which would give some liquid nitrogen at $-180°C$ and 1000 kcal h^{-1} duty, which would achieve our aim of providing refrigerant at $-40°C$ or colder. However, this would be terribly inefficient because we would be supplying refrigerant to cool a fluid to a temperature difference of at least 140°C. Therefore, we shall select a refrigerant that does condense and evaporate at the desired conditions, and that can be R22 (dichloro-difluoro ethane) or propane. For this case we shall assume propane. The cycle is depicted in Fig. 3.10(a) and its representation on the *P–H* chart is in Fig. 3.10(b). The designer again has a number of known factors which he can start with, or features that he can fix.

1. We shall assume 20°C cooling water and that propane will be condensed at about 35°C (T_2). This sets the pressure P_1 at about 1350 kPa and P_2 at 1330 kPa. Assuming all propane is condensed, then the vapour fraction $V_{f2} = 0$.
2. To evaporate propane at $-40°C$ (T_3), P_3 must be 110 kPa.
3. $T_4 = -40°C$ and $P_4 = 101$ kPa at the compressor inlet.

The designer does not know one important factor, V_{f3}, because he needs to consult the *P–H* chart or run a computer program to determine the effect of isenthalpic flash across JT1 on V_{f3}. Once this is known the portion of refrigerant available to E2 is known and the designer can calculate flow and cycle power consumption. Assuming a 75 per cent efficiency for the compressor C1, the power consumption was found to be 0.83 kW to produce 100 kcal h^{-1} refrigeration at $-40°C$. Table 3.6 compares this cycle for various refrigeration temperatures and other cycles.

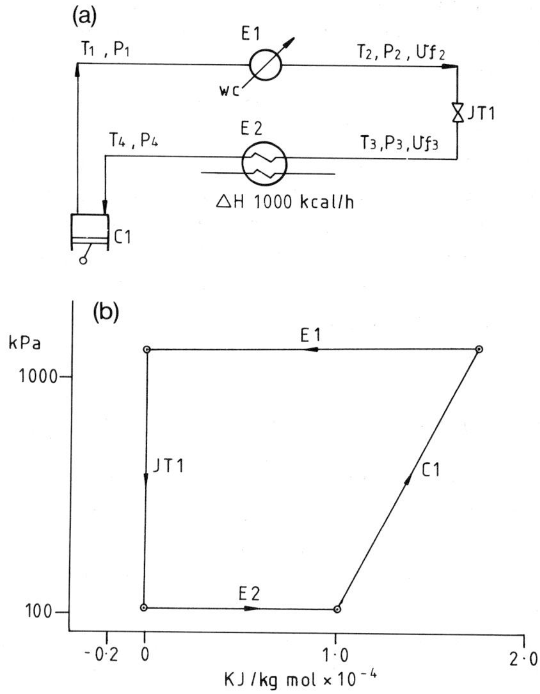

FIG. 3.10. (a) The process schematic and (b) the *P-H* chart for the propane refrigeration cycle.

3.6.3 *Expansion turbines and engines*

Expansion devices which produce work are a highly acceptable way of generating refrigeration at many temperature levels, ranging from −40°C to −269°C. Because the device converts expansion energy into work, it is considerably more efficient. Essentially we are comparing a Joule–Thomson valve where $H_1 = H_2$ and $\Delta W = 0$ to a turbo expander or expansion engine where $H_1 \gg H_2$ and available $\Delta W \gg 0$. Nitrogen is a suitable and widely used fluid for such devices and we shall use it in our theoretical example of generating 1000 kcals h^{-1} of cold at −40°C or lower. Figure 3.11(a) shows closed loop nitrogen compression expansion utilizing relatively low pressures, as encountered in practice. Figure 3.11(b) shows the various items and what duty they perform, represented on a *P–H* chart. The actual calculations for this example were done by computer program. Again, a number of parameters are known or can be fixed in order to proceed with the calculation.

$T_1 = 35$°C; $P_1 = 800$ kPa, post aftercooler
$P_2 = 770$ kPa; $P_3 = 180$ kPa

TABLE 3.6. Comparison of three refrigeration cycles. The basis is to provide 1000 kcal h^{-1} refrigeration at different levels.

Cycle type	Cycle fluid	Flowrate ($Nm^3 h^{-1}$)	Specific power consumption (kW/1000 kcal)	Required temperature of refrigerant (°C)
Joule–Thomson (3000 kPa)	N_2	895	150	−40
Expander–compressor (Comp. to 800 kPa)	N_2	60	6	−40
	N_2	75	8	−80
	N_2	95	10	−120
	N_2	130	14	−160
Propane cycle (Comp., cond., expansion, and evap.)	C_3H_8	9.1	0.83	−40
	C_3H_8	8.4	0.57	−25
	C_3H_8	8.0	0.42	−14
	C_3H_8	7.7	0.32	−4.5
	C_3H_8	7.4	0.23	+5.5

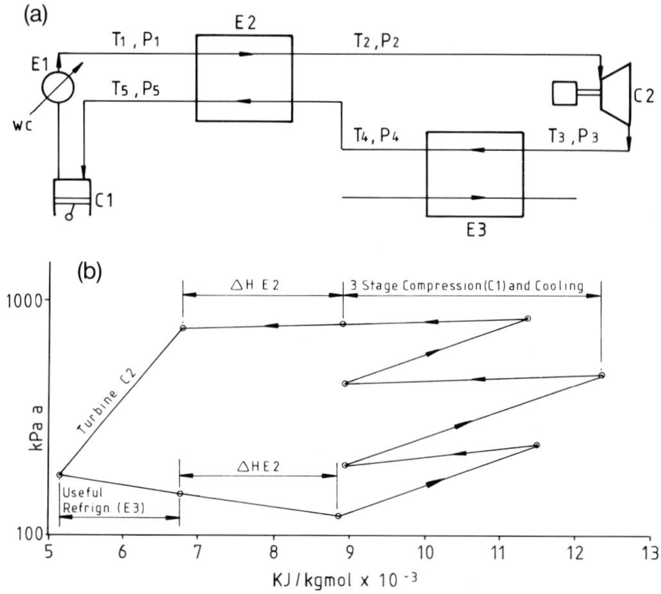

FIG. 3.11. (a) The process scheme and (b) the *P-H* diagram for a closed loop compression-expansion cycle.

$T_4 = -40°C$; $P_4 = 150$ kPa; we know the level of refrigeration we need
$T_5 = 32°C$; $P_5 = 120$ kPa at compressor inlet and 3°C warm end difference around E2
C1 efficiency 75 per cent; C2 efficiency 70 per cent; E3 duty 1000 kcal h^{-1}

The unknown parameters are T_2, T_3, cycle flow, and hence power. These can be calculated as previously described, by first doing a heat balance around E2 to find T_2, then calculating T_3 from which the load across E3 can be related to cycle flow. Thus power can be calculated.

If we compare the three schemes above we soon see which is the most efficient for this particular duty and the differences between them. Table 3.6 gives a comparison of power consumptions and clearly shows a winning condensing propane cycle.

The above three cycles are by no means fully optimized, but they are adequate for a comparison. They also illustrate how a designer proceeds in order to select the cycle type, the fluid, and working conditions. An experienced designer will start close to the optimum, but inexperience leads to a great deal of trial and error calculation.

3.6.4 *Integrated liquefiers*

In many cases it is known at the outset of the design that a given product must be predominantly in liquid form, and this feature can be built into the design of the air-separation unit. An example of such a case is a low-pressure oxygen process based on reversing heat-exchanger pretreatment, as shown in Fig. 3.12. In this process the air is compressed in C1 to about 650 kPa and after cooling passed into reversing heat exchangers E4 and E5 to remove residual water and CO_2. The clean air then passes into the base of the lower column K1. Enriched air from the base of the lower column is then reheated in E5 and passed through turbo-expander C3 to produce refrigeration, and it is then fed into the upper column K2. The double column operates in a similar fashion to that described above. The interesting point is, however, the liquefaction equipment which ties into the air-separation process. Gaseous oxygen product is withdrawn from the base of K2 reheated in E5 and E4, then compressed in C2 before being in part-routed to product usage. The remainder is cooled back down again in E1 and E2, and expanded across JT1 to form two phases which enter V1. The liquid phase becomes oxygen product for diversion into storage and the vapour returns to the base of K2 to close the liquefaction cycle. To provide the necessary liquefaction cycle cooling, part of the enriched air from the base of K1 is used to provide cold in E2 and joins up with the turbo-expander inlet. To boost the flow through the turbo-expander, C2, a flow of low pressure enriched air is reheated through E1 and passed to the main feed air compressor C1. This gas, being dry, does not raise the load on the reversing heat exchangers but does increase the cold production, thereby allowing one to make liquid.

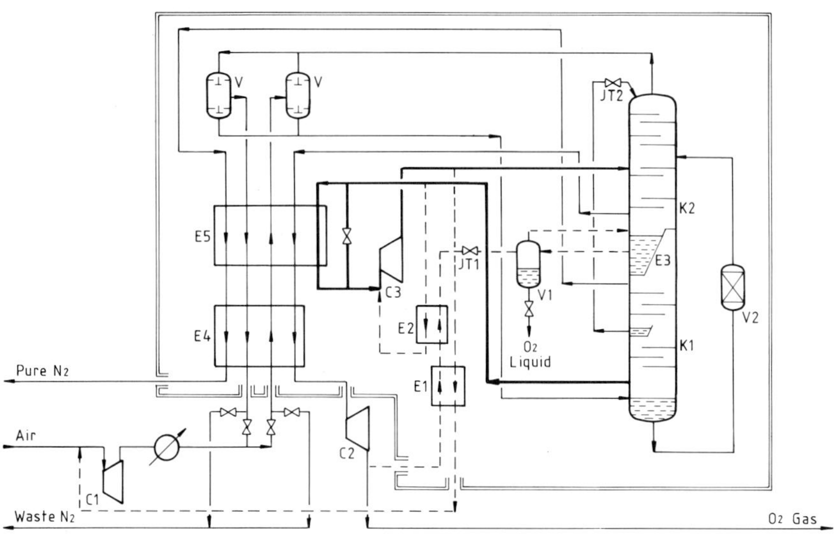

FIG. 3.12. A reversing exchanger plant with an O_2 liquefaction cycle (dashed).

This process is particularly suitable in situations where the liquid oxygen duty is variable. This liquefaction cycle can be easily turned off to revert to gas-only production. The process therefore illustrates that for a given set of circumstances an optimum integrated solution exists. However, given a constant load for liquid oxygen a more efficient process may be desirable.

3.6.5. *Simple liquefiers*

In many cases liquefiers are built separate to the main separation plant because of changing market conditions. In such cases they will admit gaseous product nitrogen, argon, or oxygen or more than one of them into a liquefier whose complexity varies depending on its capacity and frequency of use. The first of these liquefiers we shall call a 'simple' liquefier, as it utilizes only one expansion turbine to provide refrigeration. Figure 3.13 illustrates our 'simple' liquefier which actually liquefies both N_2 and O_2. The main liquefaction cycle uses nitrogen which it takes from the air separation unit as a gas at about 101 kPa, compresses it, and cools it in C1 and E4. Then combining with recycled nitrogen at about 500 kPa, it is further compressed and cooled in C2 and E5 to a pressure of about 3500 kPa. Then it is further boosted in the brake C3 of an expander compressor assembly C3–C4. The water-cooled gas from E6 enters the cryogenic section of the liquefier into E1 where it is cooled against returning medium-pressure nitrogen and against evaporating

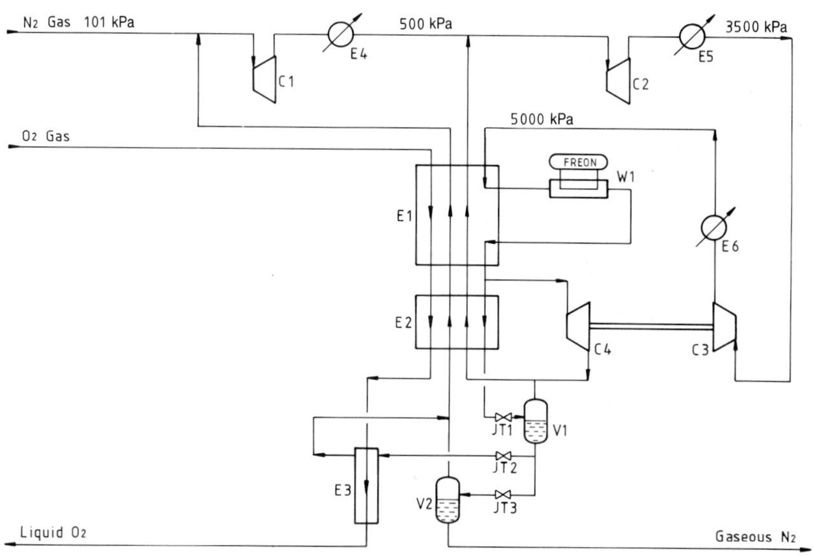

FIG. 3.13. A nitrogen and oxygen liquefier—a single stage turbine process.

Freon R22 from a packaged refrigeration unit similar to that described in Fig. 3.10. The high-pressure (HP) N_2 cools further in E1 and then some continues to be cooled supercritically in E2 while the remainder passes into the turbo-expander, which is designed to provide all of the refrigeration for the liquefaction of both the N_2 and O_2. After the HP nitrogen has been cooled below the critical temperature at above its critical pressure it is expanded to about 550 kPa in JT1. The fluid enters the subcritical regime and forms two phases: vapour and liquid. Part of the liquid is diverted via JT2 into E3 to condense oxygen product, and the remainder is flashed to low pressure in JT3, with the two resultant phases being separated in V2. The liquid N_2 from V2 is routed to storage and the low pressure flash is reheated in E2 and E1 before being recompressed with fresh feed in C1. The oxygen gas is left at relatively low pressure, cooled in E1 and E2. It is then passed into E3 which totally condenses it against liquid N_2.

It can clearly be seen that this process embodies all three of the refrigeration techniques described earlier. One interesting point is that although Joule–Thomson expansion showed itself to be very inefficient on nitrogen service in the examples shown in Fig. 3.9(a), it does show considerable merit in this case. The reason for that is because the 5000 kPa N_2 is being cooled well below its critical temperature. Then when it is again expanded in a Joule–Thomson valve it expands isenthalpically over a region where the enthalpy lines are very steep, i.e. large temperature change for small entropy rise. This basically means that in this region it is efficient to utilize Joule–Thomson expansion, while in warmer cases it is not. This point illustrates the knowledge which the designer must apply in selecting appropriate fluids and cycles.

3.6.6 *Complex liquefiers*

The liquefiers described in section 3.6.5 are relatively simple in that they use only one turbine and one Freon unit to provide refrigeration. Whilst this is highly acceptable for small to medium-sized or mixed-product liquefiers, it may not be acceptable in the cases where the capacity is large and power is expensive. In such cases it may be advantageous to add a second turbo-expander to the cycle to improve process efficiency. Addition of a second expander allows the refrigeration to be produced more where it is needed and not all at the coldest point, leading to a more efficient cycle. Figure 3.14 shows a schematic of a two-turbine liquefier on nitrogen only, although clearly oxygen could easily be incorporated in the same way as in Fig. 3.13. A third liquid expander again contributes to process efficiency as it reduces the number of Joule–Thomson valves which raise entropy isenthalpically.

The process in Fig. 3.14 resembles that of the Fig. 3.13 liquefier in that

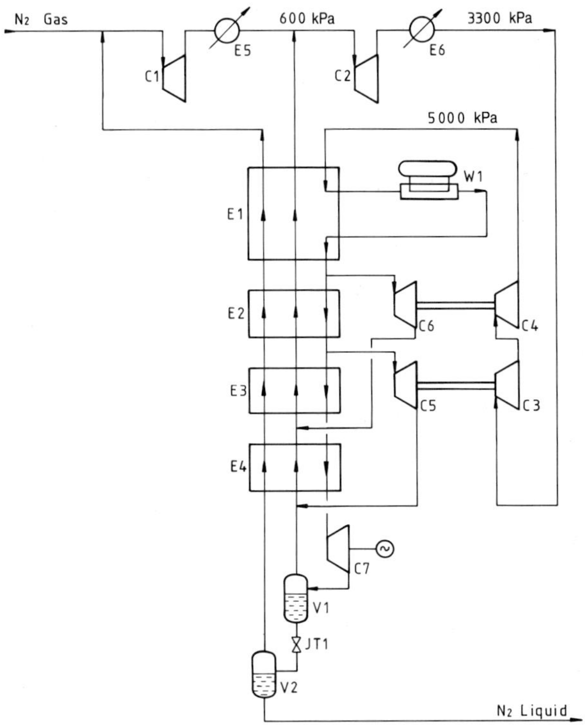

FIG. 3.14. A liquefier with two gas expanders and one liquid expansion turbine.

low pressure nitrogen makeup is compressed up to about 600 kPa in C1, mixed with medium-pressure (MP) recycle N_2, boosted to about 3300 kPa in C2 and then further boosted in two turbo-expander brakes: compressors C3 and C4. A Freon unit W1 is still used to provide refrigeration, the differences being in the number of exchanger blocks into which the cooling train is split, i.e. E1, E2, E3, and E4. After exchanger E1, a sidestream of HP N_2 is taken, expanded to MP in C6, and returned to E3 inlet for the MP return. This provides cold at an intermediate level. Further down, after E2, a second sidestream of HP nitrogen is expanded in C5 and returned to the MP reheat stream at the cold end of the cooling train at the MP N_2 inlet into E4. The now supercritical N_2, well below its critical temperature, is expanded in C7 and forms a vapour and liquid at the expander outlet. C7 duty is a little arduous but the technology is available and proven today to handle very dense 'vapours' at the inlet and condense a significant proportion at the outlet. The two phases are separated in V1 with the liquid continuing to the next stage of flash into V2, whence product liquid is formed for routing to the storage tank. The vapours from V1 and V2 are routed to the MP and low pressure (LP) N_2 reheat streams respectively.

This kind of liquefier can be easily adapted to accommodate oxygen liquefaction and is generally beneficial in larger sizes and high power cost situations.

Bibliography

BARRON, R. F. (1985). *Cryogenic Systems.* Oxford University Press.

BERNSTEIN, J. T. (1982). 'Analysis and correction of design deficiencies in a 500 T/D Oxygen Plant'. *AIChE Symposium Series,* **79**, (224), 24.

BOC (1979). The production of oxygen, nitrogen and inert gases. B.O.C., England.

DUCKETT, M., BANKS, R., and LIMB, D. (1983). 'Using nitrogen to enhance oil and gas recovery'. *World Oil,* (July), 125.

ECKLES, W. W., PRIHODA, C., and HOLDEN, W. W. (1981). 'Unique enhanced oil and gas recovery for very high pressure Wilcox sands uses cryogenic nitrogen and methane mixture'. *J. Petr. Tech.,* (June), 971.

GARDNER, J. B. (1979). 'Worldwide cryogenics—UK, cryogenics in BOC'. *Cryogenics,* (March), 131,

GAUMER, L. S. (1980). 'Small air separation plants'. *IOMA Broadcaster,* (May–June), 10.

HANDS, B. A. (1986). *Cryogenic engineering.* Academic Press, London.

HERSCH, D. J. and ABRARDO, J. M. (1977). 'Air separation plant design'. *Cryogenics,* (July), 383.

KELLER, G. E. (1986). 'New directions for an old field'. *AIChE Symposium Series,* **82** (250), 35.

LATIMER, R. E. (1967). 'Distillation of air'. *Chem. Eng. Prog.* **63** (2), 35.

SCOTT, R. B. (1966). *Cryogenic Engineering.* Van Nostrand, Princeton, N.J.

SPRINGMANN, H. (1978). 'The liquefaction of oxygen, nitrogen and argon'. *Linde Reports Sci. Tech.,* **28**, 23.

SPRINGMANN, H. (1980). 'The production of high pressure oxygen'. *Linde Reports Sci. Tech.* **31**, 3.

THOROGOOD, R. M. (1986). *Cryogenic engineering,* Chap. 16. Academic Press.

WILSON, K. (1978). 'Enhanced recovery inert gas processes compared'. *Oil and Gas J.,* (July 31), 162.

WOLFF, F., EYRE, D. V., and GRENIER, M. (1979). 'Oxygen plants: 10 years of development and operation'. *CEP,* (July), 83.

4
RARE GAS RECOVERY

4.1 Introduction

Most rare gas recovery processes are linked to other cryogenic processes, either because a major product is made or because it is necessary to do some cryogenic processing to get to the point where a rare gas can be recovered. By 'rare' we can infer that it is 'difficult to find' or in very low concentration. All of the gases considered here are present in air, and are usually recovered from it. Notice that the concentrations are indeed low: argon 0.934 per cent; neon 18 v.p.m.; krypton 1.1 v.p.m.; xenon 0.08 v.p.m. and helium 5.3 v.p.m. Only helium is generally recovered from natural gases because it is present in larger concentrations there (typically from 0.04 to 2 per cent, depending on source). Helium recovery from air separation is rarely practised and usually not economical.

Helium is, at times, the only component taken out of a natural gas mixture. However, many gases need calorific adjustment, e.g. de-nitrogenation, and those that contain helium often have a helium recovery unit attached to the de-nitrogenation process. The remaining rare gases are all recovered by adding a further treatment process to the main production unit—usually making oxygen as a basic product. Air separation is, in fact, the major source of rare gases, but the 1980s have seen an increase of rare gas recovery (particularly of argon) from ammonia synthesis purge gas. If one follows the ammonia process flow-sheet through (see Chapter 6) one finds that the rare gases are found in the purge gas as a result of air addition into the secondary reformer.

4.2 Argon recovery

4.2.1 *Argon recovery from air separation*

Argon has found favour as a high-grade inerting medium in welding and shielding in the steel industry. It is therefore a highly sought-after commodity which means that more and more air-separation complexes include recovery of argon. Argon's boiling point falls between those of nitrogen and oxygen and thus it tends to accumulate in the distillation system to concentrations reaching 15 per cent by volume. Therefore, if a sidestream is withdrawn at the appropriate point it can be treated effectively to recover pure argon. Figure 4.1 shows a typical double column of an air-separation plant to which an argon side-column has been added.

The argon take-off point is chosen so that the argon concentration is in the region of 9–15 per cent by volume and the nitrogen content is very low at about 100 v.p.m., the balance being oxygen. This off-take is fed to the base of the argon column, K3 where it is scrubbed of its main constituent—oxygen. The reflux for K3 is generated by partial re-evaporation of the oxygen-rich liquid feed in E2 prior to its entry into K2. The crude argon leaves the top of K3 at low pressure at a concentration of 95–97 per cent argon with the balance being nitrogen and oxygen. This stream is then reheated through the main process to the next stage of purification to high grade argon product.

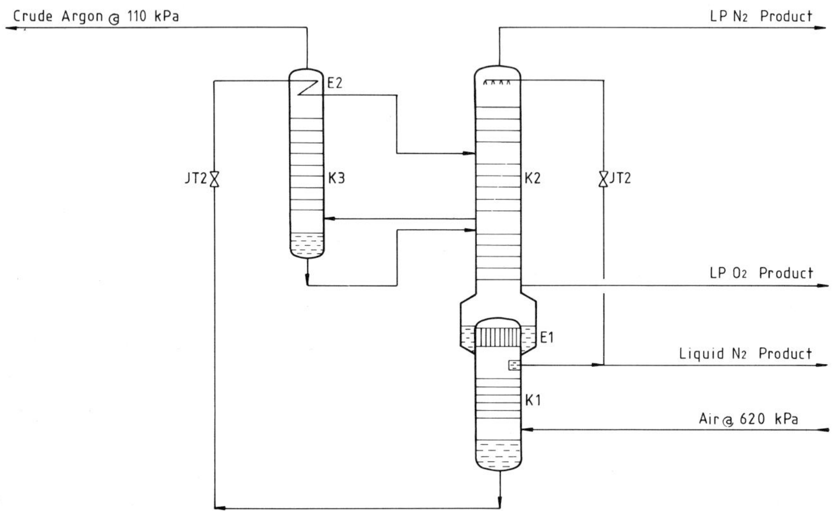

FIG. 4.1. The recovery of argon with simultaneous oxygen and nitrogen production.

There are two basic techniques for upgrading crude argon. The first is shown in Fig. 4.2 and relies on compression in C1, catalytic combustion to remove oxygen in W1, and distillation at low temperatures to make pure argon. After the combustion stage the gas is cooled in E1 to condense the majority of the water formed, dried in adsorbers V1 A/B, cooled and condensed in E2, and fed into V2. Any excess hydrogen added upstream of C1 is separated in V2 and forms the incondensibles which recycle through E3 and back to the suction of C1. The bottoms liquid from V2 will now be only argon with traces of nitrogen and no oxygen. This is then fed into K1 via JT1, distilled to remove N_2 to single figure p.p.m. levels. Liquid argon is withdrawn from the base of the

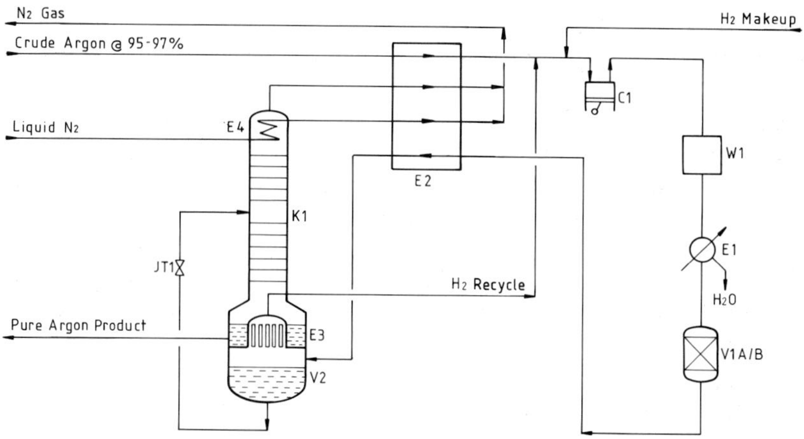

FIG. 4.2. Production of pure argon, using hydrogen combustion for oxygen removal.

column and N_2 is vented at the top. Reflux for the columns is provided by liquid nitrogen from outside, evaporated in E4, reheated in E2, and vented together with the K1 overheads nitrogen.

The second method is shown in Fig. 4.3 and in many ways differs from the combustion method described above. The crude argon is fed directly into K1 in this process and oxygen-bearing argon product is withdrawn from its base, cooled in E3, and passed into liquid cooled adsorbers (synthetic crystalline zeolite) to remove the oxygen. Two adsorbers are employed—one adsorbing, one being regenerated. The pure argon is then filtered in S1, passed into V2, and any flash is condensed totally by liquid nitrogen in E4. The total bottoms from V2 thus form the liquid argon product. Reflux and reboil are provided by liquid and gaseous nitrogen for E2 and E1 respectively from the main air-separation unit.

Each of the processes has the ability to make a high purity argon

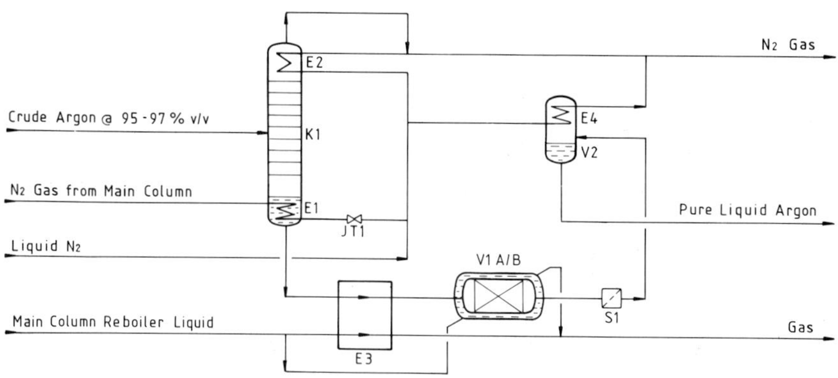

FIG. 4.3. Production of pure argon, using adsorption to remove oxygen.

product. The choice between the two is a rather subjective one, but there are some control advantages to the first route. In each case argon product of a purity of 99.999 per cent can be achieved, and, in some cases, even better.

4.2.2 *Argon recovery from ammonia purge gas*

Argon is present in air at a level of only 0.934 per cent, and is concentrated up in the main air separation process to levels of 9–15 per cent before being withdrawn for argon purification as described in Figs 4.1, 4.2, and 4.3. The conventional ammonia process yields a purge gas which contains 3–6 per cent argon which often goes into a hydrogen recovery process in the first instance. The residue gas from the hydrogen recovery can contain much higher levels of argon, typically 12–20 per cent, making it highly suitable for argon recovery. Figure 4.4 shows how argon builds up in an ammonia process which includes hydrogen recovery. The classical ammonia plant flowsheet is described in more detail in Chapter 6.

Usually the hydrogen and argon recovery are designed as a combined unit to reduce cost and increase process efficiency, and so it is not possible to isolate a separate 'fuel' gas containing 10–20 per cent argon at ambient conditions. However, the basic hydrogen recovery section produces a liquid of approximately the same composition which is then treated further to recover argon. Therefore, if through market

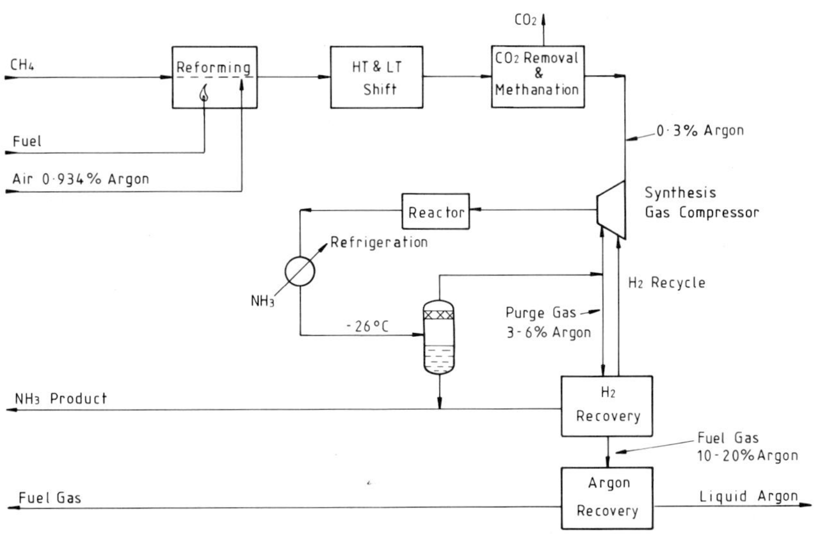

FIG. 4.4. A block diagram of the ammonia process, showing the build-up of argon.

considerations a hydrogen recovery unit already exists, it would still be practical to add an argon recovery unit. A typical argon recovery process including the hydrogen recovery is shown in Fig. 4.5. Chapters 6 and 8 give a more detailed account of pretreatment requirements for this type of gas. Table 4.1 shows a typical mass balance for the overall argon recovery process assuming an already dry and ammonia-free feed.

The dry feed gas enters E1 where it cools and partially condenses against return stream. A hydrogen-rich product vapour is separated from the liquids in V1 and this vapour returns back through E1 to the ammonia plant to improve its efficiency. The liquid in V1 contains argon at a relatively high level (in this case about 13 per cent) which has been achieved by a simple partial condensation step. This liquid is then taken

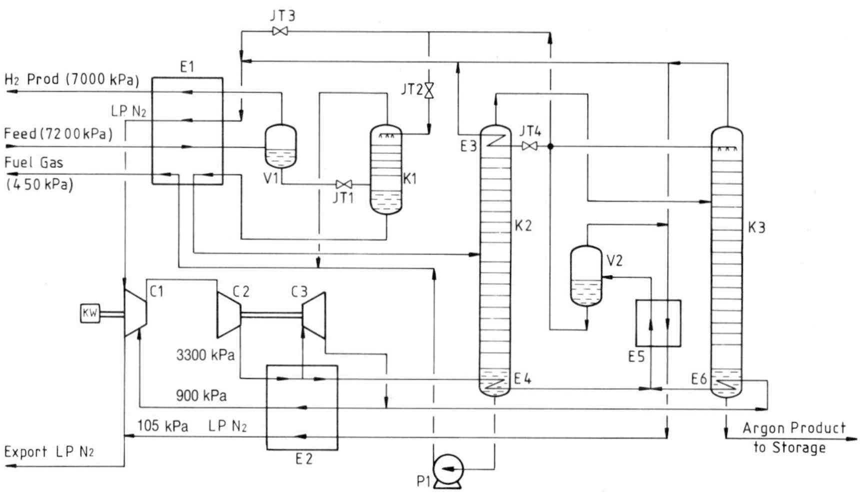

FIG. 4.5. An argon recovery process, treating ammonia purge gas.

TABLE 4.1. The overall mass balance for an argon recovery process treating NH_3 purge gas (figures are in mole per cent). Power consumption is estimated at 1600 kW.

	Feed gas	Hydrogen product	Argon product	Fuel gas	Nitrogen
H_2	62	92	—	21.7	0.5
N_2	21	7.0	—	13.0	99.0
Ar	6	0.7	100.0	1.0	0.5
CH_4	11	0.3	—	64.3	—
Flow ($Nm^3\ h^{-1}$)	18 000	10 963	732	2859	2446
Pressure (kPa)	7200	7000	140	450	110
Temperature (°C)	20	16	−181.5	16	25

to the argon recovery section of the plant. After its flashing across JT1 the argon-rich stream enters K1, where the vapour portion is washed against pure liquid nitrogen to reduce the argon content of the vapour leaving K1 to 1 per cent or less. This vapour then joins recovered methane and is reheated in E1 to form the fuel gas product.

The bottoms from K1, now containing a substantial proportion of the argon that was present in the feed, is partially re-evaporated in E1 and fed into the methane removal column K2. The object of this column is to allow *no* methane to pass in the overheads of K2 into K3. Reboil and reflux are provided to the column by 3200 kPa gaseous nitrogen and 135 kPa liquid nitrogen which condense and evaporate in the respective duties. The relatively pure methane liquid is pumped from the base of K2 in pump P1 and it joins the overheads from K1 to form the fuel gas. The overheads from K2 are fed into K3 to separate the nitrogen from the argon. At the feed to K3 the composition of the mixture is about two-thirds N_2, one-third Ar, and about 0.3–0.4 per cent H_2 with no methane. K3 is directly refluxed with nitrogen and reboiled with MP nitrogen in a separate exchanger. Liquid argon product of the correct purity is drawn from the base of K3 and either gravity-fed to horizontal storage tanks or pumped to vertical ones.

The nitrogen cycle is based on an open system where low-pressure N_2 is drawn from K3 and, together with evaporated refrigerant N_2, is reheated in the subcooler E5 and subsequently in both the cycle exchanger E2 and the main exchanger E1 before entering the LP suction of the cycle compressor C1. The nitrogen is compressed to about 900 kPa, combined with returning MP N_2 and further compressed in C1 to about 2900 kPa.

The nitrogen is then boosted to about 3300 kPa in the brake C2 of a compander set C2 and C3 (a combined compressor and expander on one rigid shaft). The dry nitrogen then cools in the cycle exchanger E2 and a side-draw of HP nitrogen is routed to turbo-expander C3 to provide refrigeration for the process. The expander outlet is at about 930 kPa and much of it reheats in E2 and back to the compressor C1. The remainder of it continues into the argon reboiler where, by providing reboil, it condenses itself and is then routed to the refrigerant accumulator V2 and subcooler E5. The HP N_2 that does not go to the turbine continues to be cooled in E2 and is then condensed in E4, the methane reboiler, whence after subcooling in E5 it passes into the liquid refrigerant reservoir V2. The reservoir serves to supply liquid refrigerant to K3 as direct reflux, to E3 via JT4, to K1 via JT2, and to E1 via JT3. Any boiled-off vapours after expanding refrigeration are collected in the spent LP N_2 header and returned through E5, E1, and E2 to the suction of the cycle compressor. Make-up to the cycle is provided by the nitrogen that is produced in K3, and any surplus is vented at the suction to C1. This type of process has a high once-through recovery, ~ 86 per cent, and when you account for the argon that is recycled to the ammonia plant in the hydrogen product, the recovery is around 95 per cent or more. An argon purity to meet the most stringent commercial specification can be produced with this process, making it a widely accepted source of argon. The USA has only a few of these units to date, with Western Europe, seeing the advantages more clearly, operating half a dozen or so. The leader in this kind of plant technology is Eastern Europe, which boasts around a dozen units in operation.

The process described above is a typical unit, and designers have introduced many minor and some more major differences from that covered above. However, the process devised by BOC Cryoplants for a plant in Oklahoma merits special mention because it involves a fundamentally different process route. Figure 4.6 shows a simplified flowscheme for this process. The first difference in this process is the treatment of the total feed gas in a wash column to produce a H_2:N_2 rich gas with an approximately 3:1 ratio. This allows the gas to be returned to the synthesis loop without a nitrogen deficiency. The second, and perhaps more fundamental, difference is that K2 yields N_2 as the top product, and Ar plus CH_4 as the bottom products, while the final column K3 separates argon from CH_4. This design has arisen perhaps by following a guideline or 'heuristic' in synthesizing distillation trains that says 'Aim to separate out the largest component first', the logic being that it reduces the flows to subsequent columns. This should theoretically give a lower power consumption. However, other things tend to counteract this—for example, the process becomes more complex as one cannot use an open cycle on the final and lowest-pressure column because it separates out Ar an CH_4. However, when carrying out a detailed analysis of both processes, little power advantage can be found

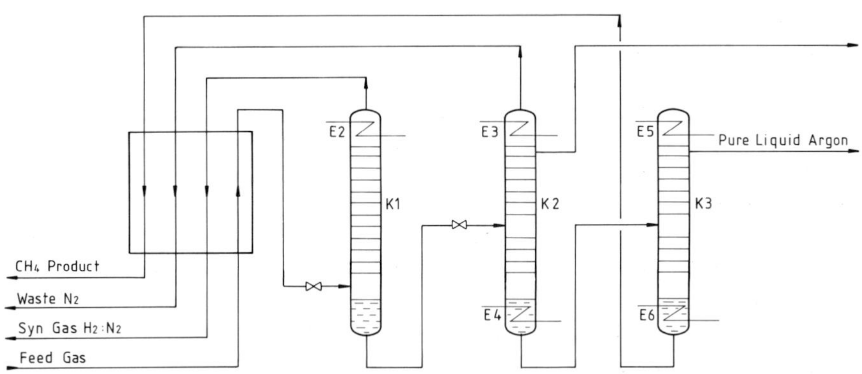

FIG. 4.6. A simplified flow-scheme of a successful variant on a typical argon recovery process.

in either. The former route has gained preference in use through experience rather than through a definitive process advantage.

4.3 Helium recovery and liquefaction

4.3.1 *Helium recovery*

All helium is extracted from natural gases by processes which are similar world-wide. In all cases the processes used are cryogenic. In most cases the natural gas also contains some nitrogen, and the primary objective of the plant is to remove the nitrogen from the hydrocarbons making the gas suitable for feeding to the national grid. The helium is extracted from the de-nitrogenation section of the plant as a purge of non-condensibles. This gas can then be further purified to produce liquid helium. Figure 4.7 shows a block diagram of a natural gas purification plant with a helium recovery unit included.

Because the main process is cryogenic, any compounds that freeze must be removed at the start of the process. Natural gas from the well-head will usually contain water and often acid gases. If the gas passes along a pipeline, heavy hydrocarbons may condense and, therefore, the first item one usually sees in the process is a condensate separator. The natural gas, which is usually at pressures between 2500 and 5500 kPa, is then treated with amine wash or another sweetening process to reduce carbon dioxide and hydrogen sulphide levels. The overheads from the

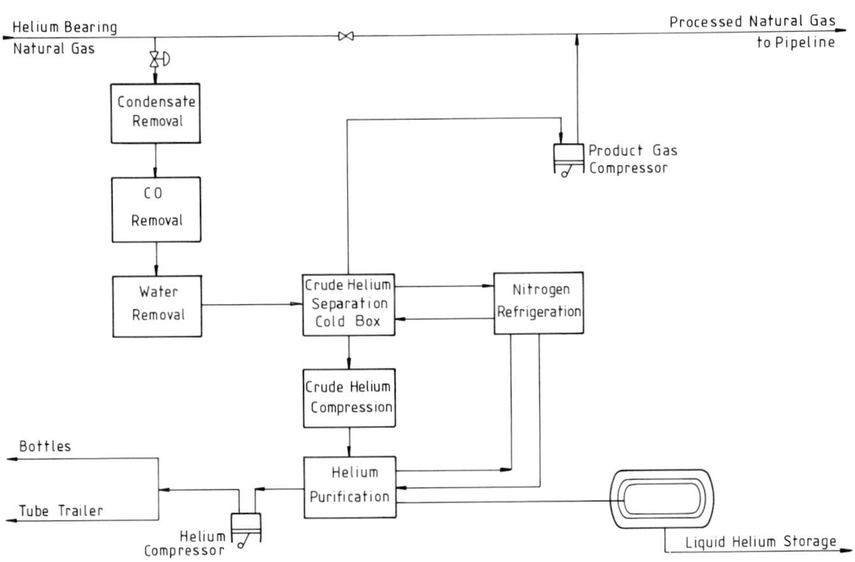

FIG. 4.7. A block diagram for the helium plant process.

sweetening unit are then dried using molecular sieves. These are often also required to remove the remaining acid gases. The clean, dry gas is then admitted to the cryogenic unit to remove nitrogen, if necessary, and, as a consequence, produce a helium-rich incondensible purge which can be further treated to recover pure helium. The purified natural gas of the correct calorific value is withdrawn from the crude helium separation 'cold-box' and compressed back into the gas grid. Depending on the composition and conditions of the raw gas, supplementary refrigeration may be required. This is supplied via a closed-loop nitrogen cycle or other compression–expansion cycle.

The crude helium stream is then compressed to a relatively high pressure and fed to the helium purification unit. The purified helium can then be liquefied or merely compressed into gas bottles for distribution to customers. The crude helium leaves the natural gas upgrading 'cold-box' at a purity of 50–80 mole per cent, with the balance being chiefly nitrogen, although methane can be found in some mixtures. A flow diagram of the process occurring in a typical crude-helium extraction unit is shown in Fig. 4.8.

The dry natural gas feed enters at a pressure of 2500–5500 kPa and is first cooled against returning product streams in a plate-fin heat exchanger E1 to condense out heavy hydrocarbons. The heavy hydrocarbons are separated out in V1 and the vapour from V1 continues being cooled in exchanger E2, and expanded to about 1400 kPa across valve JT2. The two phases resulting are separated in V2 with the vapour being sent to rectifier K1. The methane-rich liquids from V2 and K1 are

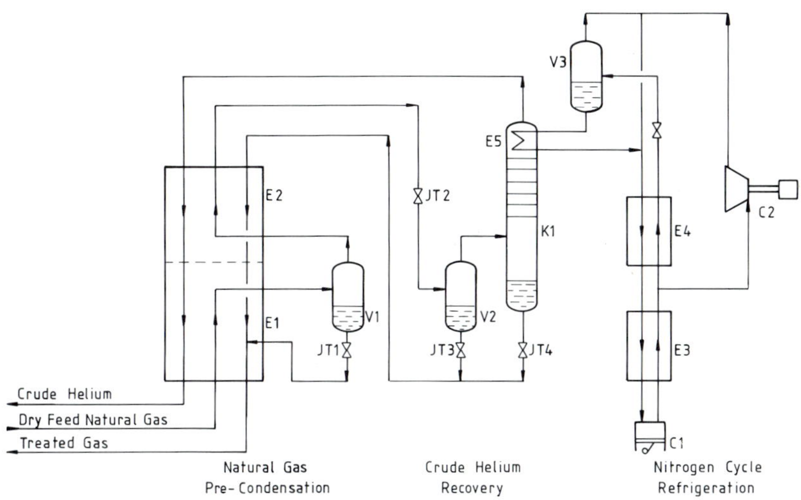

FIG. 4.8. A typical crude helium extraction process for low nitrogen content natural gas.

expanded across valves JT3 and JT4, respectively, and reheated in E2 to provide part of the refrigeration for the process. The heavy hydrocarbon liquids from V1 are expanded in valve JT1, mixed with low-pressure treated gas from E2, and further reheated in E1 to ambient temperature, forming the product reheated natural gas.

The rectification column K1 has a liquid nitrogen condenser which provides reflux for washing the vapour from V2, thereby reducing its methane content to very low values. Often the natural gas contains low levels of nitrogen and therefore does not require de-nitrogenation. Thus the mixture emanating from the top of K1 will essentially be 60–70 per cent helium and 30–40 per cent nitrogen with traces of methane. This crude helium, available at pressure a little above 800 kPa, can then be sent to a helium purification unit. In unusual cases where there is no nitrogen present in helium-bearing gas, one might use direct washing with liquid nitrogen in K1 to attain very low levels of methane. High levels of methane in crude helium may give rise to freezing problems in the final helium purification stages.

Natural gas containing substantial amounts of nitrogen needs upgrading to make it compatible with the natural gas grid. Crude helium recovery and de-nitrogenation can effectively be combined into one relatively simple process given the right feed conditions. As described in Chapter 2, Polish natural gas contains 42.75 per cent nitrogen and 0.4 per cent helium. Being available at 5500 kPa, it is ideally suited to this type of process.

Figure 4.9 shows a flowscheme for the crude helium extraction process used in this case. After removal of water and heavy hydrocarbons by adsorption, the gas is admitted into the plate-fin exchanger E1, expanded to about 2700 kPa in JT1, and is fed to the base of K1. Methane-rich liquid is taken from the base of K1, flashed to about 170 kPa in JT3, and fed into K2. Column K2 is mounted over column K1 so that the condenser, E2, for K1 is in fact the reboiler for K2, thus forming the classical double-column system. Nitrogen liquid is drawn as a side-stream from K1, flashed in JT2, and fed as reflux into K2. Liquid methane is taken from the base of K2, pumped in P1, and reheated in E1 before being routed to the natural gas grid.

A small amount of liquid nitrogen, which is drawn from the side of K1, flashed in JT4, and can be used as refrigerant in the final de-nitrogenation and crude helium recovery process, is provided by Joule–Thomson expansion across JT1, JT2, and JT3. Crude helium is withdrawn from the unit as the non-condensible purge from condenser E2, passes through a reflux exchanger E3, and runs back into the top of column K1. Then it passes to the final purification stages. At this point the helium is at a purity in excess of 85 per cent, with nitrogen being the balance. The stream is usually available at pressures of about 2000 kPa to 2500 kPa

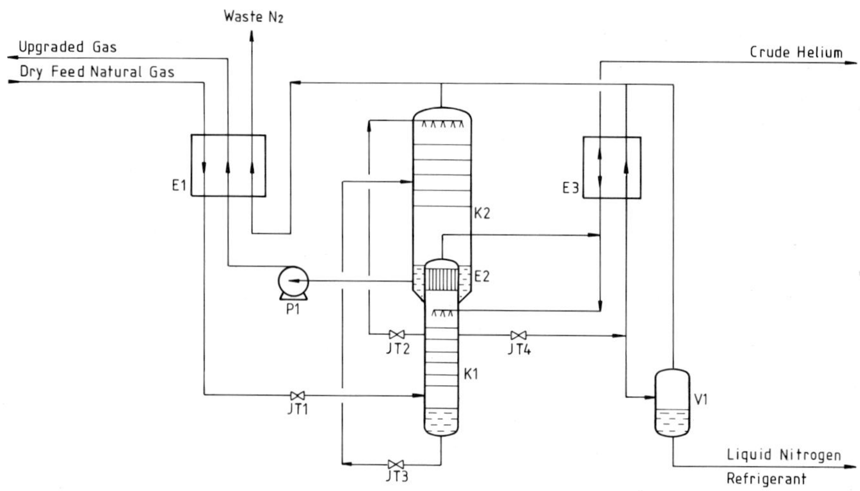

FIG. 4.9. A typical crude helium extraction process for high nitrogen content natural gas.

and about −190°C. This can then be routed at the helium purification and liquefaction section.

4.3.2 *Helium purification and liquefaction*

Two basic natural gas treatment plants are described in section 4.3.1. In the single-column process, crude helium may only be available at relatively low pressures and must be compressed in order to make its processing effective. In such cases, the crude helium is often compressed to very high pressures to make the product gas compatible with cylinder-filling requirements. Thus, by compressing the crude helium to about 18 000–20 000 kPa, the pure helium can be bottled without further compression. This is considered to be the optimum because helium compression is very inefficient and expensive. This process is shown in Fig. 4.10 and is highly suitable for gaseous helium production because of its simplicity.

Crude helium leaves the natural gas treatment plant as a dry and cold gas. Before it is admitted into the purification unit several operations are carried out. Firstly, traces of hydrogen are removed to about 0.001 per cent by catalytic oxidation at ambient temperature and elevated pressure. The gas can then be compressed, dried, and cleaned of any lubricating oil before being fed to the helium purification unit.

The dry, oil-free crude helium gas is then cooled in the main exchanger, E1, against cold return streams. Liquid nitrogen evaporating at atmospheric pressure is used in E2 to further cool this feed stream.

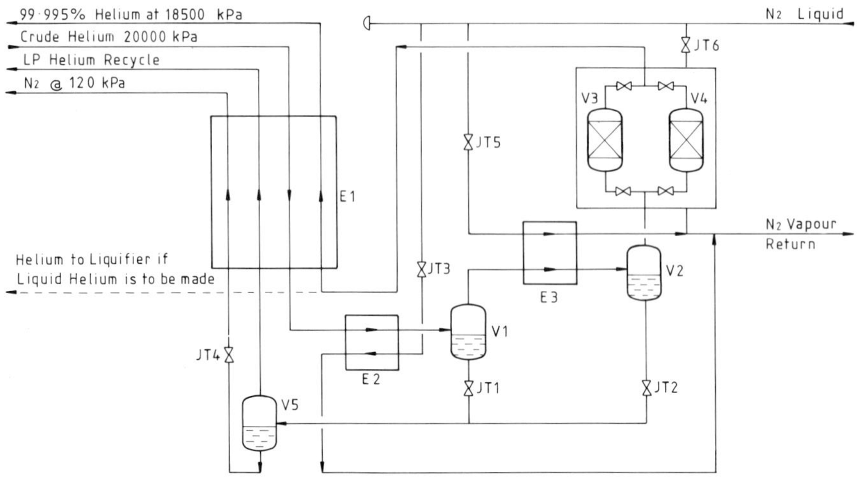

FIG. 4.10. The high-pressure helium purification process.

Nitrogen-rich liquid is condensed out and separated in V1 from the helium enriched vapour. This is further cooled against liquid nitrogen evaporating at about 20 kPa (sub-atmospheric). The resultant nitrogen-rich condensate is separated out in V2, whilst the 99.5 per cent helium-rich vapour is fed into low-temperature adsorbers V3 and V4. These adsorbers operated cyclically with one on line and the other being regenerated. They are filled with activated carbon and are maintained at very low temperatures by liquid nitrogen jacket cooling. The helium emanating from these adsorbers, usually in excess of 99.995 per cent by volume, is re-heated in E1 and bottled. The liquids from separators V1 and V2 contain typically 3–4 per cent helium dissolved in nitrogen. These are flashed in valves JT1 and JT2, releasing a helium-rich vapour which typically contains 35–40 per cent helium, and reheated in E1 before recompression and recycle back to the process. The liquid from V5, usually at about 450 kPa, is further flashed in JT4 to 130 kPa and reheated in E1 to ambient.

Nitrogen refrigeration is supplied to the process by liquid from a storage tank or closed-loop compression–expansion cycle which also makes liquid. Cycles of this type are described in more detail in Chapter 3. If a closed cycle is used, liquid nitrogen evaporated at atmospheric pressure is then returned to the suction of the cycle compressor. Liquid nitrogen evaporated at sub-atmospheric pressure is first compressed in a vacuum pump and then returned to the suction of the main cycle compressor.

The high-pressure process described above is suitable in cases where the crude helium leaves the natural purification plant at relatively low pressure. Often natural gas which contains helium also contains nitrogen which has to be removed to adjust calorific value. In such cases the process shown in Fig. 4.9 or similar is often used to de-nitrogenate the gas and at the same time extract a crude helium stream. This means that the crude helium is available at some 2500–2800 kPa and can be processed at this pressure to attain purities in excess of 99.995 per cent.

The low-pressure process lends itself to integration with a liquefaction cycle, and is shown in Fig. 4.11. The pure gaseous helium product is then admitted to the helium liquefaction cycle, which can take several forms. This cycle is described later. The intention in the purification process is to upgrade the helium from 85–95 per cent to 99.995 per cent helium. The main impurity is nitrogen, but the crude helium often also contains traces of neon and small amounts of methane and hydrogen. Because the hydrogen has a volatility quite close to that of helium, it cannot be removed by condensing it out like the methane and nitrogen, and so catalytic means are used.

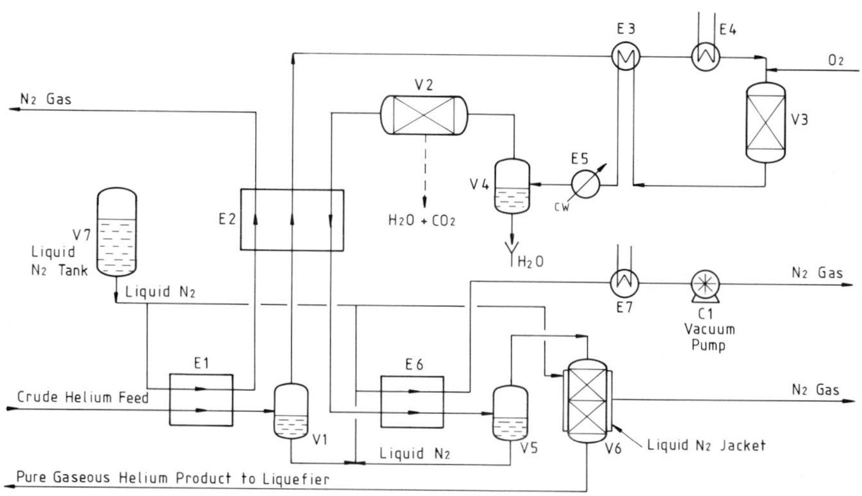

FIG. 4.11. The low-pressure helium purification process.

Taking this process step by step, the crude helium leaves the de-nitrogenation plant (see Fig. 4.9), as an inert purge stream containing approximately 85 mole per cent helium. The pressure of this stream is about 2700 kPa, at a temperature of around −160°C. The crude helium is first cooled and partially condensed against liquid nitrogen to increase the helium content. The nitrogen-rich liquid is either flashed and re-used as refrigerant, or returned to the lower column of the double-column de-nitrogenation plant. Sometimes, E1 is actually designed as a refluxing heat exchanger such that the crude helium cools upwards, allowing the condensed nitrogen to run back down the exchanger. This means that the liquid which condenses washes the vapour flowing up the exchanger, thereby reducing the methane content virtually to zero. This gives the process much more integrity as it eliminates any problems with carbon-containing molecules in the downstream process.

The two phases leaving E1 are separated in V1 and the helium-rich vapour is heated to ambient in E2 against treated crude helium. The feed stream is pre-heated in E3 and E4 before entering an oxidation reactor V3 which uses a platinum catalyst. Oxygen is added via air and the hydrogen is converted into water in an exothermic reaction. The reaction products are cooled in E3 and E5 down to about 35°C, thus condensing out water which is separated out in V4. The vapour from V4 is passed through molecular sieve adsorbers V2 to remove water and carbon dioxide. The dry gas is then cooled in E2 and further cooled and partially condensed in E6 against vacuum liquid nitrogen. The two phases are separated in V5 yielding a nitrogen-rich liquid, which can be used as coolant, and a vapour containing 98–99 per cent helium. This vapour is then passed into low-temperture adsorbers containing activated carbon and molecular sieve. The gas passes through the carbon first for bulk nitrogen removal and then the sieve for final polishing of the helium. Two adsorbers are used, one on stream and the other being regenerated. The final helium product leaves this unit in gaseous form at purities in excess of 99.995 mole per cent at low temperature and some 2300–2500 kPa.

The helium can then either be reheated in exchanger E2 to ambient and compressed into bottles, or preferably, admitted into a helium liquefaction cycle. There are many variants of helium liquefaction cycles based on combinations of Joule–Thomson refrigeration and work expansion. These have been developed by companies and individuals including Sulzer Brothers, Collins, Linde, McMahon-Gifford, etc. However, only typical schemes will be covered here.

Figure 4.12 shows a flow-scheme for a helium liquefaction unit successfully utilized on a de-nitrogenation plant in Poland. This unit takes crude helium from the lower column's inerts purge of de-nitrogenation process, as shown in Fig. 4.9, passes it through a helium purification and drying unit as per Fig. 4.11, and then admits the pure, dry helium to the liquefaction cycle described in Fig. 4.12. This liquefaction is a typical scheme applied in many plants in the USA and Europe.

The pure, dry helium enters the liquefaction section at about −191°C (81 K) and above 1650 kPa, and is reduced to about 1650 kPa in JT1 before being joined up with high-pressure (HP) cycle helium flow leaving E1 also at about −192°C (81 K). This HP helium stream continues to be cooled in E2 and E3 against colder low-pressure (LP) return helium which has been produced by expansion turbines C3 and C4. The HP

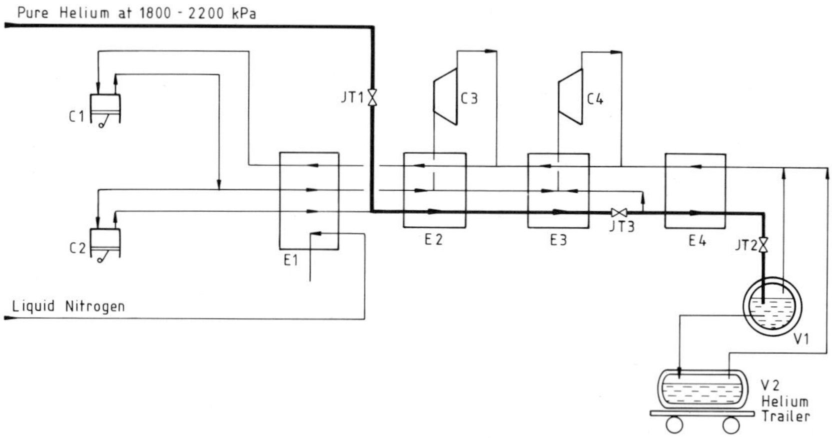

FIG. 4.12. A typical helium liquefaction process.

helium is then expanded in JT3 from about 1600 kPa and −267°C (6.1 K) to about 400 kPa and 5.9 K. Part of this medium-pressure (MP) helium is reheated in E3 and supplied to expansion turbine C4, and the remaining smaller portion is cooled further in E4 before being expanded in JT2 to about 125 kPa and 4.5 K, where it becomes a two-phase mixture (up to the inlet of JT2 it has been supercritical). The liquid is separated out in V1 and then passed to trailers V2 for shipment. The vapour flash from V1 and boil-off from V2 is returned as coolant into E4, E3, E2, and E1 before being supplied to the suction of helium compressor C1. At the LP stage, C1 of the helium compressor compresses the helium up to about 900 kPa and, after ambient cooling, part of it is admitted into the HP stage C2 and part of it is cooled in E1 and E2. The MP helium flow splits in E2, whence part of it continues to be cooled in E2 and E3 and the remainder is expanded in C3 to provide refrigeration for the process. The same situation occurs in E3 and C4, where again a significant proportion of the MP helium is expanded to provide refrigeration at about 10 K into E3. This low-pressure helium mixes with the return stream from E4, reheats through E3, and mixes with C3 outlet at about 31 K before returning through E2 and E1 into the suction of C1. Liquid nitrogen supplementary refrigeration is used in E1 to ensure that the cold-end of E1 stays at about 81 K and the cycle provides refrigeration below 81 K.

Variants of the above process are used with only one expansion turbine or engine for relatively small units where the designer has felt that the benefit of the higher process efficiency with two turbines cannot be justified. Alternatively, for larger processes one might consider expanding all of the medium-pressure stream from say 800 kPa to 400 kPa in C3 and passing the turbine outlet from C3 as medium-pressure feed to E3 at about 400 kPa. Often, the complexity or simplicity of the process, together with power cost and size, influence the designer's choice of process.

Most cryogenic units are housed in a carbon-steel plate-and-frame structure which is then filled with mineral wool or free-flowing powdered insulant such as Perlite. These units are generally purged with an inert gas such as nitrogen. Exceptions to this rule are generally hydrogen, neon, and helium liquefaction units which, because of the very low temperatures at which they operate, need some special consideration. Because of the high cost of producing refrigerant at temperatures of −268°C, the heat inleak must be reduced to an absolute minimum. This is done by encasing the helium liquefier in a similar carbon steel plate-and-frame structure but with the following conditions:

- it is filled with powdered insulant;
- the insulation cavity is evacuated to high vacuum;
- reflective foil is placed over cryogenic parts.

Each of these reduces heat inleak by convection and conduction through insulant and purge gas, as well as reducing radiation heat inleak. Even the expansion turbines and engines are mounted in similar conditions to ensure the best possible efficiency for the process.

4.4 Krypton, neon, and xenon recovery

4.4.1 *Krypton and xenon recovery*

4.4.1.1 From air-separation plants

Air contains just 1.1 v.p.m. krypton and 0.08 v.p.m. xenon. Because Kr and Xe have normal boiling points of −153.35°C and −108.1°C respectively they tend to accumulate with the oxygen product, which has a normal boiling point of −182.97°C when pure. This means that the logical place to withdraw these two components is from the base of the upper column in a double-column air separation process. Theoretically, if all of the oxygen were recovered as product the krypton and xenon concentration in it would only be 5.3 v.p.m. and 0.38 v.p.m. respectively. Therefore, some pre-concentration of krypton and xenon is carried out prior to pure product recovery. Figure 4.13 shows the arrangement of a double-column air-separation plant with a krypton and xenon enrichment section, which produces a krypton and xenon 'concentrate' containing 0.1–0.2 per cent krypton and xenon with the balance being, in the main, oxygen.

K1, K2, E1, and E2 show the typical arrangement of the double-column air-separation unit. In order to recover krypton and xenon it is

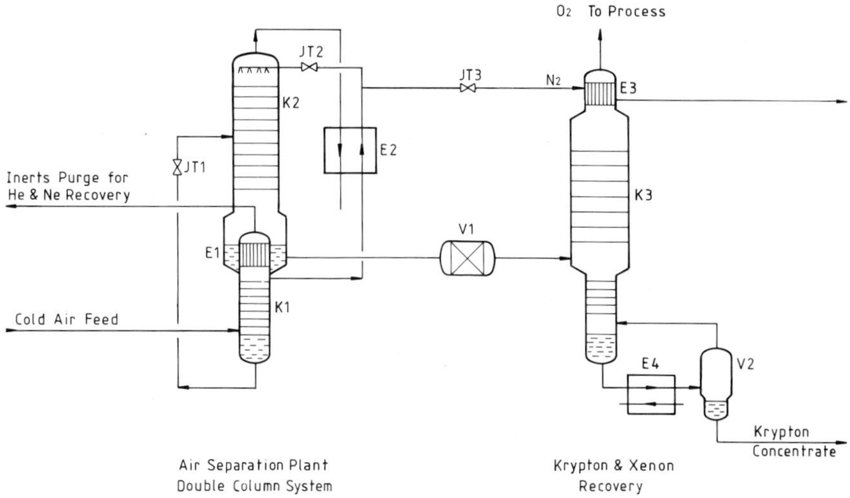

FIG. 4.13. The recovery of krypton and xenon concentrate from air.

most advantageous to withdraw a portion of the oxygen product as a liquid from the base of K2, pass it through acetylene adsorbers V1, and route it to the krypton and xenon concentrator column K3. Condensing duty is provided by liquid nitrogen in E3 and reboil by warmer process nitrogen in E4. The krypton and xenon concentrate leaves V2 at only 0.1–0.2 per cent by volume in liquid oxygen. Therefore, significant further treatment is required. This is achieved by reheating the krypton and xenon concentrate through the main heat exchanger train to ambient conditions and admitting this to the next stage of processing.

The 0.1–0.2 per cent Kr + Xe concentrate is upgraded into a product containing up to 80 per cent Kr by the process depicted in Fig. 4.14. In this process the krypton and xenon concentrate is passed into a buffer vessel, compressed to about 500 kPa in C1, and passed into a contact furnace, F1, to remove any hydrocarbons which would easily combust with the oxygen present. Many of the reactions at this stage are non-cryogenic, i.e. they occur at above-ambient temperatures. The gas is then scrubbed with caustic solution to remove carbon dioxide in K1 and dried in V2A/B. After drying the gas is cooled in E1 and admitted to column K2, where the krypton and xenon mixture is separated from oxygen which leaves the top of the column. The oxygen is then reheated against feed gas in E1 and returned to the oxygen plant from where the Kr + Xe concentrate came. The Kr + Xe product leaves the base of column K2, is treated in a furnace to remove traces of hydrocarbons, and then enters buffer tank V3 via E4. A high-pressure air cycle is used to apply reboil and condensing duty to column K2. The air, usually at a pressure greater than 10 000 kPa, is passed through E5 and then provides reboil in the

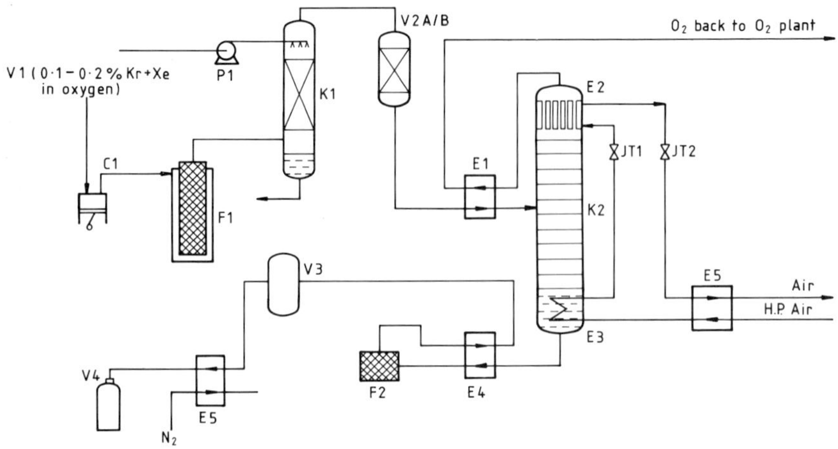

FIG. 4.14. A krypton upgrading plant producing a mixture of up to 80 per cent krypton.

base of column K2 via E3. The air mostly condenses in E3, is expanded in JT1, and is admitted as two phases into condenser E2 to provide reflux to the column. The air vapour leaving the cycle side of E2 is then further expanded across another Joule–Thomson valve JT2 before being vented after reheating in E5. Krypton is drawn from V3 batchwise into E5 which condenses the krypton and xenon product and allows it to be fed into the bottles, V4. Measured amounts are charged into the bottles allowing the pressure to rise to 4000 kPa upon vaporization through natural heat inleak. As the quantities of krypton and xenon are very small, they are generally recovered batch-wise, e.g. during a day shift—hence the inclusion of the various buffer storages V1 and V4. Krypton bottling may only be done over a few hours because of the very small volumes involved. Krypton can be separated from xenon by fractional distillation if desired, but again the process is operated batch-wise because of the very small quantities involved (the krypton and xenon mixture only contains about 7 per cent xenon).

4.4.1.2 By argon recovery units from ammonia purge gas

Krypton and xenon concentrate can also be effectively recovered from an argon recovery process, as described in section 4.2.2. In this process krypton accumulates with the heaviest of a four-component mixture of H_2, N_2, Ar, and CH_4, i.e. methane. Therefore it is the methane product which contains nearly all of the krypton and xenon that enters the ammonia plant and hence the purge gas. To operate a krypton and xenon recovery process successfully the methane product should be drawn off as liquid and gas from the base of the methane separation column in the argon recovery plant. Figure 4.15 shows the draw-off point and the initial processing of the krypton and xenon mixture.

The krypton and xenon recovery process takes methane from the base of the methane removal column of the argon plant depicted in Fig. 4.15. This methane draw-off is then processed in the continuous column K1, shown in Fig. 4.16, to enrich the Kr/Xe content to a few per cent. The Kr/Xe is further upgraded to about 80 per cent product in one of two batch columns K2A or B. This column is operated batch-wise because of

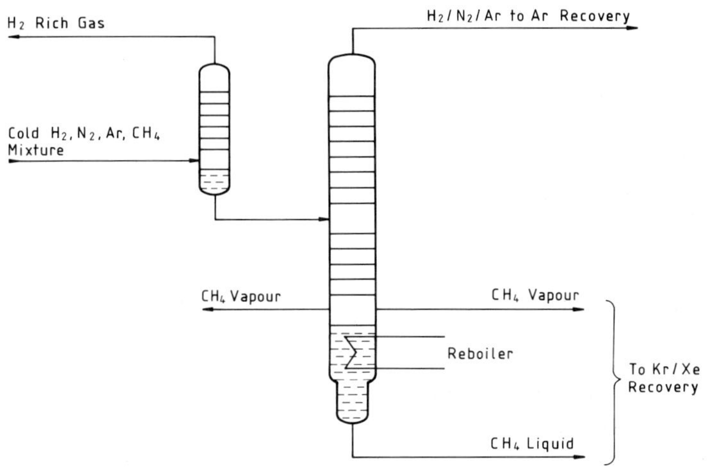

FIG. 4.15. Crude krypton and xenon recovery from ammonia purge gas.

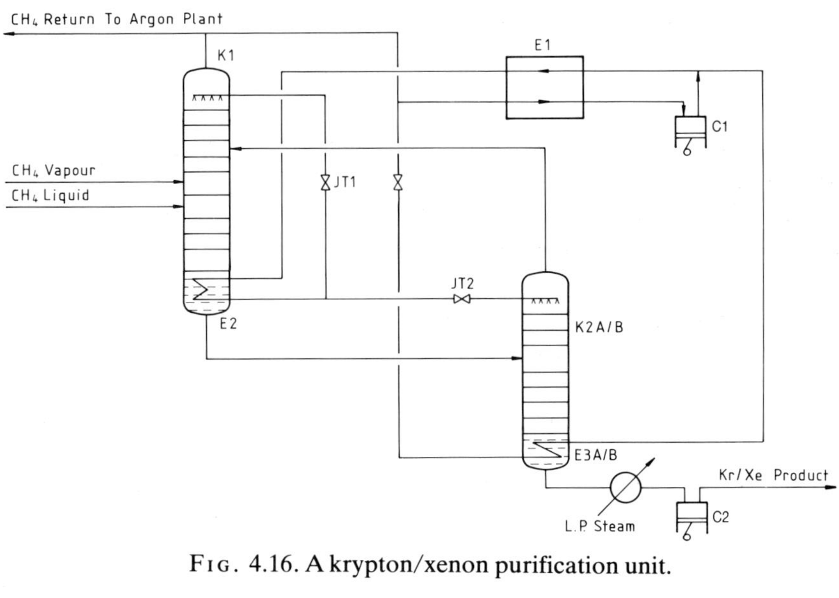

FIG. 4.16. A krypton/xenon purification unit.

the small volumes of crude Kr/Xe formed in K1. Refrigeration and reboil are provided by an open CH_4 cycle which compresses CH_4 in C1 to about 350 kPa. The CH_4 then is cooled against low-pressure return CH_4 in E1 and is used as reboil in K1. The methane itself condenses in E2, expands in JT1, and the two phase CH_4 is utilized as reflux in K1 on a continuous basis. The batch columns K2A/B are operated by taking a stream of CH_4 vapour from K1, passing it through E3A/B as reboil, and returning it through E1. Reflux is provided by using JT2 intermittently. The whole process can be self-contained in cold box and added on to existing facilities, provided tie-ins on the argon plant are in place.

4.4.3 *Neon recovery*

Neon is recovered essentially from air-separation plants, and to make it economic only larger units are considered because only 18.2 v.p.m. neon is present in air. Because its normal boiling point is much lower than N_2 or O_2, neon does not condense with any of the normal air-separation plant products. Therefore, it is drawn off as an inerts purge from the top of the condenser of the lower column, K2, of a double-column system, as shown in Fig. 4.17. Without this purge the performance of the condenser E1 would deteriorate as Ne and He built up over a period of time in its top parts. On small plants this inerts purge is withdrawn periodically, reheated through the process, and vented to atmosphere. On larger plants it becomes economic to recover neon and its commercial production began in the 1960s. The inerts purge may contain about 2.1 per cent neon, 0.8 per cent He, and the rest is N_2 with traces of H_2 and O_2. Its

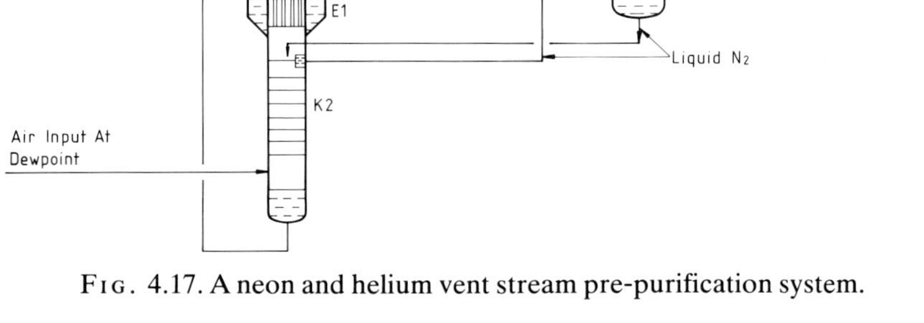

FIG. 4.17. A neon and helium vent stream pre-purification system.

flow is controlled across JT1 into V3, where it is cooled in E2 (within V3) by low pressure liquid nitrogen. Mainly nitrogen condenses out in E2, runs back into V3, and by gravity is returned into K2. By keeping the pressure of the inerts as high as possible, a significant amount of N_2 is condensed leaving a crude Ne + He mixture of the following approximate composition (v/v): 46 per cent Ne, 19 per cent He, 33 per cent N_2, 2 per cent H_2, and traces of oxygen. Liquid nitrogen refrigerant is supplied to E2 via JT2 from a side draw off-take from the main K1 reflux, prior to its subcooler E3.

The crude Ne + He mixture is then reheated and passed into the final purification and upgrading process, as shown in Fig. 4.18. In this process hydrogen is first removed by catalytic combustion with oxygen in W1, and after cooling and drying in E1 and W2 the gas is cooled by liquid N_2 in V1. A low temperature adsorber in V1 removes N_2 and excess O_2, leaving the impurities at about 1 per cent. A secondary bed (not shown) reduces these to trace amounts, leaving about 71 per cent Ne and 29 per cent He. This stream is then cooled with a closed-loop H_2 liquefier cycle W3. Initial cooling is carried out in E2 and then E3 condenses a substantial part of the neon, leaving a vapour containing only 20 per cent neon and 80 per cent helium. This can either be reheated in E2 and processed elsewhere, or simply vented. Normally it is not economic to recover helium from this vent stream as there are many natural gas streams which are highly favoured for recovering helium.

The neon product leaves the unit at a purity of about 99.99 per cent by volume as a liquid which can either be bottled at high pressure or stored in dewars. The recovery of neon in the process shown in Fig. 4.18 is over 95 per cent.

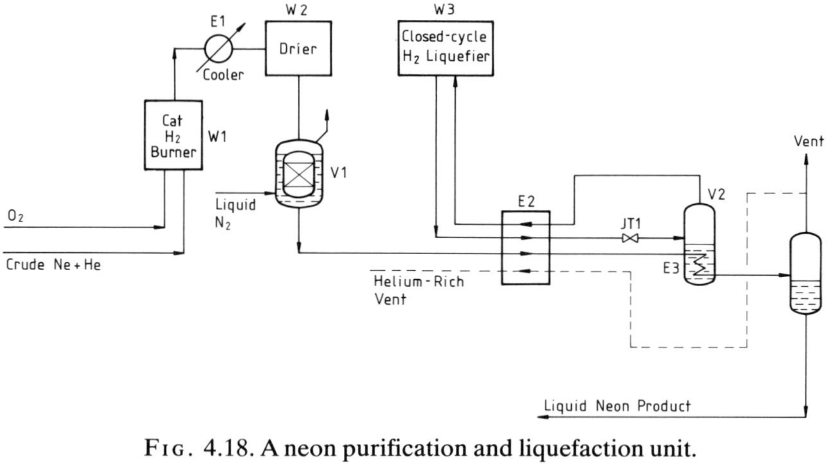

FIG. 4.18. A neon purification and liquefaction unit.

Bibliography

BARRON, R. F. (1985). *Cryogenic systems.* Oxford University Press.

BERRY, R. I. (1979). 'Argon recovery: aid to ammonia economics'. *Chem. Engng*, (July 16), 62.

BLIGH, B. R. (1983). *A process for removing CH_4 and argon from crude ammonia synthesis gas.* UK Patent Appl. 21 32 328A.

CRAWFORD, D. B. and NORENBURG, J. C. (1975). *Separation of low-boiling gas mixtures.* US Patent 3915680.

FABIAN, R. (1982). *Verfahren und Vorrichtung zum Zerlegen von Synteseabgas.* European Patent 0081849.

FABIAN, T. (1987). 'Linde development work on the NH_3–purge gas separators'. *Linde Reports Sci. Tech.* **42**, 12.

HALFON, A. (1976). *Procèdè et installation de traitement de gaz lors d'une prèparation de gas destinès à la synthèse d'ammoniac.* French Patent 2368435.

HANDS, B. A. (1986). *Cryogenic engineering.* Academic Press, London, England.

ISALSKI, W. H. (1979). 'Scope to improve NH_3 processes'. *Chemical Age*, (Nov. 30), 19.

ISALSKI, W. H. and ASHTON, G. J. (1981). 'Rare gases and the ammonia producer'. *Fertiliser Intl. Conf.*, Paper 7, p. 126.

JONES, J. K. and STACEY, J. M. (1974). *Cryogenics*, April, 198.

KREPSCHOT, R. H., BIRMINGHAM, B. W., and MANN, D. B. (1968). *Technology of liquid helium.* National Bureau of Standards Monograph 111.

LIMB, D. I. (1974). 'Poland's natural gas will fuel major helium buildup'. *Chem. Engng*, (Dec. 9), 80.

RATHMANN, U. and GRIMM, P. (1982). *Linde Reports Sci. Tech.* **34**, 14.

SPRINGMAN, H. (1982). 'Methods for argon recovery to meet increased demand on the argon market'. *A.I.Ch.E Symposium Series,* **79**(224), 12.

STREICH, M. (1971). *Removal of nitrogen and recovery of helium from Dutch Natural Gas.* International Cryogenic Conference, Brighton, England, British Cyrogenics Council.

USJUKIN, I. P. (1965). *Plant and machinery for the separation of air by low temperature methods.* Pergamon Press, Oxford.

5

REFINERY OFF-GAS PROCESSES

5.1 Introduction

A typical refinery is very complex and consists of many operations producing a variety of products and intermediates from crude oil feedstock. No two refineries are the same because of the feedstock and product varieties involved. In order to understand the areas where cryogenic processes may have an impact, it is necessary to describe a typical oil refinery and its products.

5.2 Oil refineries

Figure 5.1 shows a block flow diagram of a moderately integrated modern refinery. Crude oil is preheated and admitted into an atmospheric distillation unit where it is separated into light wet gases, butanes, the full range of unstabilized gasolines, heavy naphthas, kerosene gas oils, and topped crude. The topped crude is usually sent to a vacuum distillation tower which separated it into three basic products: vacuum gas oil, lube base stocks, and reduced crude bottoms. The reduced crude bottoms are thermally cracked in a delay coker to produce wet gas, coker gasoline, gas oil, and coke.

The atmospheric and vacuum crude unit gas oils and coker gas oil are used as feedstocks for catalytic cracking or hydrocracking units. These units crack the heavy molecules into compounds which boil in the automotive fuel and distillate fuel ranges. The products from the hydrocracker are saturated. The unsaturated catalytic cracker products are saturated and thus improved in quality by catalytic reforming or hydrotreating.

The gasoline streams from the crude tower, coker, and cracking units are fed to the catalytic reformer to improve their octane numbers. The products from the catalytic reformer are blended into automotive fuels for sale.

The wet gas streams from the crude unit, coker, and cracking units are fractionated in the vapour recovery section into fuel gas, liquefied petroleum gas (LPG), unsaturated hydrocarbons (propylene, butylenes, and pentenes), normal butane, and isobutane. The fuel gas is burned in refinery furnaces and the normal butane is blended into automotive fuel

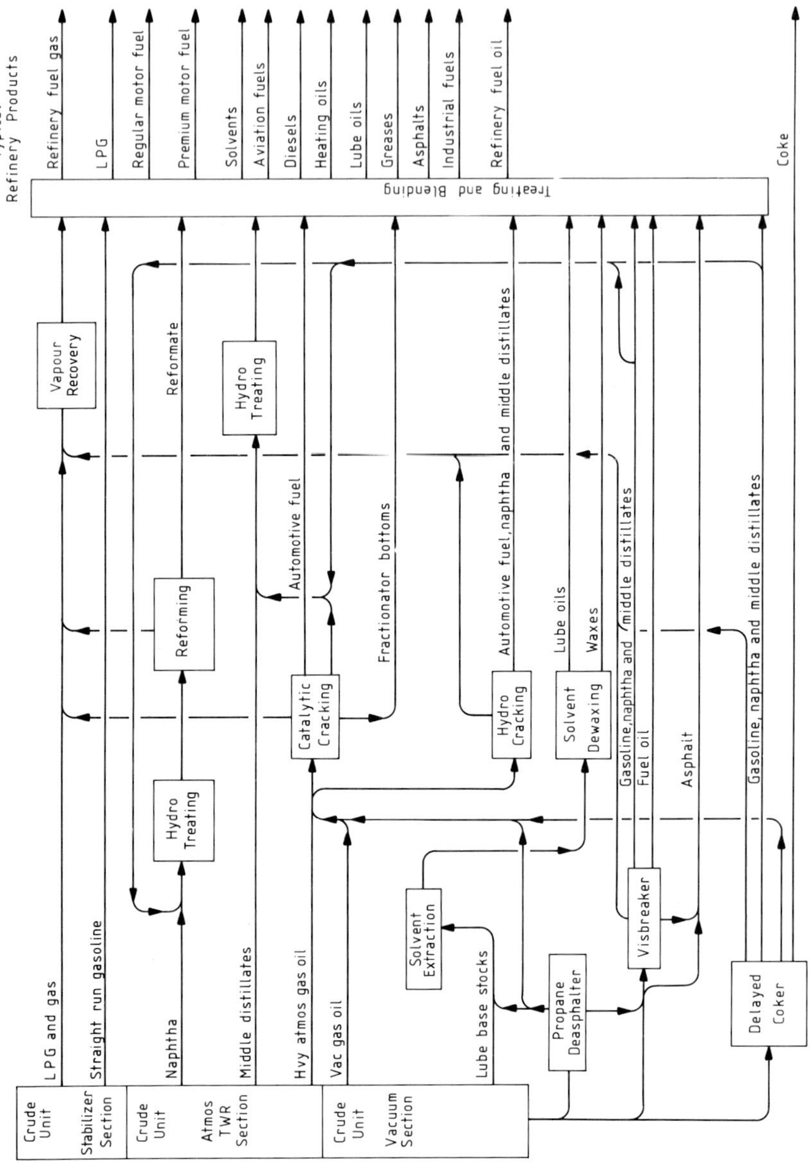

FIG. 5.1. A refinery flow diagram.

or LPG. The unsaturated hydrocarbons and isobutane are sent to the alkylation unit for processing.

The alkylation unit uses either sulphuric or hydrofluoric acid as catalyst to react olefins with isobutane to form isoparaffins which boil in the automotive fuel range. The product is called alkylate, and is a high-octane product blended into motor and aviation fuels.

The middle distillates from the crude unit, coker, and cracking units are blended into diesel and jet fuels and furnace oils.

In some refineries the heavy vacuum gas oil and reduced crude from paraffinic or naphthenic base crude oils are processed into lubricating oils. After removing the asphaltenes in a propane deasphalting unit, the reduced crude bottoms are processed with the heavy gas oils to produce lube-oil base stocks.

The heavy vacuum gas oils and deashphalted stocks are first solvent-extracted to remove the heavy aromatic compounds and then de-waxed to improve the pour point. They are then treated with special clays to improve their colour and stability before being blended into lubricating oils.

5.3 Refinery products

Products from refineries world-wide number in the thousands, and cover a far wider range than simply automotive gasoline, aviation fuel, kerosine, oils, and greases. As an example, Table 5.1 shows a breakdown of products made by the petroleum industry in the USA.

This would be fairly typical of any well-developed country or group of countries, but it is likely that developing countries will have a much narrower range of product. A typical single refinery, however, may only make 10–15 products, some of which will be the result of blending operations. There will inevitably be a small group of low molecular weight hydrocarbons, and one might already suspect that cryogenic processing may be of some assistance with such products—namely fuel gas and liquefied gases or LPGs. The low molecular weight or low-boiling products include methane, ethane, propane, and butane, along with their respective olefins, all of which are in the gaseous phase at ambient temperature and pressure. Many of these are present in the light ends and LPGs emanating from the top of a crude unit. These merit more detailed discussion as they all feature in cryogenic processes.

5.3.1 *Low boiling products*

Methane (C_1) is usually used as a refinery fuel, but can be used as a feed-stock for hydrogen production by pyrolytic cracking.

TABLE 5.1. Products made by the US petroleum industry.

Class	Number
Fuel gas	1
Liquefied gas	13
Gasolines	40
Motor	19
Aviation	9
Other (tractor, marine, etc.)	12
Gas turbine (jet) fuels	5
Kerosenes	10
Distillates (diesel fuels and light fuel oils)	27
Residual fuel oils	16
Lubricating oils	1156
White oils	100
Rust preventives	65
Transformer and cable oils	12
Greases	271
Waxes	113
Asphalts	209
Cokes	4
Carbon blacks	5
Chemicals, solvents, misc.	300
Total	2387

Ethane (C_2) can be used as refinery fuel or as a feedstock to produce hydrogen or ethylene, which are used in petrochemical processes. Ethylene and hydrogen are sometimes recovered in the refinery and sold to petrochemical plants.

Propane (C_3) is frequently used as a refinery fuel but is also sold as a liquefied petroleum gas (LPG), whose properties are specified by Natural Gas Processors Association (NGPA). Typical specifications include a maximum vapour pressure of 1550 kPa at 38°C and a 95 per cent boiling point of −38°C (235 K) or lower at 760 mmHg atmospheric pressure. In some locations, propylene is separated for sale to polypropylene manufacturers.

The butanes present in crude oils and produced by refinery processes are used as components of gasoline and in refinery processing, as well as in LPG. Because it has a lower vapour pressure than isobutane (i-C_4), normal butane (n-C_4) is usually preferred for blending into gasoline to regulate its vapour pressure and promote better starting in cold weather. On a volume basis, gasoline has a sales value higher than that of LPG; thus, it is desirable to blend as much normal butane as possible into the gasoline. Normal butane is also used as a feedstock to isomerization units to form isobutane.

5.3.2 *Automotive fuel*

This is a complex mixture of hydrocarbons which has a boiling range from 38° to 205°C. Automotive fuel contains negligible propane (C_3) but contains some butanes, pentanes, and higher hydrocarbons, along with various additives including anti-oxidants, metal deactivators, and anti-stall and anti-knock agents. The automotive fuel is often a mixture of products from the refinery blended to make the fuel of an appropriate volatility and quality for the time of year and place of use of the fuel. The products used for automotive fuel blending are made by the various operations of the refinery which are designed to maximize output of automotive or other fuels. These operations include distillation (in the crude and vacuum towers), catalytic reforming, catalytic cracking, hydrocracking, polymerization, and alkylation. Each of these operations allows the maximization of refinery output, but they also give rise to low molecular weight gas mixtures which are often only used as fuel. Many of these operations consume hydrogen and some yield low molecular weight hydrocarbons and hydrogen.

5.3.3 *Naphtha and kerosene*

Naphtha cuts are often used as catalytic reformer feedstock in order to produce high-octane reformate for automotive fuel blending and aromatics production. Catalytic reforming reactions are complex and grouped into four major areas: (1) dehydrogenation of naphthenes to aromatics, (2) dehydrocyclization of paraffins to aromatics, (3) isomerization, and (4) hydrocracking. The overall effect is that this unit produces hydrogen. Due to today's demands for high-octane fuel, catalytic reforming represents a significant source of automotive fuel.

5.3.4 *Gas oils and residuum*

The light, atmospheric, and vacuum tower gas oils are generally processed in catalytic or hydro-crackers to produce automotive aircraft and diesel fuels. The heavier ends can also be used as feedstock for lubricating oil processes. Generally processing units which admit gas oils as feedstock consume hydrogen.

The vacuum tower bottoms can be processed in a visbreaker, coker, or de-asphalter to produce heavy fuel oil, or in a cracker to form lube base feedstocks. These units often yield small amounts of low molecular weight gases which end up in the refinery fuel header.

5.4 Hydrogen and fuel balances

From the above description it becomes apparent that the design of a refinery relies on an economic balance between processes that consume and produce both hydrogen and fuel gases. Normally consumers of hydrogen include:

- hydrocrackers of residues;
- hydrotreaters of lube oils;
- hydrodealkylation units;
- hydrodesulphurizers;
- hydrogenators.

Producers of hydrogen in a typical refinery include catalytic reformers and catalytic crackers.

In addition many units yield low molecular weight gas mixtures which are not suitable for use in any reaction loop but are routed to the refinery fuel header. Some of these gases result from hydrogen-production reactions, but are unsuitable for use in hydrogen-consuming reactions due to the detrimental effect of their other components, typically mixtures of hydrogen, methane, ethane, propane, and higher hydrocarbons. These are often used as fuel by furnaces and process heaters on the refinery, leading to an overall energy balance for the refinery and minimizing the flaring or sale of hydrocarbons.

Because many refineries need more hydrogen for their 'hydrogen-consuming' reactions than their 'hydrogen-producing' (catalytic reforming) reactions make, a separate supply or relatively high-purity hydrogen is installed. This relies on steam reforming of light hydrocarbons or partial oxidation of heavy hydrocarbons or fuel oil. This is a very costly operation and requires many processing stages, namely reforming, shift conversion, gas purification, and methanation. Regardless of the high cost of the operation, it is often the only source of relatively high purity hydrogen after catalytic reforming. It is therefore highly desirable to minimize the use of this high cost hydrogen source.

With escalation of energy and crude oil costs it becomes more important to extract the maximum automotive fuel yield from every barrel of crude. With this in mind many refineries carry out energy audits and energy conservation schemes, as well as adding processing units which will yield more automotive fuels or saleable products. This often entails increasing the use of hydrogen-consuming operations, e.g. hydrocracking or fluid catalytic cracking, where the heavy ends or residual of the refinery operation are cracked into molecules of a more suitable boiling range. Therefore, some refineries become deficient in hydrogen and either need to utilize steam reforming to a greater extent or are unable to increase their operating efficiency. Also, energy conservation measures resulting in direct fuel savings can often mean that too much fuel gas is produced and fed into the fuel headers. Therefore, to make the energy conservation measures effective, the fuel gas or a proportion of it must be converted into higher value products.

One way of satisfying energy conservation measures in a refinery is to use cryogenic technology to treat the gases in a refinery fuel header. A typical refinery can have anywhere from 5 to 50 sources of fuel gas which are all supplied in a relatively constant manner into the refinery fuel main header. These gases contain varying concentrations of hydrogen, methane, propane, and higher hydrocarbons, including small concentrations of aromatics and C_{8+}. Table 5.2 gives a typical set of light gases which show the significant amounts of hydrogen and LPGs present.

The valuable components are usually:

- hydrogen for use in hydrocracking;
- propane for sale or for use in crackers;
- butanes and heavier for fractionation and blending into automotive fuels;
- ethane.

All of the above components can be recovered by the use of cryogenic technology.

5.5 Medium-purity hydrogen recovery processes

5.5.1 *Introduction*

Medium-purity hydogen is considered to be a product of 90–98 per cent purity by volume, and is often applicable to cases where the refiner is interested in rejecting hydrocarbons from a reaction loop such as hydrodealkylation. Typical refinery off-gases with high-hydrogen content include:

- hydrodealkylation (HDA) unit off-gas;
- catalytic reformer off-gas;
- hydrodesulphurizer unit off-gases;
- catalytic cracker off-gas;
- hydrotreater;
- other minor vent gases.

The process routes vary depending on the conditions of the product, i.e. purity and pressure. Basically the processes fall into three groups: autorefrigeration, via Joule–Thomson refrigeration; requiring external refrigeration; and turbo-expander processes.

TABLE 5.2. Typical low molecular weight off-gases emanating from a medium-sized refinery. Levels are given in mole per cent.

	Gas stream number							
Component	1	2	3	4	5	6	7	8
H_2	24.3	76.0	78.2	74.9	83.7	41.5	23.7	0.3
$N_2 + CO$	—	—	—	—	—	—	—	13.4
CH_4	14.0	7.8	8.0	8.6	8.5	7.8	20.1	14.2
C_2H_6	21.8	5.4	4.5	5.5	2.5	11.4	5.3	23.7
C_3H_8	15.0	3.4	4.0	5.7	3.5	8.0	45.7	10.2
$i\text{-}C_4 + n\text{-}C_4$	16.3	2.5	2.8	2.8	0.9	8.8	2.8	23.2
$i\text{-}C_5 + n\text{-}C_5$	7.4	1.6	1.3	1.4	0.2	4.7	1.3	9.4
C_{6+}	—	0.6	0.8	0.5	0.4	13.5	0.8	5.1
Aromatics	—	0.7	0.2	0.6	0.1	4.1	0.1	—
H_2S	0.2	nil	0.1	nil	0.1	0.1	0.3	0.4
H_2O	1.0	trace	0.1	trace	0.1	0.1	0.3	0.1
Typical flow-rate ($N\,m^3\,h^{-1}$)	420	9300	3300	22 000	7000	1200	2400	800
Temperature (°C)	40	35	35	35	40	40	35	35
Pressure (kPa)	1000	1250	1550	1500	2900	220	1400	300

Table 5.3 shows three gas groups which are the sources of the net refinery fuel header gas. Group 1 gases usually include hydrodealkylation unit, hydrotreater off-gases, and perhaps hydrodesulphurizer unit off-gases. Group 2 gases usually include catalytic reformer off-gas, and group 3 gases usually include fluid catalytic cracker off-gases and some of the gas plant off-gases.

TABLE 5.3. Composition ranges for hydrogen recovery units (mole per cent)

Component	Group 1	Group 2	Group 3
Hydrogen	40–65	65–90	5–20
Nitrogen	nil	nil	nil
Methane	20–35	4–10	10–50
Ethane/Ethylene	0–5	4–10	10–40
Propane/propylene	0–3	2–6	10–30
C_4	0–1	2–5	0–15
C_5	negligible	0–3	0–2
C_{6+}	negligible	0–2	0–2
Aromatics	0–1	p.p.m. levels	p.p.m. levels
H_2O, acid gases	yes	yes	saturated with H_2O

Whilst the process route may vary, the hydrogen product purity can be in the 90–98 per cent by volume range for both group 1 and 2 gases. All of these processes are able to recover a very high proportion of the hydrogen present in the feed, typically over 90 per cent. Group 3 gases do not usually lend themselves to hydrogen recovery only, and generally recovery of C_2 components and LPGs represent economic solutions.

Most of these gases contain components that become solid at low temperatures, and nearly all cryogenic units have some form of pretreatment processes to remove them. These processes can include absorption, adsorption, chilling, and solidification, and are covered in more detail in Chapter 9. Typical impurities in refinery off-gases which can give freezing problems in a cryogenic process are water, ammonia, H_2S, CO_2, aromatics, HCl and COS. A typical pretreatment process for refinery off-gases might consist of chilling the gas to remove bulk water, followed by adsorption where remaining traces of water, sulphur compounds, and aromatics are removed by adsorption onto molecular sieve and activated carbon.

Figure 5.2 shows a typical pretreatment process which could be used for group 1 gas, e.g. a pyrolysis gasoline treating unit off-gas (PGTU) which will contain some water and aromatics (C_6H_6). Feed gas from the PGTU unit enters chiller E1 where it is cooled to about 10°C (283 K),

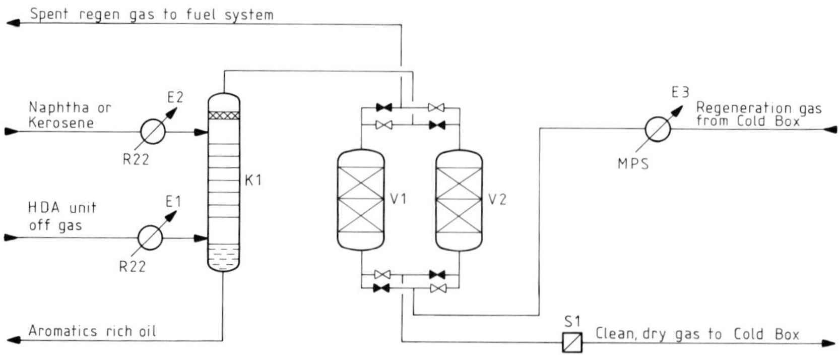

FIG. 5.2. The pretreatment process for pyrolysis gasoline treating unit (PGTU) off-gas fed to a cryogenic unit.

effecting the condensation of a significant amount of aromatics. The gas then enters column K1 where it is scrubbed with an oil of low volatility, e.g. kerosene, which washes out the bulk of aromatics and heavier hydrocarbons. This oil can be pre-chilled with Freon in E2 before entering K1. The overheads from K1, contain only small amounts of aromatics and water because of the chilling and washing of the HDA unit feed gas. The gas is then passed to vessel V1 which contains activated carbon and molecular sieve to remove aromatics and water, making the gas suitable for entry into the cold box. A fine dust filter, S1, is often installed to ensure that any activated carbon dust or sieve dust does not enter the cold box and hinder heat transfer. Regeneration is effected by heating up a portion of waste gas from the cryogenic unit using medium pressure steam or electricity, in E3, up to a temperature of 220–280°C, and passing it through vessel V2, in order to drive off the impurities adsorbed. The whole system operates on an automatically controlled cycle.

The pretreatment scheme shown in Fig. 5.2 is typical for this type of feed gas since the objective of treating PGTU unit off-gas is usually to recover both hydrogen and aromatics. As aromatics and hydrogen are present in the PGTU process loop it would be acceptable to mix the hydrogen product with the aromatics. Therefore, an alternative pretreatment method would be to eliminate the kerosene wash stage and put in much larger activated carbon beds and use hydrogen product to regenerate them. This would mean 100 per cent aromatics recovery recycled with the hydrogen product to the PGTU process. An advantage of this process is that it is very simple, but it has two small disadvantages: (i) higher consumption of steam, and (ii) fluctuating composition of aromatics in the hydrogen product due to the cyclic operation of the activated carbon adsorbers.

Yet another process option for the pretreatment of PGTU off-gas would be to dry the gas with molecular sieves first, then scrub with toluene to remove benzene, and then chill to about −40°C (233 K) using a simple R22 refrigeration unit. Again, this option would give nearly total aromatics recovery and low utility costs, but would be more complex than the two options discussed above. It becomes apparent from the above that there are many alternative pretreatments for a given feed gas. The decision to select one scheme as being better than another is made on an economic basis by comparing the cost of utilities and the importance of capital cost for a given project.

One thing that is interesting when considering pretreatment schemes is that often they produce a secondary product which enhances the plant profitability as well as being a necessary step in making the downstream cryogenic process work. The skill of the design engineer lies in devising a suitable pretreatment scheme that is truly in synergy with the cryogenic unit, rather than merely a necessary step to prevent freezing. The PGTU unit example above is one such synergy where both aromatics and hydrogen are recovered.

5.5.2 *Auto-refrigerated cryogenic units*

This type of cryogenic unit, or cold box, is one which relies totally on the expansion of gases or liquids across a valve to give a drop in temperature and hence refrigeration for the process. This phenomenon is known as the Joule–Thomson effect, and is often used in the pretreatment of refinery off-gases.

Figure 5.3 shows a typical process flowsheet for the cryogenic section of a group 1 gas separation plant, i.e. PGTU or HDA unit off-gases. The unit comprises of E4 and E5, which are multi-stream plate-fin heat exchangers, and separators V3 and V4. The feed gas enters exchangers E4 where it is cooled and partially condensed against warm hydrogen product and fuel waste gases. The condensate is separated in V3, while the vapour further cools and partially condenses in heat exchanger E5. The final condensate is separated in V4 giving relatively pure hydrogen product vapour, which is reheated through E5 and E4 before delivery to the HDA unit as recycle.

Refrigeration for the process is achieved by reheating the hydrogen product and fuel stream in exchangers E5 and E4. A positive temperature difference is achieved by expansion of fuel gas to a lower pressure in valves JT1 and JT2, and with the resultant Joule–Thomson effect there is a fall in temperature after expansion. This gives the necessary driving force for the cooling and partial condensate process. The gases quoted in group 1 usually can be processed using the Joule–Thomson effect alone for hydrogen purities up to 95–96 per cent. The process shown in Fig.

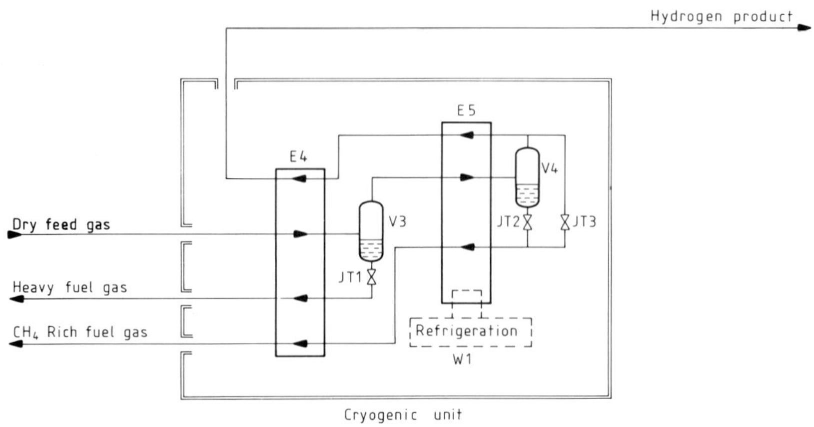

FIG. 5.3. A simplified flow-scheme for hydrogen recovery from refinery off-gas.

5.3 uses two stages of cooling and partial condensation with two phase separations in V3 and V4. Depending on the nature of the feed gas it may be advantageous to use three stages, or for simple mixtures only one stage. The choice depends on the quantity of propane, butanes, and heavier hydrocarbons present. If the mixture contained substantial amounts of C_6 and higher hydrocarbons it might be advantageous to use three separation stages in order to prevent the hydrocarbons passing to cold sections of the plant and giving problems.

5.5.3 *Processes with external refrigeration*

In order to attain hydrogen product purities of 97–98 per cent, an external refrigeration cycle is required, even with group 1 feed gases and almost certainly with group 2 gas mixtures. This external cycle may take several forms and in some cases a combination of two cycles may be used. Possible cycles are:

(1) a closed loop nitrogen cycle, utilizing a turbo-expander to provide refrigeration;
(2) a process gas expansion turbine expanding, say, hydrogen product thereby producing refrigeration;
(3) a Freon R22 or propane compression, condensation, expansion, and evaporation cycle;
(4) a cascade of two or more systems like (3), e.g. propane and ethylene;
(5) a mixed refrigerant cascade cycle.

The choice of routes (1) to (5) depends on the required product pressures and can only be made when an optimization study is done which takes account of the desired product conditions and qualities.

Inevitably, group 2 gases usually require a higher capital investment and a higher operating cost, but even these do not outweigh the advantages of hydrogen recovery against reforming more hydrogen to make up the deficiency in the refinery.

A typical example of a group 2 gas is catalytic reformer off-gas, which is often processed to recover hydrogen. Catalytic reformer off-gas will vary in composition depending on catalyst activity and type of process used, but the example in Table 5.4 is considered to be a typical case. The cryogenic process used to treat this gas can be very similar to that shown in Fig. 5.3, except that additional refrigeration is required. The pre-treatment for this feed gas is different from that in Fig. 5.2 in that no kerosene wash stage is required, and only molecular sieve driers are installed. The gas is dried in molecular sieves and passed to a two-stage partial condensation and separation unit, as shown in Fig. 5.4. The process requires external refrigeration which can be supplied by a closed loop nitrogen cycle. Several units of this type are in use, giving very flexible operability.

Catalytic reformer off-gas, normally available at ambient temperature and about 2500 kPa, enters adsorber V2 while adsorber V1 is being thermally regenerated. The adsorber dries the gas down to less than 1 v.p.m. water and the feed gas enters the cryogenic unit via the filter S1.

TABLE 5.4. Hydrogen recovery from refinery off-gas—mass balance.

	Composition (mol.%)			
	Feedgas catalytic reformer off-gas	Hydrogen to hydro-treaters	Reject fuel gas	LPG C_{3+} product
Hydrogen	77.7	95.5	43.3	1.1
Methane	5.2	4.4	11.5	1.4
Ethane	7.2	0.1	33.5	18.0
Propane	5.0	trace	10.0	33.4
Butane	2.8		1.7	25.1
Pentane +	1.1			10.9
C_{6+}	0.9			9.2
H_2S	2.0 (max.)			
H_2O (p.p.m.)	50.0 (max.)			
HCl (p.p.m.)	0.1			0.9
Flow-rate ($Nm^3\ h^{-1}$)	10 000	7411	1580	1009
Pressure (kPa)	2500	2420	500	1500
Temperature (°C) (approx.)	40	36	40	36

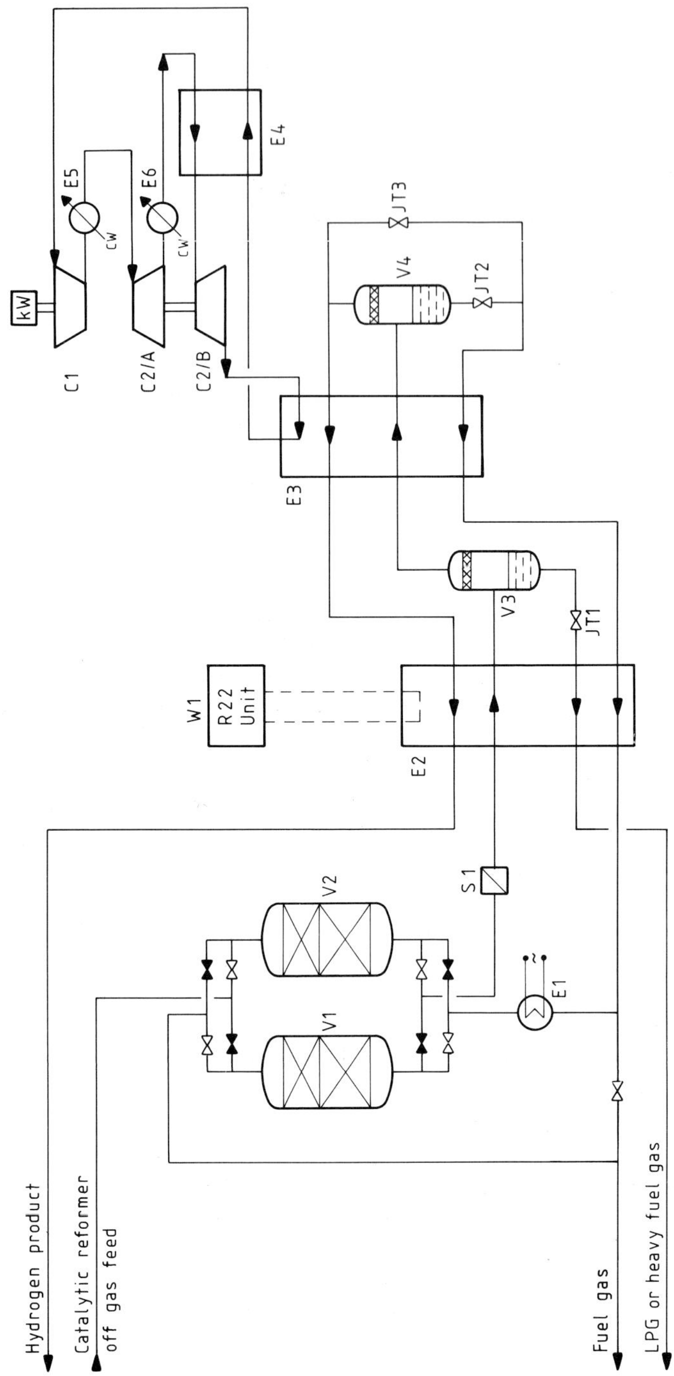

FIG. 5.4. A flow-sheet of a hydrogen recovery unit treating catalytic reformer off-gas.

The feed gas is cooled and partially condensed in the plate-fin heat exchanger E2, and the resulting liquid phase is separated from the hydrogen-rich vapour in V3. The vapour from V3 passes into a second heat exchanger E3, where it is cooled and partially condensed against returning hydrogen product, fuel gas, and nitrogen refrigerant. The liquid phase containing mainly CH_4 is separated from the vapour in the feed gas in V4 and upon flashing across valve JT2, is reheated through exchangers E3 and E2 before being used as regeneration gas and passed to fuel. The vapour from the separator V4 is hydrogen of the correct purity at 95.5 mole per cent. This is reheated in exchangers E3 and E2 before being passed to battery limit as product. A small portion of hydrogen product is injected into the fuel gas via expansion valve JT3 and it joins the liquids from V4 after they have expanded through JT2. This aids the evaporation of methane and thus lowers the temperature further than by expansion of the liquid stream alone.

The fuel gas mixture thus formed by the hydrogen injection and the methane from JT2 now evaporates through E3 giving part of the necessary cold and driving force to cool down the plant to the desired operating temperatures. The fuel gas then reheats further through E2 before being passed to regenerate the adsorbers and the fuel system.

The remainder of the refrigeration requirements are supplied by a closed-loop nitrogen cycle. This cycle usually operates by compression of low-pressure nitrogen to about 800 kPa in C1, cooling in E5, boosting in the brake, C2/A, of the expansion turbine, and further cooling in E6 before passing to the low temperature section.

The nitrogen, now at about 950 kPa, is cooled against low-pressure return nitrogen in E4 before entering the expansion turbine. Work is produced in the expander C2/B which is transmitted by a stiff shaft to the brake, C2/A. The outlet gas from C2/B is at a much lower temperature than at the inlet, due to the transmitted work. For example, expanding from 900 kPa to 150 kPa with an inlet temperature of $-123°C$ (150 K), the outlet temperature may be as low as 95 K, which gives a large potential refrigeration load to be dissipated by E3. The gas leaves E3 at about 130 kPa and 140 K. The low pressure nitrogen then returns through E4 into the suction of the cycle compressor C1.

Sometimes, LPG recovery is required from catalytic reformer off-gas whilst still recovery hydrogen. In these situations, a closed-loop refrigeration unit is supplied operating with Freon 22 compressed to 1400 kPa and condensed against cooling water before being expanded to 120 kPa and evaporated at 233 K to give refrigeration in E2, so that C_3 rich liquid product can be withdrawn through E2 or directly from V3 to the LPG recovery system on the refinery. Table 5.4 gives a typical balance for this type of unit.

This material balance shows a typical hydrogen recovery of 91 per cent. Higher recovery is possible by either additional capital investment or by lowering the fuel gas pressure, when it is possible to increase the hydrogen recovery to above 97 per cent. Note also that a separate C_{3+}-rich stream is also recovered and passed to the refinery LPG recovery unit. The addition of other hydrogen-rich gas streams sent directly to the fuel header can provide an additional incentive for the installation of this unit. The utility requirements are relatively small for the processing of catalytic reformer off-gas alone—typically approximately 450 kW of electricity and 50 $m^3\ h^{-1}$ of cooling water at 10°C rise in compressor aftercoolers are consumed.

5.6 LPG recovery from catalytic reformer off-gas

As can be seen from Table 5.4, the hydrogen recovery process can be easily adapted to produce a crude LPG product recovering a substantial proportion of C_{3+}. Some refineries do not have a shortage of hydrogen although they do have a market for LPGs. These cases are particularly suitable for the application of an expander process where the energy in the hydrogen product or light gas is utilized in a turbo-expander to provide reflux in an LPG stabilization column. This process is illustrated in its simplest form in Fig. 5.5. The feed gas is first dried in molecular

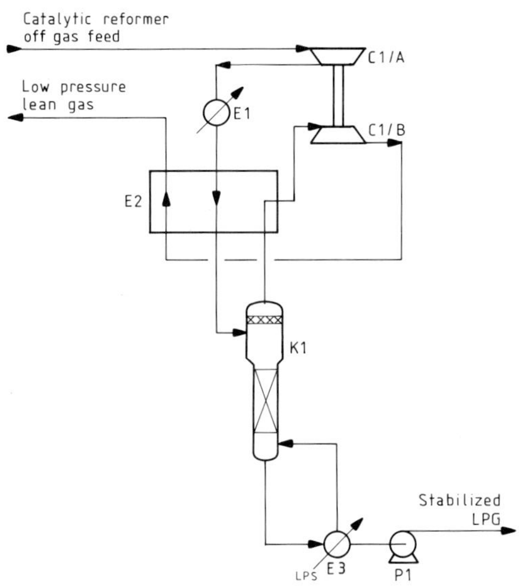

FIG. 5.5. A process flow-sheet for basic LPG recovery unit using a turboexpander process.

sieve adsorbers before passing to the LPG recovery unit. The feed gas is then compressed in the brake C1/A of a turbo-expander-compressor set, and cooled in water cooler E1. If the feed gas pressure is high enough, this compression is not required and the plant then actually *generates* electricity as well as LPGs because the expander is coupled to a generator. The feed gas is then cooled against returning streams in a plate-fin heat exchanger E2. Partial condensation occurs and the total stream is passed to the top of a distillation column, K1. The liquid resulting from the condensation in the plate fin exchanger provides the reflux to the column. Reboil is provided by a small low-pressure steam reboiler, E3, at the base of the column. This stabilizes the LPG product to the desired degree of stability making it suitable for pumping by P1. All of the required refrigeration is provided by the hydrogen-rich portion of the gas which exits from the top of the distillation column, is slightly reheated in the plate-fin heat exchanger E2, and then enters the expansion turbine C1/B. The pressure of this stream is reduced, converting this energy into work, and thus providing the necessary refrigeration. The low-pressure gas then is at a much colder temperature whence it reheats through the plate-fin heat exchanger again and passes to the fuel system of the refinery.

Catalytic reformer off-gas is highly suitable for this type of application where the feed gas to the cryogenic unit contains from 75 to 95 per cent hydrogen. Very high recoveries of C_{3+} can be achieved. Table 5.5. shows a typical mass balance around a Costain Engineering LPG recovery unit. Over 94 per cent of C_{3+} are recovered in this process, which is typical of the high efficiencies attainable through the use of expanders and plate-fin heat exchangers.

TABLE 5.5. LPG recovery from catalytic reformer off-gas.

Component	Feed (mol.%)	LPG (mol.%)	Fuel gas (mol.%)
Hydrogen	94.1	—	96.6
Methane	2.1	—	2.2
Ethane	1.1	0.6	1.1
Propane	0.8	28.1	0.1
Butane	0.5	18.4	—
Pentane	0.2	8.6	—
C_{6+}	1.2	44.3	—
Flow-rate ($Nm^3\ h^{-1}$)	20 000	510	19 490
Pressure (kPa)	1100	1550	450
Temperature (°C)	40	90	36

5.7 Ethane–ethylene recovery

Many refiners install fluid catalytic cracker units (FCCU) to crack heavier crude residual hydrocarbons to improve the economics of their refinery. These FCCU units also have a significant vent gas which is routed to fuel. This gas is generally low in hydrogen content but has many saturated and unsaturated hydrocarbons which are potentially valuable products. Table 5.6 shows a typical breakdown of components present in this type of gas.

TABLE 5.6. Typical FCCU off-gas.

Component	Range (mol.%)
Hydrogen	8–20
Nitrogen	2–15
Carbon monoxide	0–2
Methane	25–45
Ethylene	8–15
Ethane	15–25
Propylene	1–6
Propane	1–6
Butanes	0.5–2
Pentanes	0.5–2
C_6	0–0.5
C_{7+}	0–0.5
Flow-rate ($Nm^3\ h^{-1}$)	10 000–40 000
Pressure (kPa)	1000–1500
Temperature (°C)	40–50

In addition to these hydrocarbons, a typical FCCU off-gas contains many impurities including H_2O, CO_2, H_2S, COS, AsH_3, CS_2, and NH_3. Many of these components are present at trace levels but need careful consideration because of product specifications. Cryogenic technology is available to process FCCU off-gas and several such units are already in operation.

There are many processes that can be applied to FCCU off-gas, and depending on what products, purities, and recoveries are sought, the complexity of the process route changes. Although the pretreatment or removal of trace freezable impurities such as water is common to most of the processes, the cryogenic sections differ greatly, depending on what product is sought. One of the simplest processes used is to recover a low-pressure product stream rich in ethane and ethylene which is then fed to an ethylene plant cooling and fractionation train to produce pure ethylene products.

The integration of the process is as shown in Fig. 5.6. The process itself consists of the traditional pretreatment selected to reduce impurities to a level that presents no freezing problems for the cryogenic unit.

This process, shown in Fig. 5.7 consists of chilling in E1 to remove the bulk of water and some heavy hydrocarbons which are separated in V1. This is followed by molecular sieve adsorption in V2 or V3 to remove the remainder of the water and any other undesirable impurities, e.g. excessive amounts of H_2S and CO_2. These adsorption units are then regenerated by heating dry fuel gas in E2 and passing it through the off-stream bed (either V2 or V3).

The dry feed gas then enters the cryogenic unit which may contain one or more plate-fin heat exchangers, E3, and separators, V4 and V5. The feed gas is cooled and partially condensed to recover the maximum amounts of C_2 and C_3 components. The product stream and the fuel gas stream are used to provide cooling for the feed stream. Additional

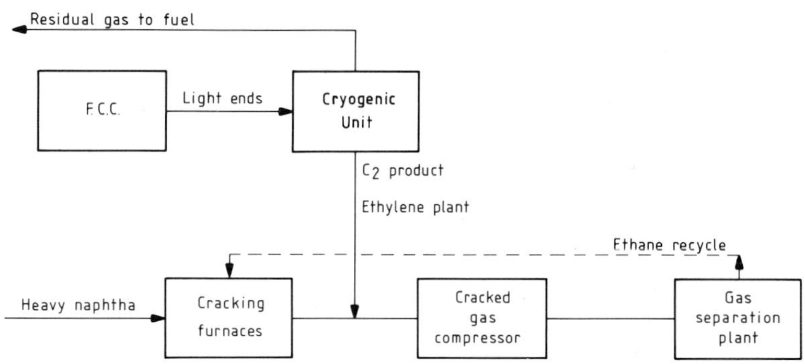

FIG. 5.6. A typical application—ethylene recovery from FCCU off-gas.

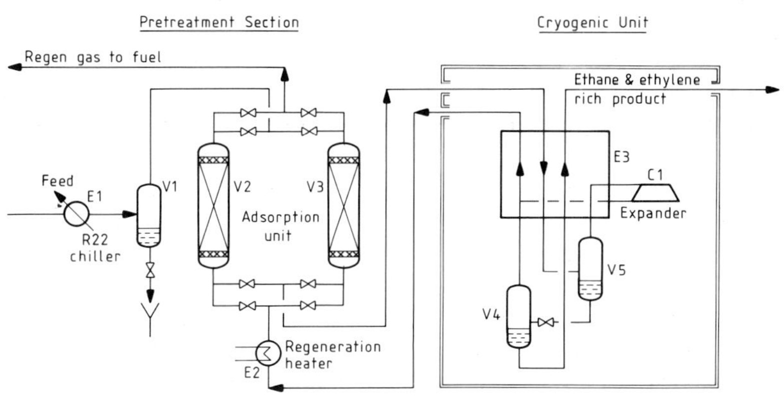

FIG. 5.7. A typical ethylene and ethane recovery system processing FCCU off-gas.

refrigeration is supplied by using a turbo-expander, C1, on either the fuel gas stream or the C_2 rich stream, depending on product and feed conditions. Usually considerable optimization is required to match the customer's best conditions. Table 5.7 shows an actual mass balance for an operating unit designed for treating FCCU off-gas.

This unit is very simple and easy to operate, but the proximity of a large petrochemical complex to a refinery is not too frequent an occurrence. Often, FCCU off-gas has to be processed differently to gain the most benefit.

TABLE 5.7. Typical mass balance for cryogenic unit process FCCU off-gas.

	Stream composition (mol.%)		
	FCC light ends	Fuel gas	C_2 product
Hydrogen	15	24	—
Nitrogen	16	26	—
CO	1	2	—
CH_4	33	46	15
C_2H_4	9	1	20
C_2H_6	20	1	50
C_3H_6	2	—	6
C_3H_8	3	—	7
C_4	0.5	—	1
C_5	0.5	—	1
C_{6+}	—	—	—
H_2O	saturated	—	—
H_2S	300 v.p.m.	—	—
CO_2	300 v.p.m.	—	—
Average molecular wt	22	17	30
Flow-rate ($Nm^3\,h^{-1}$)	12 200	7440	4780
Pressure (kPa)	1000	450	150
Temperature (°C)	45	30	30

5.8 Butanes and higher molecular weight component recovery

Butanes and higher molecular weight components are high-value components for a refinery because they require only a little processing to make them suitable for blending into automotive fuel. It is possible to use relatively simple cryogenic technology very effectively to recover butanes, C_4, and heavier from FCCU or other hydrogen and hydrocarbon-containing gases. Again this process adopts a relatively simple pretreatment route similar to that shown in Fig. 5.7, and although the cryogenic section is also very simple it has some fundamental differences from the C_2 recovery process. The C_4 and heavier recovery process cryogenic section is shown in Fig. 5.8.

In the C_4 and heavier component recovery process the dry feed gas is admitted to the cryogenic unit at about 20°C. The feed gas is cooled in exchanger E2 against warming fuel gas and in some cases Freon (R22) refrigerant. The feed gas then enters the product separator, V1, in which any C_4-rich product which condensed out in an exchanger E2 is separated out and joins condensate returning from reflux exchanger E3. The vapour from separator V1 contains a substantial amount of C_4 and heavier components which the reflux exchanger, E3, is intended to recover. This vapour cools and partially condenses up reflux exchanger, E3, which is sized so that the feed cooling passages are large enough to allow the condensed liquid to run back into separator V1. The temperature of the feed stream at the top of the reflux exchanger is about −56°C (217 K). This means that most of the C_4 and heavier components are no longer present in the vapour leaving the top of the reflux exchanger. Power is generated in the brake end (C1/A) of the turbo-expander assembly, C1/A,B. This vapour is then reheated in the reflux exchanger, E3, and work-expanded in the expansion turbine end (B) of the turbo-expander assembly, C1. The outlet gas is then substantially colder at about −73°C (200 K) and is reheated through the exchanger, E3, to provide the necessary refrigeration to make the reflux exchanger work. The C_4-rich product can then be routed to the refinery for fractionation into individual components and blending into automotive fuels. A typical mass balance for this type of process is shown in Table 5.8. This process is suitable for recovering unsaturated hydrocarbons (C_4) for feedstock into MTBE (methyl tertiary butyl ether), a lead-free anti-knock gasoline additive.

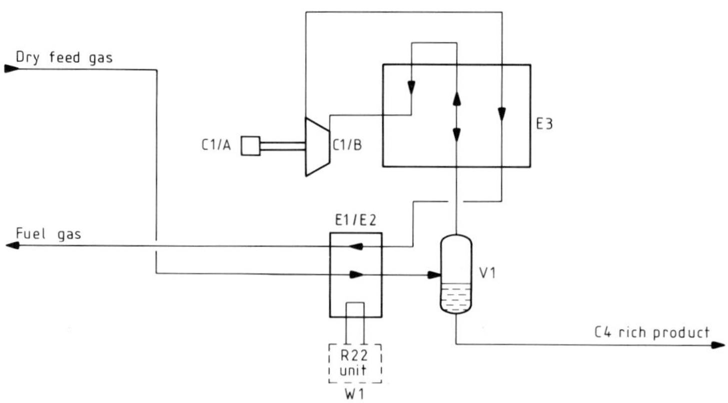

FIG. 5.8. The recovery of butanes and heavier components from refinery off-gas.

TABLE 5.8. The mass balance for the reflux exchanger C_{4+} recovery process.

	Feed	Fuel	Product
H_2	71.9	81.4	1.8
N_2+CO	12.1	13.6	0.7
CH_4	2.6	2.9	0.5
C_2H_4	0.2	0.2	0.1
C_2H_6	0.2	0.2	0.2
C_3H_6	1.2	0.9	3.6
C_3H_8	1.1	0.7	3.9
i-C_4 and $C_4^=$	10.3	0.1	85.6
n-C_{4+}	0.4	—	3.6
Flow-rate (kg mol h^{-1})	1200	1056	144
Pressure (kPa)	3100	760	3000
Approx. temperature (°C)	21	10	20

A simple alternative process for recovering C_4 and heavier components from the same feed gas would be by refrigeration and partial condensation, as shown in Fig. 5.9. In this process, power is generated in C1/A, as shown in Fig. 5.8 for the reflux exchanger process. However, in this process the feed is cooled in a conventional plate-fin exchanger, E1/E2, and the C_4-rich liquids are partially condensed and separated out in V1. The vapour from V1 is reheated in E2 and then work-expanded in C1/B to provide refrigeration for the process. This refrigeration is given up in exchanger E2, whereupon the gas exits warm and is vented to the refinery fuel header. Table 5.9 shows a typical mass balance for this type of process.

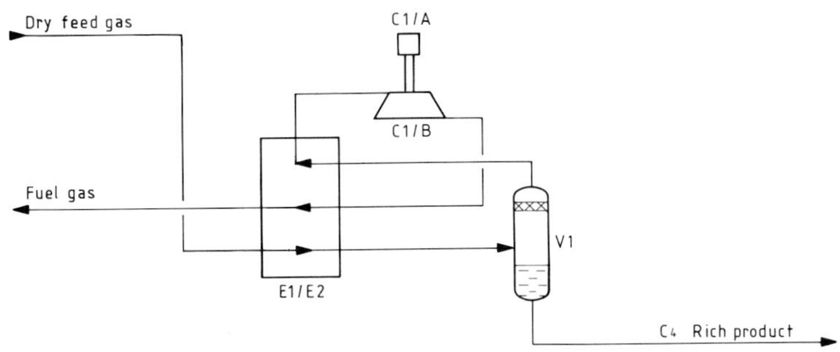

FIG. 5.9. Cooling and partial condensation to recover C_4 and heavier components.

TABLE 5.9. The mass balance for the conventional exchanger C_{4+} recovery process.

	Feed	Fuel	Product
H_2	71.9	82.9	1.1
N_2+CO	12.1	13.8	0.9
CH_4	2.6	2.8	1.4
C_2H_4	0.2	0.1	0.6
C_2H_6	0.2	0.1	1.0
C_3H_6	1.2	0.1	8.3
C_3H_8	1.1	0.1	7.5
i-C_4 and $C_4^=$	10.3	0.1	76.0
n-C_{4+}	0.4	—	3.2
Flow-rate (kg mol h^{-1})	1200	1038	162
Pressure (kPa)	3100	760	3000
Approx. temperature (°C)	21	10	−75

It is worth mentioning at this point the advantages that may be gained from processes similar to that shown in Fig. 5.8, i.e. using refluxing exchangers, compared with those using conventional heat exchangers as in Fig. 5.9. One advantage can be seen very clearly when comparing the mass balances in Tables 5.8 and 5.9: the levels of C_3 and lighter are much lower in the reflux exchanger process. The reason for this is that the equilibrium constant for propane and lighter components is much higher at the warmer temperatures which are produced in the reflux exchanger process (compare +20°C and −75°C in the separator V1 for the reflux and conventional exchanger process respectively). The equilibrium constants for some of the relevant compounds are given for the two temperatures at 3000 kPa in Table 5.10. The Peng–Robinson equation of state was used as the basis of calculation.

TABLE 5.10. Comparison of equilibrium constants versus temperature at 3000 kPa.

	Temperature (°C)	
	−75	+20
K-value C_2H_4	0.17	1.21
K-value C_2H_6	0.084	0.90
K-value C_3H_6	0.011	0.24
K-value C_3H_8	0.008	0.17

5.9 Application of loss analysis

Another reason for the superiority of the reflux exchanger process is that it is generally more efficient. Loss or exergy analysis is an accepted technique used by cryogenic designers, and increasingly so amongst other branches of engineering, for process efficiency assessement. By applying this technique one may ascertain why the processes are different.

Exergy is a thermodynamic function of state and can be calculated from thermodynamic principles. The efficiency of a given process can be defined in many ways—hence it is important to define a common basis to allow a proper comparative analysis. Therefore, typical analyses can use the feed temperature and pressure as the reference point. The efficiency of a process can be defined as a ratio of the minimum work of separation and the 'actual exergy input' into the process to bring the products back to inlet conditions.

In exergy terms, 'minimum work of separation' of feed into products at feed conditions of temperature and pressure is the exergy difference between the feed and products. The 'actual' work is the exergy change incurred when bringing the actual products back from their actual conditions up to the feed temperature and pressure.

If external energy is put into the process in the form of a refrigeration cycle, it should be added as an exergy input to the 'actual' work. Thus by comparing process efficiencies, the designer can gain a rough appreciation of his design against an alternative. By carrying out an exergy balance around each item of equipment, the designer can identify where the largest 'losses' occur. Thus it becomes an extremely valuable tool for process concept optimization. Individual item exergy balances on a typical cryogenic process show that most of the losses occur in the heat exchangers. A typical breakdown of losses for Joule–Thomson refrigerated processes can be 40–70 per cent for heat exchangers, with gas and liquid expansion valves taking the remainder. Liquid expansion valves tend to be the larger of the two expansion losses in well designed Joule–Thomson processes. Table 5.11 gives a comparison of some of the processes discussed in this chapter.

The results of the analysis presented in Table 5.11 indicate that the reflux exchanger process shows considerable benefit over the conventional process. The product recoveries are the same, yet the power input is much larger. Therefore, one can use the 'yardstick' of efficiency effectively when comparing two process alternatives to achieve the same feed to product separation. When we use an absolute value for efficiency the 'yardstick' is less clear and hence much less useful (compare processes 1, 2, and 3 with 4).

TABLE 5.11. A comparison of processes for hydrocarbon recovery

	Process			
	Refluxing	Conventional	Integrated	Vapour products
Product recovered	C_4	C_4	C_3	C_2
Process used	Fig. 5.8, Table 5.8	Fig. 5.9, Table 5.9	Fig. 5.10, Table 5.12	Fig. 5.7, Table 5.7
Minimum work of separation (kcal kg^{-1} mol^{-1} feed)	141	147	132	295
Actual work (kcal kg^{-1} mol^{-1} feed)	807	1500	627	708
Efficiency (%)	16	10	21	42
Power input into process (kW kg^{-1} mol^{-1} feed)	0.07	0.9	negl.	negl.
Product ratio of feed	0.12	0.135	0.0725	0.31
Product state	liquid	liquid	liquid	vapour

A two turbine, multi-separator process with an LPG stabilization column recovering C_3 and heavier components (Fig. 5.10) is shown in Table 5.11 (process 3). This is a highly integrated process where considerable effort has been spent in its optimization, and yet it has only half the efficiency of process 4 and only 5 per cent more than the relatively simple refluxing process 1. When breaking down the exergy losses into exchangers, valves, and machinery one can piece together a picture of where the inefficiencies lie for a given process route. This information can be used to improve the process, but absolute process efficiency values should always be treated with caution.

5.10 C_3 and heavier hydrocarbon recovery from refinery off-gases

5.10.1 *Introduction*

When the LPG market is buoyant many refiners look to recovering propylene and propane components as well as C_4 components. FCCU unit off-gas is one of the few refinery off-gases that contains unsaturated hydrocarbons and often gains special attention in terms of their recovery. In some cases a simple process such as that shown in Figs 5.8 and 5.9 can be used with perhaps addition of Freon R22 refrigeration into exchangers E2 in each process. This yields a relatively low recovery of propylene and propane and may satisfy some refiners but not others. This product is also likely to contain light ends such as methane (C_1), ethylene, and ethane (C_2). These then find their way into the refinery gas separation train and reduce the efficiency or cause unnecessary bottlenecks.

5.10.2 *Turbine processes*

The cryogenic designer often has to look for ways of reducing the levels of C_1 and C_2 components. One way may be to use reflux exchangers as shown in Fig. 5.8, but at lower temperatures in order to recover C_3 components. A more acceptable way is to use a distillation column to stabilise the C_{3+} (LPG) product because it gives a potentially higher recovery. This process technique, illustrated in Fig. 5.10, treats the same FCCU unit off-gas, but maximizes the recovery of propylene and propane and ensures that the C_{3+} product has less than 0.5 per cent of ethane and lighter components.

The process is relatively complex as it involves a number of features which enhance the recovery of propylene and propane without the addition of external refrigeration. It adopts the basic concept shown in Fig. 5.5 for catalytic reformer (CRU) off-gas processing, but is considerably modified to recover high quantities of the more volatile propylene as well as propane which are both present in FCCU off-gas (CRU off-gas contains saturated hydrocarbons only).

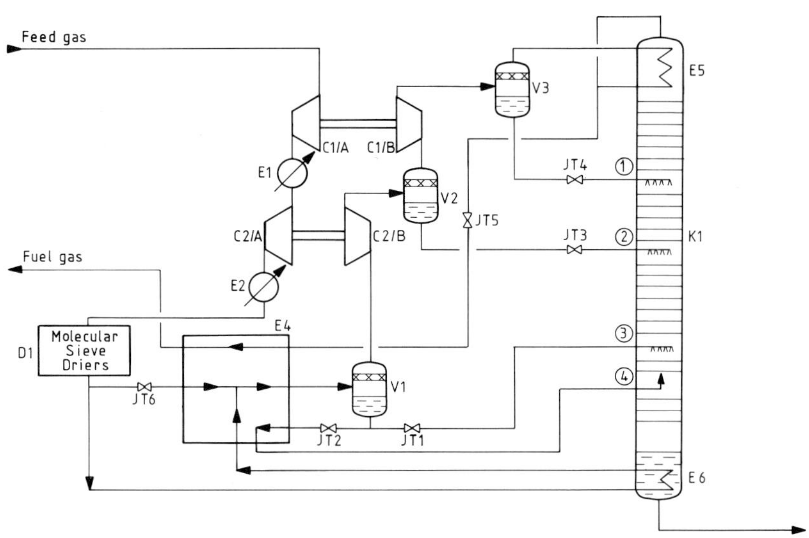

FIG. 5.10. An integrated process for stabilized LPG recovery from refinery off-gases.

In this process, the feed gas at about 1050 kPa is passed to the brake and C1/A of the first of two turbo-expander assemblies, C1 and C2. The feed gas is compressed in C1/A to about 1400 kPa, and cooled in a water cooler, E1, to about 35°C. Any heavy hydrocarbons that may have condensed are knocked out in a separator upstream of the second brake, C2/A (the separator is not shown). The gas then enters into the second brake, C2/A, where it is further compressed to about 1750 kPa and is subsequently cooled to about 20°C in cooler E2. Any hydrocarbon liquids condensing out in E2 are knocked out in a separator before the gas enters the driers D1. The knockout separator is not shown. The dry feed gas then passes partly to the reboiler, E6, and partly to the main exchanger, E4, where it cools to about −45°C and partially condenses. The liquids are separated from the vapour in separator V1 and form part of the feeds to column K1. Some of the liquid from separator V1 is flashed across a valve JT1 and is passed to the stabilizer column directly, whilst the remainder of the V1 liquids are flashed across JT2, reheated in the main exchanger E4, and then fed to the column. The aim of this reheated stream is to provide some of the refrigeration in the main exchanger, E4, and also to provide some intermediate reboil in the stabilizer column.

The vapour from separator, V1, is passed to the first expansion turbine C2/B where it is expanded to about 1000 kPa. As a result of the work expansion the temperature falls to about 212 K, and some liquid forms in the turbine outlet which is separated from the vapour phase in separator, V2. This liquid is rich in C_3 and heavier components and is therefore fed to an intermediate point in the distillation column, K1, via expansion valve, JT3. The vapour is further expanded in C1/B and the resultant work is given up into the corresponding brake, C1/A, in the same way as for the turbine brake set C2/A–C2/B. The outlet pressure from C1/B is about 370 kPa and the temperature reaches approximately 187 K, resulting in some more liquid being formed in the turbine outlet. This liquid is separated from the vapour phase in separator, V3, and fed to the distillation column K1 via valve JT4. The vapour from the separator V3 is partly used to provide refrigeration in the column overheads condenser, E5, and the remainder is passed via valve JT5 to join the warmed low pressure gas from E5. This vapour, together with ethane and ethylene-rich overheads fom K1 form the residue gas which is passed to the refinery fuel system via the main exchanger, E4. This fuel gas represents some 91.5 per cent of the feed gas flow that enters the plant, and thus provides nearly all of the refrigeration for the process. This refrigeration has been provided by work expansion in expansion turbine C1/B and C2/B. A very small fraction of refrigeration has been provided by Joule–Thomson expansion in valves JT1, JT2, and JT3. The stabilized LPG is withdrawn from the base of the column K1 and can be pumped and/or heated before supply into the refinery fractionation and blending equipment.

5.10.3 *Two turbine operation*

This process has two features which increase the overall process efficiency and hence improve the product recovery. The first is the use of two expanders C2/B and C1/B to work-expand the gas from 1700 to 370 kPa.

The reason for using two expanders in series is the improvement in isentropic efficiency when using two wheels to work-expand over a pressure ratio of 4.6 instead of using one wheel. The efficiencies attainable in practice for expanding FCCU off-gas in C2/B and C2/A are 86 and 85 per cent respectively. If a single expansion wheel were to be used then the efficiency attainable would only be ~ 80–81 per cent. A similar comparison can be made for the compressor or brake wheels C2/A and C2/B attached to each expansion wheel. This gives several percent extra cold production for the process, and hence a better potential for recovery of C_{3+} products. Typically, a single turbine option can only recover 55–60 per cent of the propylene and propane in the feed. The mass balance for a two-turbine process as described above is given in Table 5.12.

TABLE 5.12. The mass balance for a C_{3+} recovery unit treating FCCU off-gas using the process route shown in Fig. 5.10.

	Feed (mol.%)	Product (mol.%)	Fuel (mol.%)
H_2	15.1	0.0	16.3
N_2+CO	1.5	0.0	1.6
CH_4	45.6	0.0	49.1
C_2H_4	6.0	0.0	6.4
C_2H_6	23.9	1.2	25.7
C_3H_6	1.6	18.5	0.3
C_3H_8	4.8	59.8	0.6
C_4	0.7	9.5	0.0
C_5	0.5	7.0	0.0
C_6	0.2	2.0	0.0
C_7	0.1	2.0	0.0
H_2S+CO_2	traces	traces	traces
H_2O	satd	10 v.p.m.	variable
Pressure (kPa)	1050	1480	340
Temperature (°C)	27	21	24
Flow-rate ($Nm^3\,h^{-1}$)	44 700	3240	41 460

5.10.4 *Multiple column feeds*

A second feature of this process worthy of mention is the multiplicity of feeds to the column, K1. This gives a considerable improvement in efficiency within the column. The reason for this can be explained with the aid of a McCabe–Thiele diagram. The aim of the process is to recover C_3 and heavier components and to reduce C_2 and lighter components to a minimum: hence the 'key' components are ethane and propylene in a pseudo-binary mixture. Thus given a fixed sum of feeds into the column via valves JT1, JT2, JT3, and JT4 (shown in Fig. 5.10) one could put all of them on one tray or on different trays. Let us assume we put them all on one tray—then the McCabe–Thiele plot using a fixed approach to minimum reflux is illustrated in Fig. 5.11.

Figure 5.12 represents something close to the highest efficiency attainable for the column, but clearly with the many feeds shown this becomes both too complex and uneconomic. Therefore something of a compromise is desirable and this is depicted in Fig. 5.13, which represents the true case as applied in a number of units recovering LPG from refinery off-gases.

The total enthalpy of feeds 1–4 is the same as for the feed in Fig. 5.11. The product LPG and the C_2-rich gas overheads are also the same as for Fig. 5.11, and thus a comparison can be made between the options in Figs 5.11, 5.12, and 5.13 in order to show how different the efficiencies are by doing the same kind of exergy balance as above for Fig. 5.11.

By visual inspection, we can see that the slope of the top operating line (TOL) above feed 1 is lower than the slope for the TOL in the single feed case. This means that the overheads condenser load is smaller,

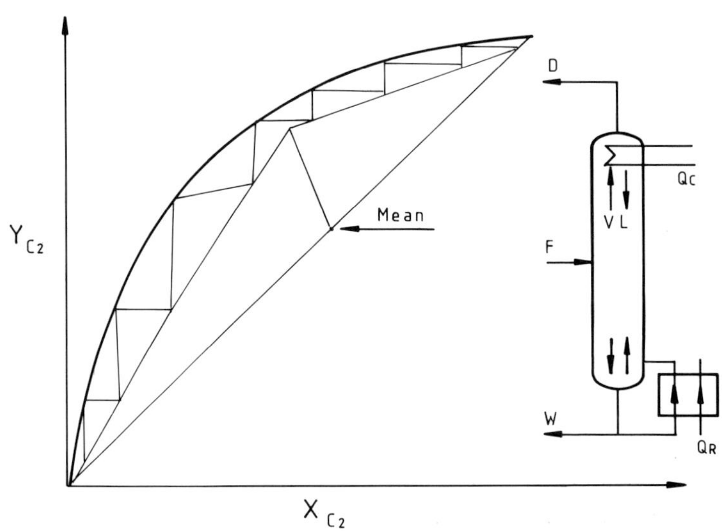

FIG. 5.11. The McCabe–Thiele plot of the de-ethanizer K1 with single feed.

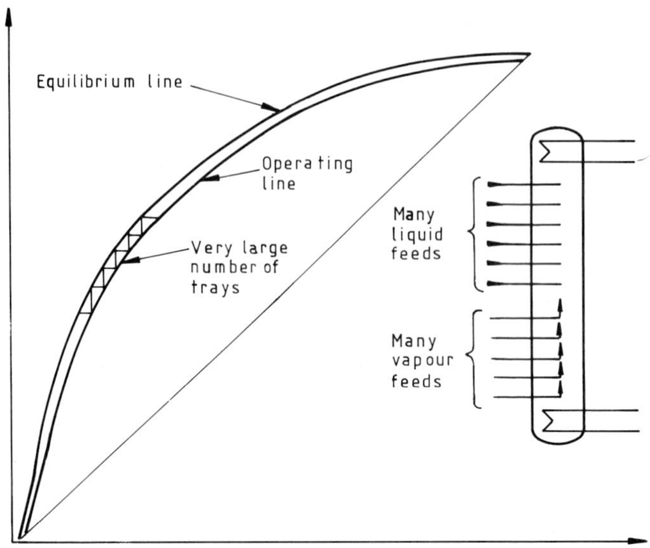

FIG. 5.12. A highly efficient configuration, theorctically, with many feeds and an operating line very close to equilibrium.

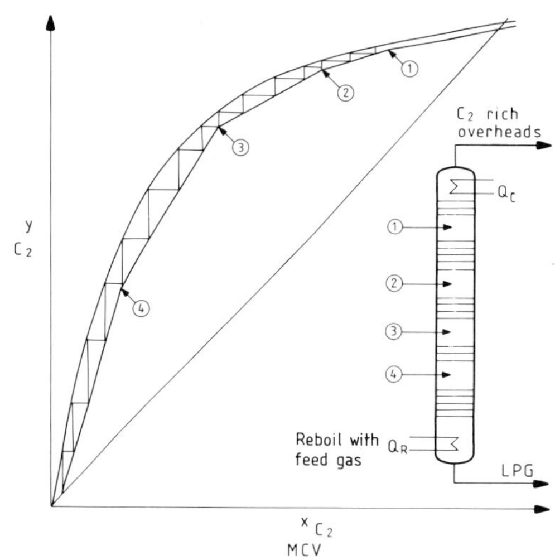

FIG. 5.13. The actual configuration of a column with four feeds.

making the process more efficient. Also we can see that the bottom operating line (BOL) is steeper for Fig. 5.13 (below feed 4) than for Fig. 5.11 (the single feed case). This in turn gives a lower reboiler load for Fig. 5.13 than for Fig. 5.11. One can see the logic of this when we see that feed 1 is colder than the single feed case (Fig. 5.11), and thus provides some refrigeration for the upper part of the column. Conversely feed 4 in Fig. 5.13 is hotter (or more vaporized) than the feed in the single feed case. This provides 'intermediate' reboil in the lower section of the column. Looking at Fig. 5.10 we see that feed 4 is vaporized after JT2 in E4 before being fed to the main column. This significantly reduces reboil in E6. This kind of integration is widely used in gas separation technology by any good cryogenics design engineer.

5.11 Ethane and ethylene recovery using C_{3+} recovery processes

Refineries often look for great flexibility in their processes to meet the wide variations in demand for C_3 and heavier components as well as ethane and ethylene.

The process shown in Fig. 5.10 can easily be modified to operate as a C_2 recovery unit, and it is this kind of flexibility that refiners often seek. By the addition of a simple propane, propylene, or R22 refrigeration unit the process can be made to operate as an ethane and ethylene recovery unit. Figure 5.14 shows the extent of modifications required. Often these modifications can be planned into the plant at the design stage and the refrigeration unit can be supplied later or at the same time, depending on projected demands for C_2s and/or C_3s.

The result is that the refrigeration unit W1 provides the majority of refrigeration required at temperatures of −40°C and warmer, thus allowing separators, V1, V2, and V3 to operate much colder which results in knockout of ethylene and ethane into the column K1. The turbine C1/B and C2/B also operate colder and thus generate less power than for C_{3+} recovery alone. However, as W1 is supplying a large portion of refrigeration to condense out C_{3+}, less needs to be supplied to

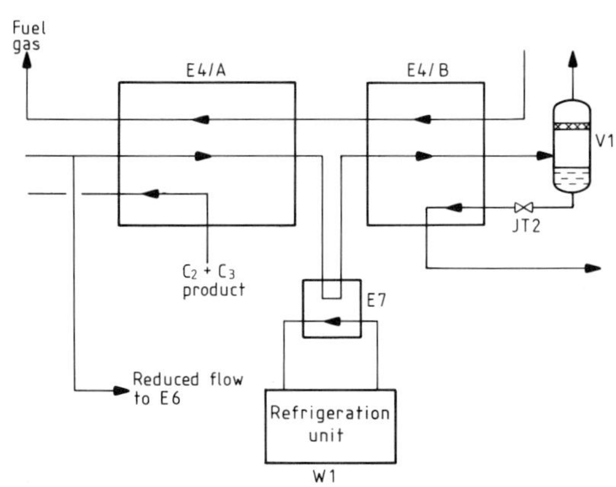

FIG. 5.14. Ethane–ethylene recovery using a basic C_3 recovery process.

condense out reasonable amounts of C_2 components. For very high recoveries of C_2 components, feed gas compression may be required because the brakes C1/A and C2/A may not boost the feed gas to a high enough pressure. The actual process conditions required vary significantly depending on the extent of C_2 recovery sought, but recoveries of 95 per cent ethane and ethylene are possible with compression of feed gas in a separate machine to only about 2000–2500 kPa, whilst the remaining compression duty is handled by the brakes C1/A and C2/A.

There are many variations of the basic concepts shown in Figs 5.8 to 5.12 offered by design companies, depending on the product requirements and nature of the feedstock. Other variables that often affect process route selection include the designer's experience, his traditional design attitude, and the availability of certain kinds of equipment. The operating company can also judge the design of the unit by its utility costs, finance availability, degree of operator skills, familiarity with cryogenics, and many other reasons which may force the designer away from what is technically the most efficient solution. This chapter has considered processes which for various reasons have been found acceptable to operating companies. Typical examples of the wide variety of processes used to recover C_3 components from refinery off-gases are patented and published by Higgins, Jordan *et al.*, Rowell, Bernhard *et al.*, and others.

Many of the processes shown in Figs 5.1 to 5.11 are relatively simple and little can be done to change them dramatically. However, one might consider three cooling and separation stages instead of the two shown in Figs 5.3 and 5.4. This might be done to remove heavier hydrocarbons or acid gases at the warmer end of the plant in order to prevent them from freezing in the final separator.

The complex C_3 recovery process shown in Fig. 5.10 could be modified in many ways, as shown in literature and patents. For example, the turbine brake could be utilized on the fuel gas side of the process and not at the feed gas side. This might give an advantage of efficiency for the compression of low-pressure fuel gas compared with higher-pressure feed gas, as shown in Fig. 5.10. However, it gives the disadvantage that in order to liquefy the same amount of C_{3+} components one has to reach colder temperatures, although the C_2 light ends content is likely to be lower at the lower pressure.

The use of a pre-cooler before turbine C1/B and separator, V1 for process 5.10 might improve process efficiency because it can recover cold from the LP return gas from V3 and E5, but it adds complexity to the process.

The process shown in Fig. 5.10 can also be modified to operate the distillation column, K1, at medium or high pressure with the turbine

C2/B feed coming from the column K1 overheads. This has the advantage that the overheads temperature is much warmer to recover the same amount of C_3 compounds. However, there is a disadvantage in that the relative volatilities reduce at higher pressures. Also low-pressure steam would most probably be required in reboiler E6 at higher column pressures, thereby increasing utility costs.

Theoreticaly one could postulate the use of three turbines and one might come up with a more efficient process, but the route becomes too complex for even the most experienced operators.

It is often found that if many designers are considering a plant for the same products from the same feed gas, the process routes vary depending on experience, equipment availability, and capital cost constraints. However, the process efficiencies can often be similar, thus making the selection of the 'optimum' route very difficult.

Bibliography

BANKS, R. and ISALSKI, W. H. (1987). Excess fuel gas? Recover H_2/LPG. *Hydrocarbon processing,* (Oct.), 47.

BAUER, H. (1985). 'Extraction of high-utility substances obtained from refinery fuel gases'. *Linde Reports Sci. Tech.* **40**, 36.

BERNHARD, D. P., ROWLES, H. C., and TEICHMAN, R. L. (1984). *Process for recovering C_{4+} hydrocarbons using a dephlagmator.* European Patent Application No. 0126309.

CORNELL, H. G. and HEINZELMANN, F. J. (1980). 'Hydrogen for future refining'. *Hydrocarbon Processing.* (Aug.), 85.

ELSHOUT, R. V. (1982). 'Retrofitting for energy savings'. *Hydrocarbon Processing,* (July) 109.

GARY, J. H. and HANDWERK, G. E. (1975). *Petroleum refining technology and economics.* Dekker, New York.

HIGGINS, R. D. (1981). *Cryogenic recovery of liquid hydrocarbons from hydrogen-rich gas.* U.S. Patent 4272270.

JOHNSON, J. A. and HILDER, G. K. (1984). *Dehydrocyclodimerisation, converting LPG to aromatics.* NPRA Annual Meeting, San Antonio, Texas.

JORDAN, C. H. and SAMPAT, A. (1978). *Recovery of light hydrocarbons from refinery gas.* U.S. Patent 4115086.

KARAFIAN, M. (1983). *Refinery gas utilisation—hydrogen generation and management.* Hydrocarbon Processing Symposium, Italy, April 1983.

LAM, W. K., *et al.* (1986). 'Recover valuable off-gases by the Braun ROE process'. *Energy Progress,* **6**(4), 205.

LIMB, D. I., and HEALY, M. J. (1984). 'Loss analysis aids process selection'. *Hydrocarbon processing,* (Aug.), 87.

OVERAL, J. (1969). 'Conoco's Humber refinery'. *Petroleum review,* (Dec.), 333.

ROWELL, R. L. (1982). 'New low-investment process to recover liquids from refinery fuel gas being used in Texas'. *Oil and Gas Journal* (May), 127.

TOMLINSON, T. R. and ISALSKI, W. H. (1979). *Developments in hydrogen recovery processes.* XV Intl. Congress of Refrigeration, Venice, paper A3/26, IIF, Paris, France.

WANG, S. I., *et al.* (1984). 'Analysis dictates H_2 purification process'. *Oil & Gas Journal (Technology),* (Feb.), 111.

6

RECOVERY OF INDUSTRIAL GASES FROM PETROCHEMICAL AND FERTILIZER PLANTS

6.1 Introduction

Oil refineries treat crude oil with a variety of distillation and catalytic treating stages to make aviation, automotive, and industrial fuels, naphtha, and a variety of lubricating oils and gases. Some of the refinery products such as LPGs and naphtha can be used as feedstocks for petrochemical processes. In today's integrated world of refinery and petrochemical complexes, it is often difficult to draw a clean line dividing the two areas. However, for our purposes we shall define refineries as a collection of unit operations aimed at maximizing the production of suitable grades of fuels, LPGs, oils, and greases from crude oil. Petrochemical plants shall be defined as a group of unit operations which treat a refinery product or group of products, making an intermediate compound which can then be converted into a consumer useable product. Fertilizer plants may use naphtha as feedstock to make ammonia and downstream chemicals. This chapter will consider gas processing associated with these technologies.

6.2 Hydrogen recovery from petrochemical plant off-gases

An example of a petrochemical process is the conversion of propane and/or butane into aromatics including benzene, toluene, and xylene (BTX). Benzene can be considered as an intermediate product which can then be converted into styrene and polystyrene. Figure 6.1 shows the divisions between refinery, intermediate petrochemical, and end product processes.

The reactions for the various stages in the petrochemical processes are:

Unit 1 (1) $2C_3H_8 \xrightarrow[\text{catalyst}]{} C_6H_6 + 5H_2$

(2) $7C_3H_8 \xrightarrow[\text{catalyst}]{} 3C_7H_8 + 16H_2$

(3) $8C_3H_8 \xrightarrow[\text{catalyst}]{} 3C_8H_{10} + 17H_2$

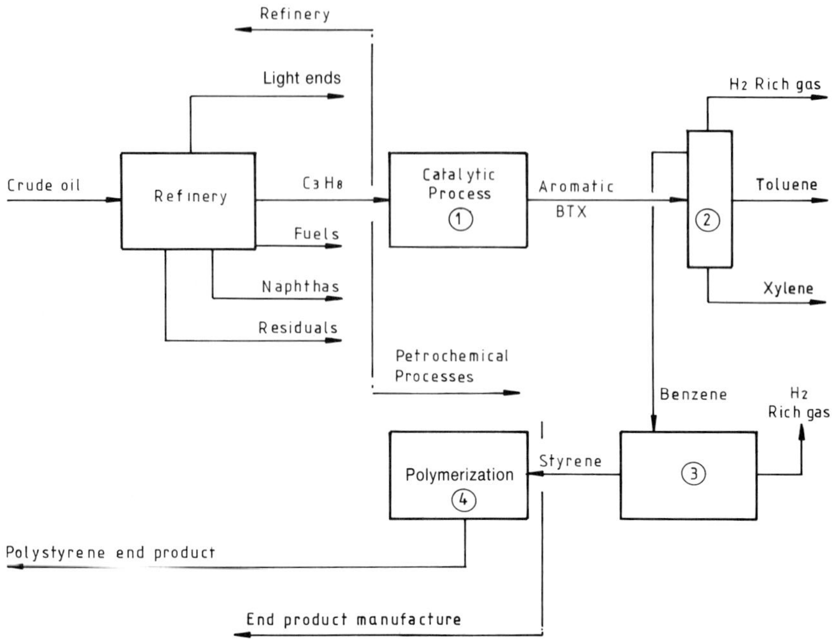

FIG. 6.1. Crude oil to polystyrene.

Unit 2 Separation process recovery of benzene, toluene, and xylene.

Benzene is an important building block in petrochemical processing and there are several ways to produce it.

Unit 3 Taking benzene.

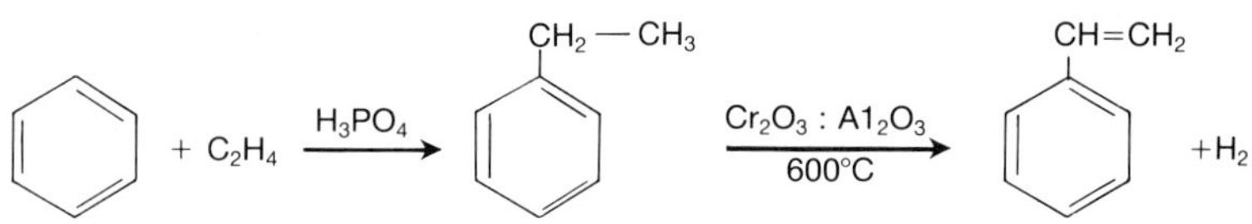

Unit 4

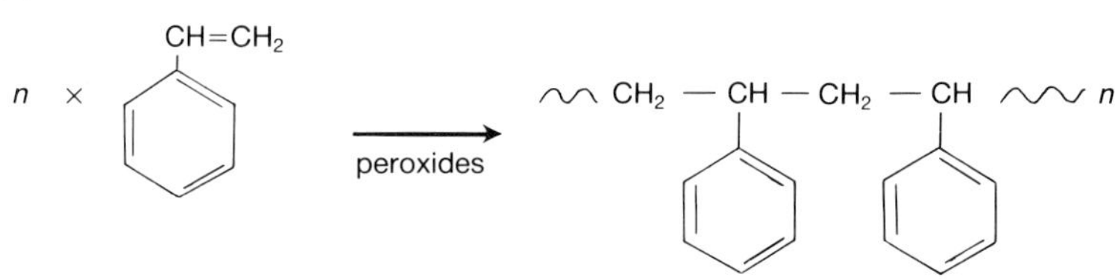

It can be seen that the above reactions produce hydrogen which could be useful in other areas of a petrochemical complex or back at the refinery. Cryogenic technology can be used to recover this hydrogen at a suitable purity for use elsewhere.

An example of a consumer of hydrogen in a petrochemical complex is the hydrodealkylation of toluene to make benzene. This reaction occurs as follows:

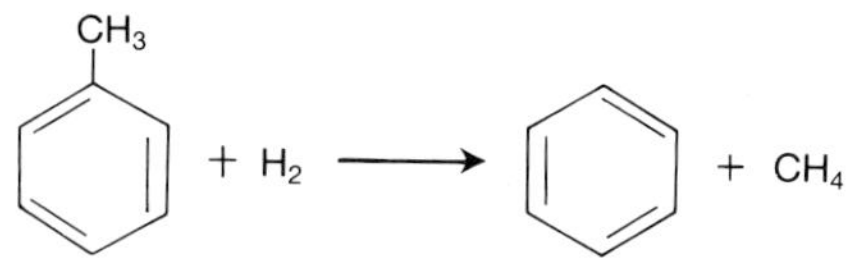

Toluene and heavier aromatics hydrodealkylation (HDA) is an accepted technique in industry for making benzene, and being a hydrogen consumer needs an outside source of hydrogen, usually via reforming.

Since reformed hydrogen is expensive it is desirable to maximize its use in the reaction stages. A typical aromatics production facility is shown in Fig. 6.2. A cryogenic unit is shown linked into the system showing how hydrogen is recovered to boost the efficiency of the overall scheme.

The cryogenic unit required for this process is virtually identical to that shown in Fig. 5.3, but without the use of external refrigeration. This type of unit is very simple and does not use rotating machinery, thus making it a highly reliable source of hydrogen product. A typical mass balance for this kind of unit is shown in Table 6.1. A 90 per cent hydrogen purity is assumed in this case which is usually acceptable, although purities in excess of 95 per cent can be attained depending on the feed gas and fuel gas pressures, and hence on the Joule–Thomson refrigeration available.

There are a few other examples of proprietary petrochemical processes which emit hydrogen-rich gases which are currently routed to fuel but could be recovered via cryogenic technology. These include the very similar unit for treating pyrolysis gasoline to increase benzene yields. The feed gas to the cryogenic process from this kind of unit is very similar and therefore warrants a very similar approach in terms of

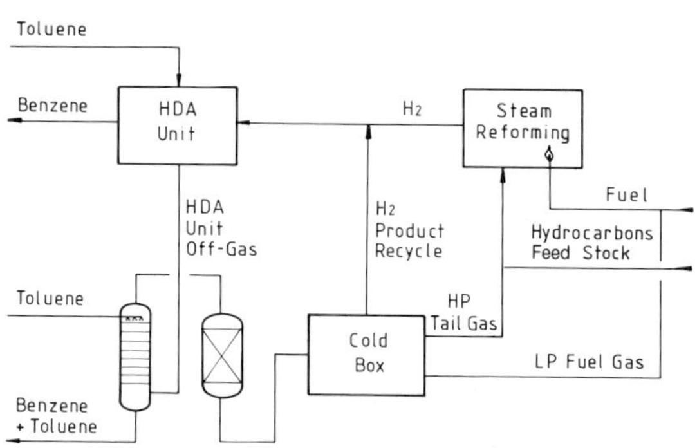

FIG. 6.2. The integration of hydrogen recovery and an aromatics plant.

TABLE 6.1. The mass balance for a typical HDA off-gas H_2 recovery unit (mole per cent).

	Feed gas	Lean wash solution	Rich solution	Hydrogen product	Fuel gas
H_2	60	—	traces	90	15
CH_4	34	—	traces	10	74
C_2H_6	3	—	traces	traces	8
C_3H_8	1	—	traces	—	3
Benzene	2	—	10	—	traces
Toluene	—	100	90	—	traces
Flow-rate ($Nm^3\ h^{-1}$)	19 000	3420	3800	11 475	7145

cryogenic process route. Table 6.2 shows a typical range of feedstock composition.

The flowrate depends, of course, on the size of the treater unit, and there are other impurities present in the feed gas which need removing prior to cryogenic unit treatment. These includc H_2S and H_2O which are usually removed by caustic scrubbing and drying respectively.

In both of the above cases the economics favour the use of a cryogenic unit to recover hydrogen, because one must compare the cost of recovering this hydrogen against reforming hydrocarbons to produce the hydrogen. Hydrogen produced by the reforming process costs some four to six times that of hydrogen recovered via cryogenic technology. The annual benefits from recovering hydrogen from, say, HDA unit off-gas as in Table 6.1 represent a simple payback of only a few months, making this type of process very attractive in most circumstances.

TABLE 6.2. Pyrolysis gasoline treater off-gases—typical range.

Gas	Level (mol.%)
H_2	57–64
CH_4	25–33
C_2H_6	9–10
C_3H_8	0–2
C_4H_{10}	0–0.1
C_{5+}	traces
C_6H_6	0–300 v.p.m.
Toluene +	500–2000 v.p.m.
Flow-rate ($Nm^3\ h^{-1}$)	5000–35 000
Pressure (kPa)	3500–5300
Temperature (°C)	15–40

6.3 Hydrogen recovery from ammonia synthesis purge gas

The majority of fertilizer production worldwide involves the fixation of nitrogen in order to produce a fertilizer that can be administered onto barren ground safely and efficiently. Highly favoured nitrogen-containing fertilizers are ammonium nitrate (NH_4NO_3) and urea (NH_2CONH_2). Both of these compounds are formed from ammonia in the following manner:

1. Ammonium nitrate

$$NH_3 + HNO_3 \rightleftharpoons NH_4NO_3$$

2. Urea

$$2NH_3 + CO_2 \rightleftharpoons NH_4COONH_2 \rightleftharpoons NH_2CONH_2 + H_2O$$

Therefore most larger fertilizer complexes have a large ammonia production facility as the hub of their activities.

The majority of larger capacity single-train ammonia-producing units adopt the classical technique of hydrocarbon-reforming synthesis-gas production to provide feedstock for the synthesis of ammonia.

A typical ammonia plant using this route is depicted in Fig. 6.3. In this process raw material gas which is usually available at pressures greater than 3000 kPa is used as feedstock. The first stage in the process is the removal of acid components (usually H_2S) from the natural gas. This is achieved by hydrodesulphurization which uses a small quantity of hydrogen-rich synthesis gas or ammonia loop purge gas. The natural gas is admitted to the primary reformer, where part of it is used as feedstock, together with steam to make hydrogen. The remainder of it is used as fuel gas to fire the reformer. Air is added to the secondary reformer where further reactions take place involving oxygen. Nitrogen is introduced into the secondary reformer with the air and this flow is controlled so that the synthesis gas ultimately entering the synthesis contains a ratio of 3:1 hydrogen:nitrogen. Based on natural gas as feedstock and fuel, the reforming reactions are as follows:

$$C_nH_m + nH_2O \rightarrow nCO + (n + \tfrac{1}{2}m)H_2 \quad (1)$$

$$CO + H_2O \rightleftharpoons CO_2 + H_2 \quad (2)$$

$$CO + 3H_2 \rightleftharpoons CH_4 + H_2O \quad (3)$$

Reaction (1) goes essentially to completion whilst (2) and (3) are reversible.

In the secondary reformer, oxygen is used up in the oxidation of some residual natural gas as well as combination with CO and H_2 forming CO_2 and H_2O. The next stages are high and low temperature shift conversion reactions which reduce CO levels by conversion with water into CO_2 and

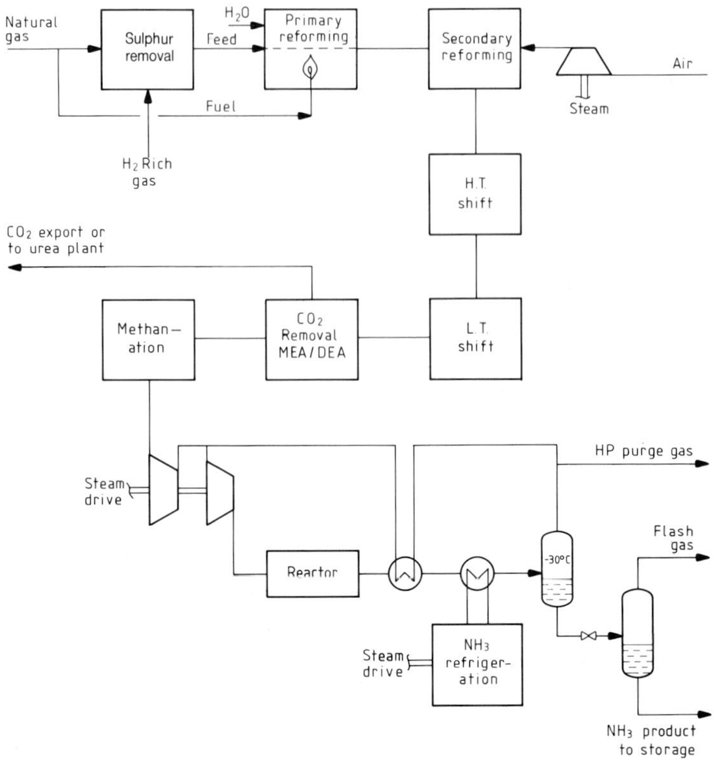

FIG. 6.3. An ammonia plant block flow-scheme.

hydrogen. CO_2 is then removed by working with a variety of possibilities including mono-ethanolamine (MEA), diethanolamine (DEA), Sulphinol, Benfield or Selexol. These are all proprietory processes which are the accepted techniques for removing CO_2 from synthesis gas. The final stage before compression of the synthesis gas is to ensure that no carbon oxides are present before entering the synthesis loop. This is achieved by methanation of residual CO and CO_2 into methane as follows:

$$CO + 3H_2 \rightarrow CH_4 + H_2O \tag{3}$$

$$CO_2 + 4H_2 \rightarrow CH_4 + 2H_2O \tag{4}$$

Reactions (3) and (4) are essentially a reversal of reaction (1) and (2) above, but this is an effective way of making the synthesis gas suitable for conversion to ammonia. The gas at the inlet of the synthesis gas compressor usually contains approximately 74.5 per cent H_2, 24.4 per cent N_2, 0.4 per cent Ar, and 0.7 per cent CH_4 at 2700 kPa and 40°C.

The gas is then compressed into the ammonia synthesis circuit to about 14 500 kPa, where ammonia is produced by the reaction

$$3H_2 + N_2 \rightleftharpoons 2NH_3 \quad (5)$$

The reaction requires moderate temperatures, high pressures, and an iron-based catalyst to ensure reasonable conversion in the reactor. The ammonia product is usually condensed out of the gas circulating around the synthesis loop by cooling at the outlet of the reactor down to about −30°C using process fluids. The ammonia product is separated at high pressure, and flashed to low pressure to remove dissolved H_2, N_2, Ar, and CH_4 before passing to storage.

The synthesis gas entering the synthesis loop contains 1.0–1.5 per cent Ar + CH_4 which do not react, acting as inerts which build up to significant levels which could inhibit the reaction. In order to keep the Ar and CH_4 at an acceptable level a purge is taken off the loop which represents 4–6 per cent of the total synthesis gas supplied into the unit (the flash gas flow above is too small to purge off large amounts of argon and methane). The early 1980s saw a rapid growth in the process which recovered the hydrogen from the ammonia purge gas, recycling it back to the synthesis loop in order to allow the ammonia plant operator to make 4–5 per cent more ammonia or to save 2–3 per cent of his energy consumption for the same ammonia output. Cryogenic technology is a very common method for recovering hydrogen from ammonia purge, although permeation and, to a lesser extent, adsorption are less efficient alternatives which are sometimes adopted where limited capital is available and flexibility is not as important.

One can see why cryogenic technology lends itself very effectively to treatment of ammonia purge gas when one considers the wide difference in the relative volatility of hydrogen with respect to the other components N_2, Ar, and CH_4. Table 6.3 shows a typical ammonia plant purge gas showing the differences in volatility between the components.

The composition of the purge gas spans the expected variations from plant to plant and shows variations due to catalyst activity in the reforming and shift stages of synthesis gas preparation. The hydrogen

TABLE 6.3. Loop purge gas composition.

Component	Normal (mol.%)	Range (mol.%)	Volatility relative to methane at 7300 kPa at low temperature
Hydrogen	63	60–65	305
Nitrogen	21	19–22	9.1
Argon	4.5	3–5	4.3
Methane	8.5	6–14	1.0

recovery plant is capable of operating over the whole range of the given compositions whilst maintaining the overall recovery of hydrogen.

The relative volatility of components with respect to methane are for a pressure of 7300 kPa and about −183 to −173°C. Clearly, with such a high volatility of hydrogen, cryogenic processes involving cooling and partial condensation of heavier components, namely argon and methane, will show considerable potential.

Cryogenic technology has been applied to ammonia purge gas treatment for some 25 years, although the processes that gained recognition in the late 1970s were considerably more efficient than their predecessors. Early processes used wound-coil heat exchangers and could operate at virtually any pressure. However, most of the processes built in the 1960s utilized plate-fin heat exchangers with a maximum allowable working pressure of about 4500 kPa. Convenient levels for returning hydrogen product were 2500–3000 kPa and 6500–7500 kPa. The product could only be returned at the lower pressure levels into the suction of the synthesis gas compressor.

Ammonia freezes at −77°C, and since the cryogenic processes generally operated at −188 to −173°C at the plant cold end, ammonia had to be removed. This was done in two ways: ammonia condensation by chilling followed by adsorption; and waterwash of the purge gas followed by adsorption.

6.3.1 *Ammonia removal by chilling and adsorption*

The first method had the advantage of being fairly simple as there was no introduction of water into the purge gas, and the whole process was essentially cryogenic and well integrated with the hydrogen recovery parts.

An early process utilized a propane compression, condensation, and evaporation cycle to provide refrigeration to condense ammonia from the purge gas. High-pressure ammonia purge gas enters the plant at 13 500 kPa and ~ −25°C, and is expanded to about 4500 kPa before it enters the plant if plate-fin exchangers are used.

The feed gas is cooled and most of the ammonia is condensed. The hydrogen is produced by partial condensation following adsorption of ammonia. The process was similar to that shown in Fig. 6.4, but was limited by its low operating pressure and 2600 kPa H_2.

This type of process has been successfully implemented at higher pressures in 1978 by the IMC Corporation, on a 1043 MTPD ammonia plant in Sterlington, Louisiana, USA. This unit is based on wound-coil copper heat-exchanger techology and operates at pressures up to about 13 000 kPa. This type of heat exchanger is considerably more expensive than plate-fin heat exchangers, but it gives the advantage of returning

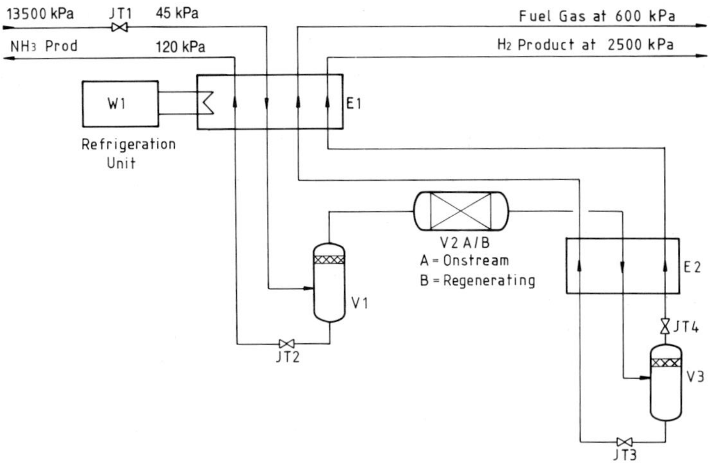

FIG. 6.4. Ammonia purge gas recovery (total cryogenic option).

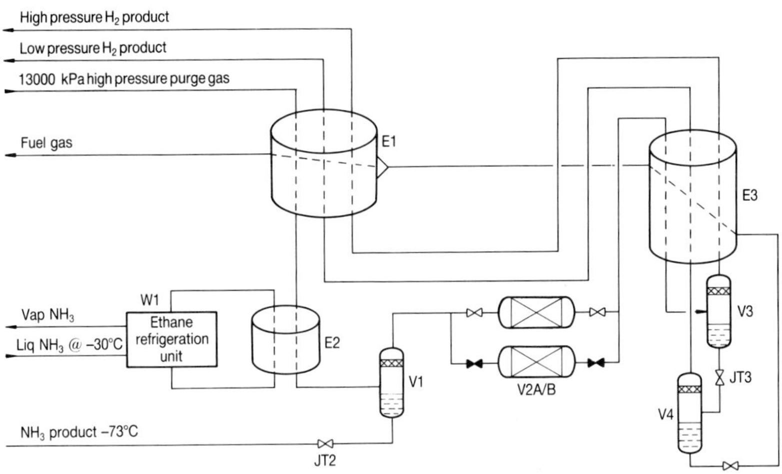

FIG. 6.5. The high-pressure purge gas process.

substantial amounts of hydrogen product at 12 400 kPa, thereby saving considerable amounts of energy in the ammonia plant. Figure 6.5 shows the basic flow-scheme for this high-pressure process.

In this process, high-pressure purge gas enters the plant at about 13 000 kPa and −23°C, and is cooled in E1 and E2 to −73°C by returning hydrogen product, fuel gas, and ethane refrigerant respectively. As a result, the majority of the ammonia is condensed and then separated in V1. Liquid ammonia is withdrawn from V1 and returned to the ammonia plant as product. The vapour from V1 is then passed into molecular sieve adsorbers to remove the final traces of ammonia at

12 800 kPa, before being cooled down further to −173°C in order to condense out the argon and methane. The vapour forms the high-pressure product and is reheated through E3 and E1 before being fed into the ammonia synthesis loop. The liquids are separated out from hydrogen-rich vapour in V3 at about 12 400 kPa. Because this pressure is very high a considerable amount of hydrogen is dissolved in the liquids leaving V3.

In order to increase the recovery of hydrogen to 96 or 97 per cent it is necessary to flash the liquid across JT3 to an intermediate pressure of 2760 kPa which releases a large proportion of this hydrogen from the liquid leaving V3. This hydrogen is then separated from the flashed liquid in V4 and allowed to reheat through E3, E2, and E1, and forms low-pressure product suitable for feeding the suction of the ammonia plant synthesis compressor. The liquids from V4 are further flashed to 360 kPa across JT4, re-evaporated in E3, and reheated back to ambient conditions in E3 and E1 before being routed to the ammonia plant fuel system. All of the necessary refrigeration for exchanger E1 is provided by the high and low pressure hydrogen products and fuel gas. To achieve a temperatrue of −73°C in separator V1, ethane refrigerant is evaporated in E2 and the remaining refrigeration is provided by reheating fuel gas and hydrogen products in E1. An overall hydrogen recovery of 97 per cent is achieved with 90 per cent being available at 12 400 kPa and the remainder at 2670 kPa. In addition, about 40 per cent of the nitrogen is also recovered which is of benefit to the ammonia plant operator. The argon and methane levels are relatively low in both products, i.e. about 85 per cent of the argon and methane is rejected in the fuel gas. This unit handles between 9400 and 18 000 $Nm^3\ h^{-1}$ of ammonia purge gas, depending on how the ammonia plant operator wishes to run and can accommodate reasonably wide variations in purge gas composition. The refrigeration unit W1 consumes about 85 kW of power and 590 kg h^{-1} of liquid ammonia refrigerant.

The IMC unit has one advantage in that it can produce a substantial amount of hydrogen product at very high pressure and indeed the cryogenic section to provide the hydrogen product is relatively efficient. However, the cost of removal of the ammonia is very high because in addition to the refrigeration requirements given above, the unit consumes a further 700 kg h^{-1} of steam at 10 300 kPa to regenerate the molecular sieve adsorbers. Moreover, the use of wound-coil heat exchangers at these high pressures makes the unit very expensive to buy, and thus there are few of these types of units in operation today. Some operators are reluctant to use copper on ammonia service, even though conditions are dry and the risk of corrosion is negligible. Other expensive alternatives are wound-coil stainless-steel exchanger designs or the aluminium paired-tube exchangers.

6.3.2 *Ammonia removal by water wash*

This method of ammonia removal from ammonia purge gas adopts water as a wash medium at high pressure and therefore requires the use of molecular sieve driers to remove water as well as traces of ammonia. At first this process rate appears more complex because it needs more stages of operation, but a closer look shows great advantages in reliability and lower operating costs. Figure 6.6 shows the process flow-sheet for the water wash route as a pretreatment method which is currently applied on most hydrogen recovery units for ammonia purge gas.

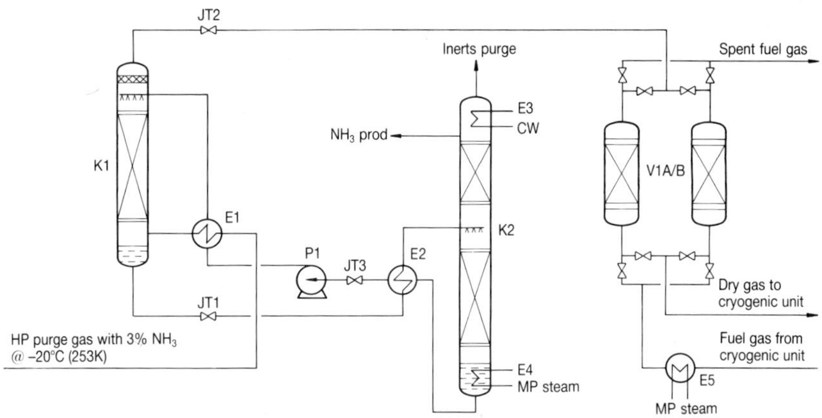

FIG. 6.6. Ammonia removal and the drying of ammonia purge gas using the water-wash route.

There are minor variations and additional equipment in many cases depending on the design philosophy, whether any other gases (e.g. low-pressure ammonia purge gas) are processed, and conditions of the high-pressure purge gas which vary widely depending on which ammonia synthesis process is being used.

In the typical water-wash ammonia removal process, high-pressure feed gas enters the plant at anything from 13 000 kPa to 20 000 kPa and −25°C to 20°C containing from 1.3 per cent to as much as 7 per cent of ammonia. These variations depend on the conditions and arrangement in the ammonia synthesis loop. The gas first enters a lean solution chiller, E1, which cools the absorber lean solution to about 5°C. The feed gas then enters the absorber, K1, where it is contacted with lean solution which takes out the bulk of the ammonia. The gas leaving the absorber is now relatively free from ammonia but is also saturated with water. The gas is then expanded across JT2 to about 7500 kPa before being dried in V1A/B on molecular sieves. These molecular sieves remove both the water and the ammonia down to less than 1 v.p.m. Two adsorbers are normally employed using a fixed time cycle where one is adsorbing and the other is being regenerated by hot, dry fuel gas which has been drawn from the cryogenic unit and heated in E5.

The rich solution leaving the base of K1 is flashed down to about 2800 kPa under level control before it is preheated in E2 and fed to the ammonia recovery column. The ammonia recovery column is either a packed or trayed tower, depending on turndown requirements, and utilizes a conventional condenser and reboiler system to drive the vapour and liquid flows up and down the column. The overheads condenser E3 uses water or air cooling whilst the reboiler uses medium pressure steam at about 3800 kPa. Both of these utilities are readily available on the ammonia plant, which ensures good integration. The hot lean solution is withdrawn from the base of the column at about 220°C and contains less than 1 per cent ammonia. It is then cooled against incoming rich solution before it is again pumped in P1 to the operating pressure of K1.

There are several beneficial features of this process which have earned it considerable popularity. One of these is the usc of the cold feed gas in chilling the lean solution in E1 which cools the gas overheads from K1. This is advantageous due to the high heat of absorption of ammonia in water which is passed to a larger extent into the liquid phase. Secondly a high pressure in K1 favours the production of high-strength rich solution and minimizes ammonia *and* water content in the overheads gas. In some cases 30 per cent by weight aqueous ammonia solution can be made which minimizes the flow of rich solution to the distillation column K2 and reduces utilities required to drive it, namely steam and cooling water. Thirdly, pumps P1 are duplicated giving very high plant reliability as there are no other machines involved. Fourthly, ammonia product is drawn off at purities of up to 99.95 per cent by weight as liquid with an inerts vapour purge for the top of the condenser which can be recycled to the ammonia plant and also recovered. The result is a reliable unit with low running costs giving ammonia recoveries in the high nineties.

6.3.3 *Modern cryogenic units treating ammonia purge gas*

Improvements were continually being made in the design and manufacture of plate-fin heat exchangers with a specific aim at increasing the design pressure, and in the mid 1970s a design pressure of 7700 kPa was attainable. As a result, a new era for cryogenic units developed for treating ammonia purge gas where the product could be returned at about 7100 kPa to the ammonia plant, saving considerable energy for the ammonia plant operator. Moreover operating at this pressure, the available Joule–Thomson refrigeration was much higher than before.

The cryogenic section of the process has therefore become very simple, as shown in Fig. 6.7.

The dry, ammonia-free purge gas from the adsorbers shown in Fig. 6.6 enters the cold box, W1, where the gas is cooled in a cold box exchanger, E1, against revert hydrogen product and fuel gas streams together with a small quantity of supplementary refrigeration. The level and amount of supplementary refrigeration is determined by the feed pressure into the cryogenic unit and the fuel pressure leaving the unit. A vapour/liquid phase separation in a feed gas separator, V1, is carried out to produce a hydrogen-rich vapour phase and a liquid phase containing some nitrogen, a large proportion of argon and methane, and a small proportion of dissolved hydrogen.

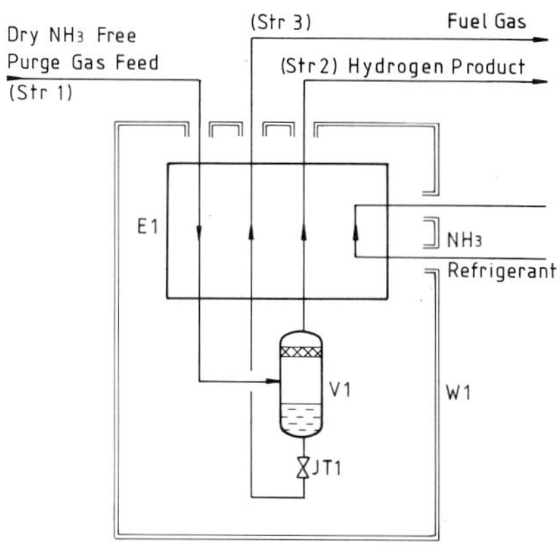

FIG. 6.7. A modern cryogenic unit treating ammonia purge gas.

The hydrogen product stream is warmed to approximately ambient temperature by heat exchange with the incoming feed gas and delivered to battery limits. The liquid phase from the separator is expanded to fuel gas pressure and also evaporated and re-heated to approximately ambient temperature.

Fuel gas leaving the cold box heat exchanger is split, a side-stream being used for adsorber regeneration and the remainder by-passed and mixed with the spent regeneration gas. The combined fuel gas product stream is then delivered to the ammonia plant at the required pressure for use in the reformer burners.

A typical mass balance for a unit operating in two ammonia plants in USA for the Agrico Company is given in Table 6.4.

For the process described above to operate to the mass balance in Table 6.4, only a small amount of ammonia refrigeration is required. This is because there is a large Joule–Thomson effect between 7300 kPa

TABLE 6.4. The performance of a cryogenic unit treating ammonia purge gas.

	Composition (mol.%)		
	Feed gas to cryogenic unit	H_2 product	Fuel gas
Hydrogen	63.0	86.7	10.8
Nitrogen	21.4	10.7	44.2
Argon	5.6	1.6	14.2
Methane	10.1	1.0	30.8
Temperature (°C)	10	7	7
Pressure (kPa)	7300	7000	500
Flow-rate ($Nm^3\ h^{-1}$)	16 539	11 373	5166
($kg\ mol\ h^{-1}$)	737.9	507.4	230.5

and 500 kPa for the liquids expanding across JT1, thus supplying enough refrigeration to drive the process.

The process is widely used for purge gas treatment because of its simplicity and reliability. The pretreatment adopted is nearly always the water-wash route described in section 6.3.2.

6.3.4 *Calculation of cold balances for cryogenic units*

Most cryogenic equipment is housed in a carbon steel plate-and-frame structure called the cold box. The cold box in a modern ammonia purge gas unit contains E1, V1, and JT1. The interior space is then filled with free-flowing Perlite insulation or fibre-glass wool and purged with nitrogen to keep the enclosure dry and inert. Using the NH_3 purge gas cryogenic unit as an example, the heat balance calculation basis can be illustrated. Figure 6.8 shows the overall inputs and outputs into a cryogenic unit, as described in section 6.3.3. These calculations are a fundamental part of cryogenic unit design.

An early assumption that is made in cold-box heat-balance calculations is the approach temperature differences between the incoming feed gas and the return product streams. This gives the amount of refrigeration that must be supplied to the process but does *not* intimate the *level*

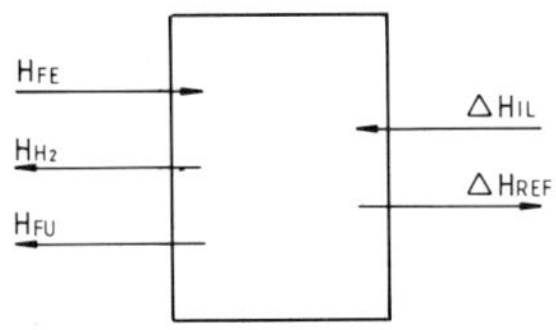

FIG. 6.8. The overall cold box balance.

at which it must be supplied. This information is gained by a different type of calculation which is illustrated a little later in this section. As plate-fin heat exchangers are highly efficient, current design often use a 3°C approach temperature at the warm end for this type of plant. Therefore, with a 10°C feed gas, the products leave at 7°C. Thus, for a mass balance in Table 6.4 we can calculate heat flows in and out of the cold box by multiplying the flowrate and the enthalpy. Heat inleak to a cold box is based largely on operating experience or a very complex calculation, since an exact thickness of insulation is always difficult to establish. Also varying atmospheric conditions can change the heat transfer coefficient between the cold box outer wall and the atmosphere. For this type of plant we shall assume an overall heat inleak $H_{IL} = 4000$ kcal h^{-1}. Therefore, the heat balance becomes:

$$(H_{FE} \times 737.9) + 4000 = (H_{H_2} \times 507.4) + (H_{FU} \times 230.5) + H_{REF}$$

where H_{FE} = feed gas enthalpy at 7300 kPa and 10°C = 1745 kcal kg^{-1} mol^{-1}

H_{H_2} = hydrogen product enthalpy at 7000 kPa and 7°C = 1687 kcal kg^{-1} mol^{-1}

H_{FU} = fuel gas enthalpy at 500 kPa and 70°C = 1887 kcal kg^{-1} mol^{-1}

Flows are in kg mol h^{-1}; enthalpies in kcal kg^{-1} mol^{-1}.

Therefore ΔH_{REF} = 5350 kcal h^{-1}, which is the amount of refrigeration that we need to supply to the process, but we do not know at what temperature. The next stage of the calculation is to determine the level required for the refrigeration which can be done by plotting cooling curves for each of the streams to determine the temperature differences all the way up the heat exchanger E1. This is done by calculating a series of enthalpies for different temperatures for each of the streams and normalizing them to zero at the cold end of the plant. The enthalpies are then multiplied by the flow rate for each of the streams and the heat inleak is added to the feed gas. The sum of the heat inleak plus feed gas cooling is plotted on a graph of heat load versus temperature; the sum of the warming streams (in this case H_2 product and fuel gas) is also plotted on the same graph. Figure 6.9 shows typical curves for the unit described in Table 6.4. The Kelvin scale is shown for convenience. Temperature differences are greater than 3 K over temperature ranges from 90 to 253 K (−183 to −30°C). At warmer temperatures ΔT falls below 3 K and thus refrigeration is required at about −30°C to bring all ΔT to above 3 K.

This calculation can be repeated for feed gas pressures of 4000 and 9500 kPa to see what effect the feed pressure has on the Joule–Thomson effect and how it affects temperature differences at both the warm and

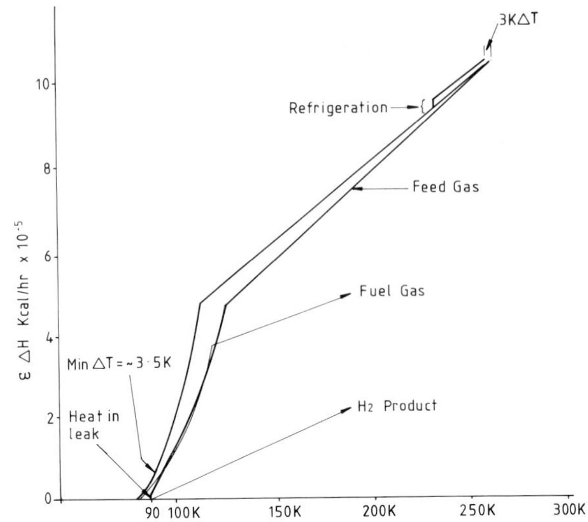

FIG. 6.9. Cooling curves for a hydrogen recovery unit.

cold ends of the exchanger E1. Table 6.5 shows the result of this calculation, which assumes the same fuel gas pressure and allows for changes in equilibrium compositions as a result of operating at different pressures.

One can see from Table 6.5 that not only does the warm end temperature difference reduce with reducing feed gas pressure, but so do the cold end temperature differences. In extreme cases, cold end temperature differences may become unacceptably small and the designer would have to reduce the hydrogen purity (design the cold end separator V1 for a warmer temperature) in order to produce an acceptable cold end temperature difference. This is the usual procedure the skilled designer of a cryogenic unit adopts when optimizing a unit of any type and complexity to suit product recovery and purity requirements.

If, instead of varying feed gas pressures, one varies the fuel gas pressure one gets a similar effect which can be taken advantage of in a few cases. Ammonia purge gas treatment is an example where sometimes the ammonia plant operator wants to recover a poorer purity hydrogen

TABLE 6.5. The effect of feed pressure on the Joule–Thomson effect.

Feed gas pressure (kPa)	Warm end ΔT (K)	Min. cold end ΔT (K)	Refrigeration level (°C)	Refrigeration load (kcal h^{-1})
4000	0.3	> 3	−73	14 000
7000	2	> 3	about −30	5350
9500	> 3	> 3	none required	zero

product knowing that he will get more nitrogen product, which is also beneficial to the maximization of ammonia output. This can be done by progressively raising fuel gas pressure to reduce the Joule–Thomson effect and reducing the purity.

Another point worth mentioning is that Table 6.5 shows that all of the process options can operate using aluminium plate-fin heat exchangers which currently have a design limitation of about 10 000 kPa. The third option also uses wound-coil heat exchangers and this can be done in copper, stainless steel, or aluminium. A considerable amount of experience was gained in the 1960s and 1970s by Chemie Anlagenbau, Dresden, GDR in using copper wound-coil heat exchangers for cryogenic processes treating ammonia purge gas. Stainless steel wound-coil heat exchangers have been little used in this application due to their expense, but aluminum paired-tube wound-coil heat exchangers have been applied with some success in a novel design of ammonia plant which started up in Redwater, Canada in 1984. This process returns hydrogen product at synthesis loop pressure in excess of 14 000 kPa. Another novel ammonia synthesis process has been designed by ICI and called the AMV process, which proved itself successfully for the first time at CIL at Sarnia, Ontario, Canada in 1985. This process also uses a cryogenic process with aluminium-tubed exchangers to recover hydrogen from ammonia purge gas returning it at synthesis loop pressure, again minimizing compression energy losses. This ammonia synthesis loop operates at approximately 8000 to 10 000 kPa. Although these pressures are tantalizingly low, they are not low enough to be comfortably within aluminium plate-fin heat exchanger design pressures.

In general, the new ammonia processes that emerged onto the market in the early 1980s mostly used some form of cryogenic recovery unit to recover ammonia purge gas or, in some cases, to purify synthesis gas. Moreover, many of these put a greater reliance on the cryogenic unit by putting larger flows to it, because its high efficiency was beneficial to the ammonia process.

6.4 Ammonia synthesis gas purification

6.4.1 *General process overview*

An alternative to the 'conventional' ammonia process described in section 6.2 and Fig. 6.3 is the partial oxidation route which successfully operates on many ammonia production sites where fuel oil, heavy hydrocarbons, or even coal are used as feedstocks.

This process requires extensive use of cryogenic technology in two areas:

(1) Production of oxygen and nitrogen for the partial oxidation of hydrocarbon feedstock and synthesis of ammonia respectively;
(2) To purify the synthesis gas before entry into the synthesis loop.

The process is shown in schematic form in Fig. 6.10 and is particularly suitable for use with heavier hydrocarbons as feedstock.

Oxygen from an air-separation unit and steam are used to achieve partial oxidation of the heavy hydrocarbon feedstock into hydrogen, CO, CO_2 and unconverted hydrocarbons—mainly methane. This process is made to operate at pressures in the range of 2500–8000 kPa, often higher than those favoured for steam reforming of natural gas. The air separation process is similar to those described in Chapter 3, and for this particular application a high pressure cycle is likely to be used where oxygen at relatively high pressures is obtained in order to avoid gaseous O_2 compression. After the reforming stage, CO shift conversion is achieved on chrome–molybdenum catalysts to ensure that nearly all CO converts to CO_2. The gas then contains large amounts of CO_2 which are removed by a Rectisol unit which washes out the CO_2 with an alcohol at sub-ambient temperatures and elevated pressures. The hydrogen-rich gas then enters the cryogenic unit which both adds the nitrogen to ensure a 3:1 hydrogen:nitrogen ratio, and scrubs out all of the CO down to less than 1 v.p.m. A typical cryogenic unit for this duty is shown in Fig. 6.11.

The feed to this process has already been passed through a Rectisol wash unit, is now relatively cold at $-50°C$, and is relatively free from CO_2 and water. The final removal of traces of CO_2, H_2O, and methanol is carried out on molecular sieves V1 and V2. Vessel V1 is onstream whilst vessel V2 is being regenerated and after a pre-set time, vessel V1 is regenerated and V2 is onstream. The clean gas leaving the molecular sieves will usually be well over 90 per cent H_2, containing some CO and methane. The gas is then cooled against returning streams in E2 to

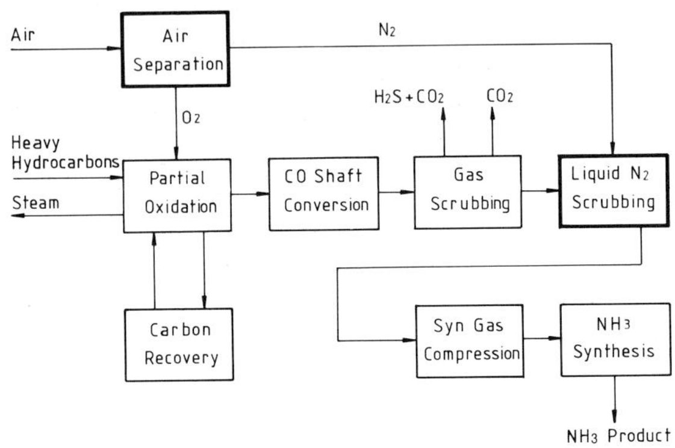

FIG. 6.10. The partial oxidation route to ammonia synthesis.

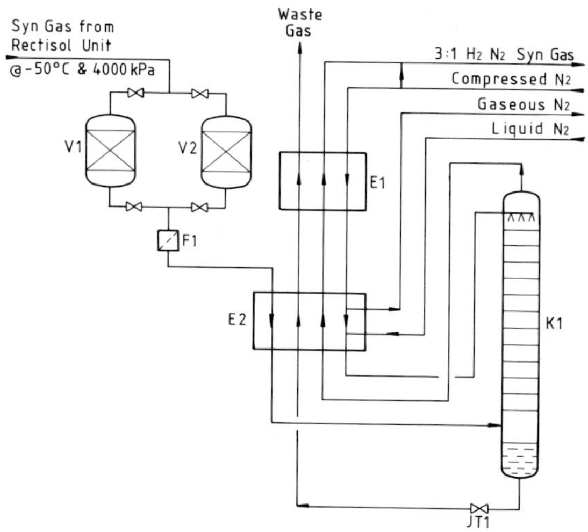

FIG. 6.11. Ammonia synthesis gas purification using liquid nitrogen wash.

around $-183°C$, a little below its dewpoint. The gas is then passed into the wash column, K1, where is it scrubbed with pure liquid nitrogen in order to remove methane and carbon monoxide to levels below 5 v.p.m. The overheads from the column are essentially pure hydrogen and nitrogen, which has now been added via the column reflux. This stream is reheated in E2 and after further make-up with N_2, forms the synthesis gas for ammonia production.

The liquid withdrawn from the base of the column K1 is mainly nitrogen, but contains virtually all of the CO and CH_4 that was present in the feed gas that entered K1. This liquid is flashed across valve JT1 to a low pressure, re-evaporated in E2, and reheated to ambient temperature in E1. The nitrogen required to make up the H_2:N_2 ratio to 3:1 is taken from an air-separation plant in both gaseous and liquid form. Some of this nitrogen is required as liquid to provide the necessary refrigeration for the process. The vapour nitrogen is admitted at high pressure, cooled in E1 and E2, mixed with liquid nitrogen in E2, and fed as a two-phase mixture into K1 as reflux. At higher pressures this nitrogen may be supercritical in E2, but as soon as it reaches the hydrogen in the column K1 it becomes subcritical again and thus allows mass transfer to take place within the column. There is no machinery associated with the unit as all the necesary compression and pumping equipment is near the less hazardous air-separation unit.

6.4.2 *The Braun purifier*

A variant of the process shown in Fig. 6.10 and 6.11 is the famous Braun purifier process, which has been in operation since the mid 1960s. This process has been modified and improved to reduce energy consumption over the years and the unit in operation today competes favourably in the ammonia technology market. The process uses a cryogenic unit to purify the synthesis gas in a similar fashion to that shown in Figs 6.10 and 6.11. The Braun process is based on the steam reforming route for producing hydrogen and uses excess air fed into the secondary reforming to provide the nitrogen for the synthesis of ammonia. The cryogenic unit is added after the methanator which converts any CO and CO_2 into CH_4. The Braun purifier process is shown in block diagram form in Fig. 6.12, whilst the cryogenic unit is shown in Fig. 6.13.

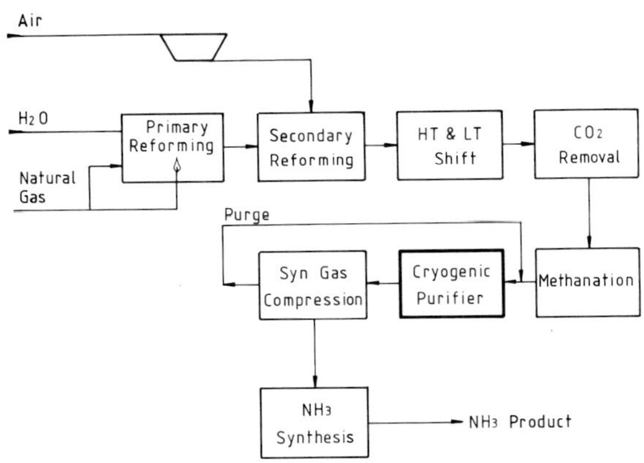

FIG. 6.12. The Braun purifier ammonia synthesis process.

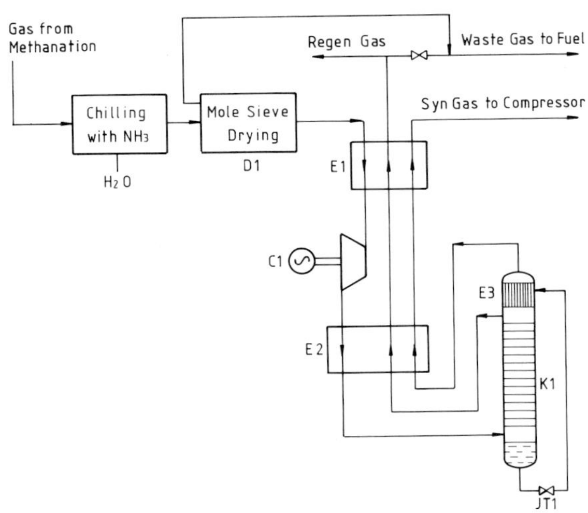

FIG. 6.13. The Braun purifier cryogenic unit.

The gas entering this cryogenic purifier is saturated with water as it leaves the methanator and chilled with ammonia refrigerant to reduce water content. The gas is then dried in molecular sieve adsorbers D1 before entering the cold box which houses the cryogenic equipment. The feed gas entering the cold box has a hydrogen:nitrogen ratio lower than 3:1, along with argon and methane. The cryogenic unit achieves two things: it reduces the level of argon and methane to a very low value, and adjusts the H_2:N_2 ratio to 3:1 before compression into the ammonia synthesis loop. The dry feed gas enters the first exchanger at typically 2800 to 3300 kPa and is cooled before entering the expansion turbine C1 which work expands the synthesis gas generating electricity in a generator coupled to the expander. The outlet gas from C1 is not much colder and at about 2600 kPa, and is further cooled and partially condensed in E2. The two-phase mixture passes into the base of column K1 where the synthesis gas is scrubbed from argon and methane by reflux produced in overheads condenser E3. The liquid collected at the base of column K1 is flashed in JT1 to ~ 300–700 kPa and supplied to E3 to provide the necessary refrigeration which produces the reflux in column K1. The low-pressure vapour leaving the condenser E3 is reheated through E2 and E1, used in part to regenerate the driers, and is then fed to the ammonia plant fuel header. The column K1 overheads, now essentially 3:1 H_2:N_2 and relatively free from argon and methane, are reheated in E2 and E1 before being compressed from about 2400 kPa into the ammonia synthesis section. The amount of purge gas taken from the synthesis loop is much smaller than for the conventional ammonia process described in the early parts of this chapter, and is recycled to the upstream side of the cryogenic unit.

6.4.3 *Simple nitrogen adjustment processes*

New ammonia synthesis processes are continually emerging in an ever-competitive market. One such process was patented in 1983 by Foster Wheeler and is based on the partial oxidation (PO) of heavy hydrocarbons to produce hydrogen using air, both as an oxidation medium and as the source of nitrogen for ammonia manufacture. A block flow diagram for this process is shown in Fig. 6.14. The outlet gases from the PO unit pass into a CO shift reactor to produce CO_2 and are then treated with a sweetening process such as MEA/DEA or Selexol to remove H_2S and CO_2 from the gas. The gas then enters the cryogenic unit, which is sometimes termed a 'nitrogen adjustment plant'. This unit, shown in Fig. 6.15, comprises a pretreatment to remove water and CO_2 and a low temperature section which produces a 3:1 H_2:N_2 ratio product which has only small amounts of argon and methane. The gas is then methanated to convert any CO into CH_4.

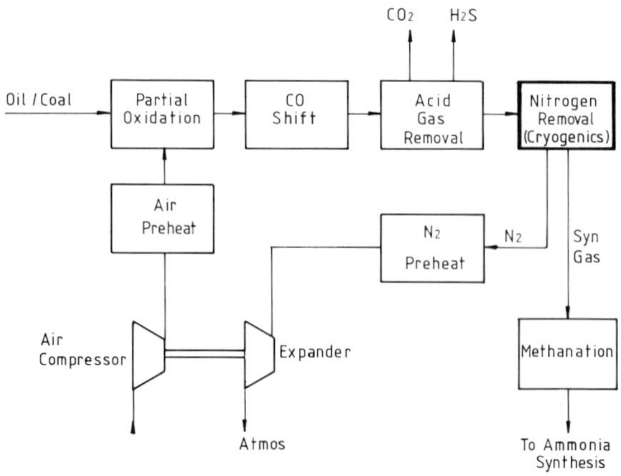

FIG. 6.14. The air partial-oxidation process for ammonia manufacture.

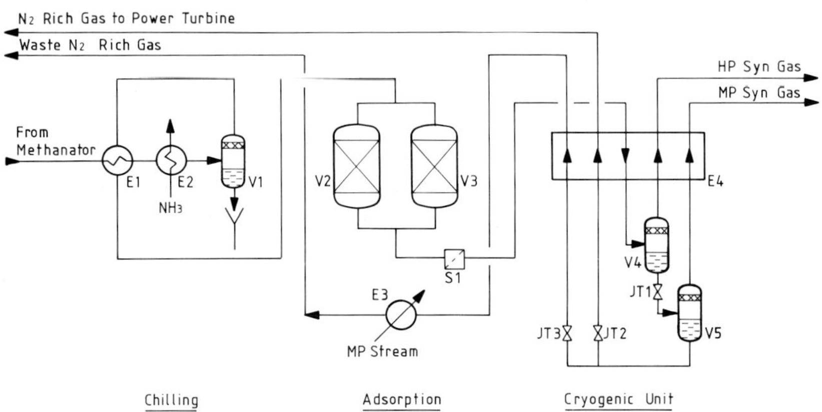

FIG. 6.15. A nitrogen adjustment plant.

This cryogenic unit treats the whole of the gas, chilling out the water using ammonia refrigerant and reheating the gas in E1 and E2. After separating out the water in V1 the gas is reheated in E1 and dried in one of two adsorbers V2 and V3. Regeneration of V2 or V3 is carried out cyclically by passing dry low pressure N_2 waste gas through regeneration heater E3 and through the vessel requiring regeneration. After the dry synthesis gas passes through the filter, S1, it is admitted to the low temperature section of the unit, first passing into plate-fin heat exchanger assembly E4. The feed gas leaving E4 is a two-phase mixture, the vapour phase having approximately 3:1 H_2:N_2 and small amounts of argon and methane, and the liquid being mainly N_2 with some argon and CH_4. V4 separates the two phases, whereupon the liquid is flashed to a lower pressure and the hydrogen-rich phase is separated out in V5. This hydrogen-rich vapour, a much smaller flow than the vapour from V4, is reheated in E1 before being compressed into the HP main flow of the synthesis gas. The vapour from V4 is reheated in exchanger E1 before being sent first to the methanator and then to the main ammonia synthesis gas compressor, after mixing with the small flow of reheated and compressed V5 vapour. Expansion across JT2 is only to about 700 kPa and, upon reheat in E1, this portion of the gas is sent to a preheater and then into a gas turbine which drives the main air compressor on the plant. The remainder of the V5 liquids are expanded in JT3, reheated in E1, and partly used to regenerate the driers V2 and V3. A typical mass balance for this type of cryogenic unit is shown in Table 6.6, and it can clearly be seen that this unit is very large in capacity and a mainstream item in the process. A variant on this process is one which could have a wash column to remove CO using a nitrogen stream generated from the feed gas. The type of process used depends on feed conditions and on the complex economics involved.

An alternative scheme to that shown in Fig. 6.13 would be to methanate the raw synthesis gas immediately after the acid gas removal stage. This would mean that the pretreatment would not have to remove CO_2 and the cryogenic unit would produce CO-free synthesis gas. The result is a slightly cheaper cryogenic unit because of the reduced pretreatment cost. Alternatively, a Braun-type cryogenic unit could also be used for this process where part of the feed gas could be methanated before the cryogenic unit and the remainder allowed to enter the cryogenic unit un-methanated. The methanated feed gas could be cooled and partially condensed to the same temperature as required by the

TABLE 6.6. The mass balance for a typical cryogenic N_2 adjustment plant (mole per cent).

	Feed gas to plant	HP H_2 product	MP H_2 product	MP waste	LP waste
H_2	49.36	76.34	50.01	0.90	0.90
N_2	48.42	23.13	48.93	95.36	95.36
CO	0.87	0.35	0.71	1.87	1.87
Ar	0.56	0.16	0.32	1.32	1.32
CH_4	0.20	0.02	0.03	0.55	0.55
H_2O	0.49	—	—	—	Variable
CO_2	(200 v.p.m.)	—	—	—	Variable
Flow-rate ($Nm^3\ h^{-1}$)	194 000	119 980	7 120	46 030	20 870 (ave.)
Pressure (kPa)	2750	2550	400–1400	600–800	120
Temperature (°C)	50–60	40	40	40	40

process shown in Fig. 6.15. However, instead of V4, a column is used taking methanated feed at the top and CO-containing feed at the base—both two-phase. The liquid part fed to the top washes out CO from the vapour part fed to the column base. The liquid from the column sump is flashed in JT1 into V5 to provide the refrigeration needed in E1 to drive the process via JT2 and JT3. The MP synthesis gas would have to be recycled to the front end to remove CO.

This shows how versatile cryogenic technology can be when different duties are presented to the designer. It is this versatility that has kept such mature technology in the forefront of process developments.

6.5 Ethylene plants

6.5.1 *Ethylene production facilities*

Ethylene is one of the fundamental building blocks in the petrochemical industry. Chapter 2 describes the variety of uses that commercially produced ethylene has been put to, ranging from acrylonitrile, vinyl acetate through to PVC and polyethylene. The 1950s and early 1960s saw a rapid growth in the petrochemical industry. During the decade since the mid-1950s, the ethylene capacity increased by 300 per cent in Western Europe, by 1000 per cent in the USA and by a staggering 5000 per cent in Japan. As a result, many petrochemical complexes had a small- to medium-sized ethylene plant. These plants ranged in size from 50 000 to 100 000 tonnes per year in the 1950s and early 1960s.

Worldwide ethylene capacity continued its rapid growth in the late 1960s. This period saw several significant technology changes in the ethylene plant itself. First, the size of a typical single train ethylene plant increased to 200 000 tonnes per year, with some designs as large as 450 000 tonnes per year. A typical ethylene manufacturing process is shown in Fig. 6.16 and is based on naphtha feedstock. To appreciate the areas where the ethylene process has changed, it is necessary to understand its flowsheet in a naphtha-based process. The naphtha is first admitted to a cracking furnace along with steam. The result is a mixture of hydrogen, olefins, aliphatic compounds ranging from C1 to C4 hydrocarbons, some heavier hydrocarbons, and some aromatics. A typical naphtha feedstock is given in Table 6.7 below.

The cracking furnace operates at relatively low pressures and high temperatures in the region of 727–770°C. Often designers of the larger ethylene plants had to consider the changing market situation and ensure that the cracking furnace could not only handle different feedstocks but also crack at different severities. This meant that the operator could make different amounts of valuable by-products such as propylene,

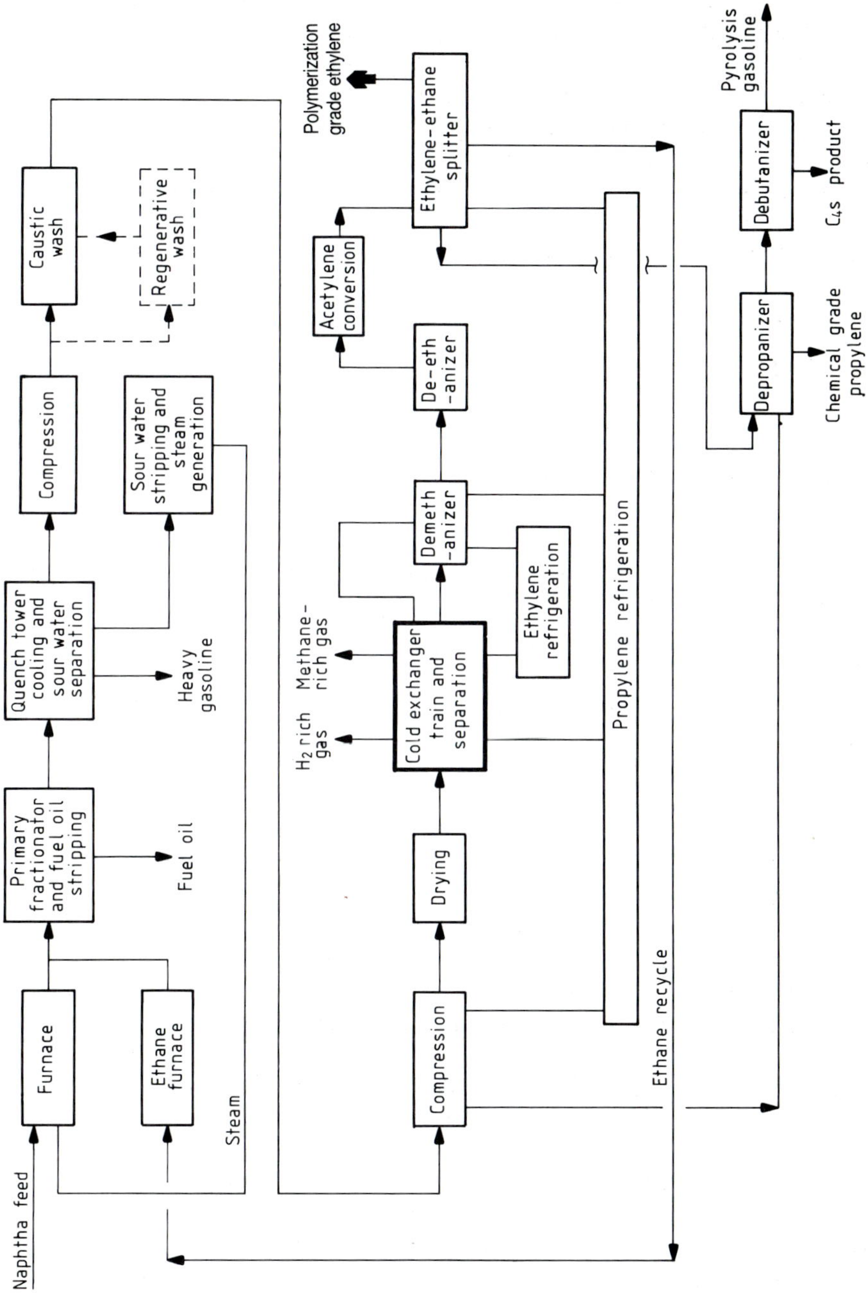

Fig. 6.16. An ethylene plant based on naphtha feed.

TABLE 6.7. Typical naphtha feedstock.

Specific gravity at 15°C	0.69
Boiling range (°C)	
IBP	41
50%	65
EP	154
Composition (mol.%)	
Paraffins	83
Naphthenes	14
Aromatics	3

butadiene, cracked gasoline, and fuel oil. This gave him the necessary flexibility to accommodate the changing market and stay profitable. A typical yield pattern for a cracking furnace fed with light naphtha is given in Table 6.8.

Flexibility was recognized to be an important factor in the 1970s and at that time the Union Carbide Corporation embarked on a development program to test an advanced cracking reactor (ACR) which ran from 1979 to 1981. Results published in 1983 showed that liquid feedstocks such as naphtha and atmospheric and vacuum gas oils could be cracked using oxygen and steam to produce very high yields of ethylene. Typically, 5–10 per cent more ethylene results from the ACR process compared with the conventional process treating naphtha or atmospheric or vacuum gas oils. Following these trials, commercial interest varied due to uncertainties in prices of oil and hydrocarbon feedstocks.

Rapid increases in chemical plant equipment manufacturing costs in the 1950s raised the capital cost of the ethylene plant, which had its

TABLE 6.8. Cracker yields for naphtha feedstock (mole per cent).

	Cracking severity		
	Mild	Moderate	Severe
Hydrogen	0.4	0.6	0.7
Methane	9.4	15.5	18.4
Ethylene	16.1	22.5	25.3
Ethane	4.5	5.2	4.8
Propylene	14.5	15.8	15.5
Propane	0.6	0.9	0.9
Butadiene	2.4	3.1	3.1
Other C_4s	8.0	7.1	6.1
Cracked gasoline	40.6	25.0	20.2
Fuel oil	3.5	4.3	5.0

effect on the product selling price. This made designers consider much larger units incorporating larger single-train furnaces and centrifugal compressors. This, along with judicious design, resulted in more efficient and much larger single train ethylene plants approaching 450 000 tonnes per annum in the 1960s and early 1970s. The down-turn in the chemical industry in the 1970s meant that little new plant was constructed for the chemical industry. The late 1970s and early 1980s saw a rationalization in the petrochemical industry with a few, relatively large and flexible ethylene plants coming into operation.

A good example of such a large ethylene plant was started up in October 1985 at Mossmorran in Scotland. This unit has a capacity of 500 000 tonnes per annum of ethylene. Similarly sized units also came on-stream in Saudi Arabia in 1985 and 1986 at Al Jubail and Yanbu. The ethylene plant at Yanbu was designed for a capacity of 450 000 tonnes per year, whilst the Al Jubail plant was capable of a staggering 656 000 tonnes per year. The process route for these plants was similar to that shown in Fig. 6.16 with a few important differences. The main feedstock for the above units is ethane which is admitted to several cracking furnaces in parallel. For the Al Jubail plant seven furnaces are required, which means that the design has departed considerably from the true single-train concept. The second improvement is that the gas sweetening/caustic wash is being carried out at higher pressures downstream of the cracked gas compressor.

The cryogenic part of these large ethylene plants also had some important additions to improve process efficiency. For example, the Al Jubail plant utilized a turbo-expander on the demethanizer overheads to provide refrigeration at levels of $-160°C$. This allowed the production of relatively pure hydrogen by partial condensation of methane and heavier components from the demethanizer overheads. This stream was further upgraded to pure hydrogen by using pressure swing adsorption (PSA) technology.

The ethylene plant of the 1980s, therefore, retains most of the features of its predecessors of the 1950s and 1960s. The basic process consists first of cracking, which may use naphtha, ethane, propane, butanes, or gas oils as a feedstock. Often, ethane recycle crackers will be operated alongside naphtha crackers to increase yield. In some cases, dual feedstock capability is built into the design, or the unit is revamped in order to maximize the plant profitability over rapidly changing feedstock prices. The next phase of the process for a naphtha fed plant is primary fractionation and fuel oil stripping, followed by quench cooling and sour-water separation. These steps essentially remove the heavier hydrocarbons and water from the mixture before compression. Centrifugal compressors are used on modern plants, often using five or more stages with cooling and caustic wash in between the stages to remove acid gases.

6.5.2 *Cryogenic processing in ethylene plants*

The resultant mixture contains hydrogen, methane, ethylene, ethane, propylene, propane, butanes, and heavier hydrocarbons. However, hydrocarbons of carbon number greater than five are present in relatively low concentrations. As the above mixture is made of low molecular weight components, it is an excellent example for the application of cryogenic technology. As a result the next phase of the process is to dry the gas using molecular sieve adsorbers to remove water.

The cryogenic section of an ethylene plant is a highly integrated example which uses shell-and-tube heat exchangers as well as plate-fin heat exchangers for the application of refrigeration and cold recovery. The work of separation is relatively high for this process because of the multitude of pure products. This requires a large refrigeration cycle which uses both ethylene and propylene as refrigerants.

Figure 6.17 shows the 'cooling-train' and associated low-temperature equipment of a typical ethylene plant. After compression in the cracked-gas compressor the feed gas is dried in molecular sieves before it enters the cryogenic unit at about 3500 kPa and is cooled to about −10°C. This cooling is often done with cold available from the ethylene fractionation column reboiler. Ethane produced in the reactions earlier and separated in the ethylene–ethane splitter is reheated, recycled, and cracked to produce more ethylene which is mixed with the feed gas.

The main feed gas enters the first shell-and-tube heat exchanger E10, where it is cooled to about −20°C and then cooled and partially condensed further to about −37°C using propylene refrigerant in E11.

At the outlet of E11, the condensed fraction is removed in separator, V1, and sent to the demethanizer column as a feed stream, while the vapour fraction is sent to the cold box plate-fin exchanger E2. About 10 per cent of the charge gas condenses in E2 against the return hydrogen and fuel gas streams. The two-phase stream then leaves the cold box and enters exchangers E12 and E13, where two stages of ethylene refrigeration at −50°C and −75°C chill and further condense about 30 per cent of the remainder of the feed. At the exit of E13, the condensate is again separated in V2 and the vapour portion returned to the cold box. At the second separator, the majority (almost 75 per cent) of the original feed stream has been condensed. The cold-box exchanger E3 condenses about 25 per cent of the remaining vapour portion against the hydrogen and fuel gas streams, and the partly condensed stream is sent to the final chilling train exchanger E14, where ethylene refrigeration at −100°C. is used.

Again, the vapour portion is returned to the cold box, chilled, and partially condensed against the hydrogen and fuel streams to about

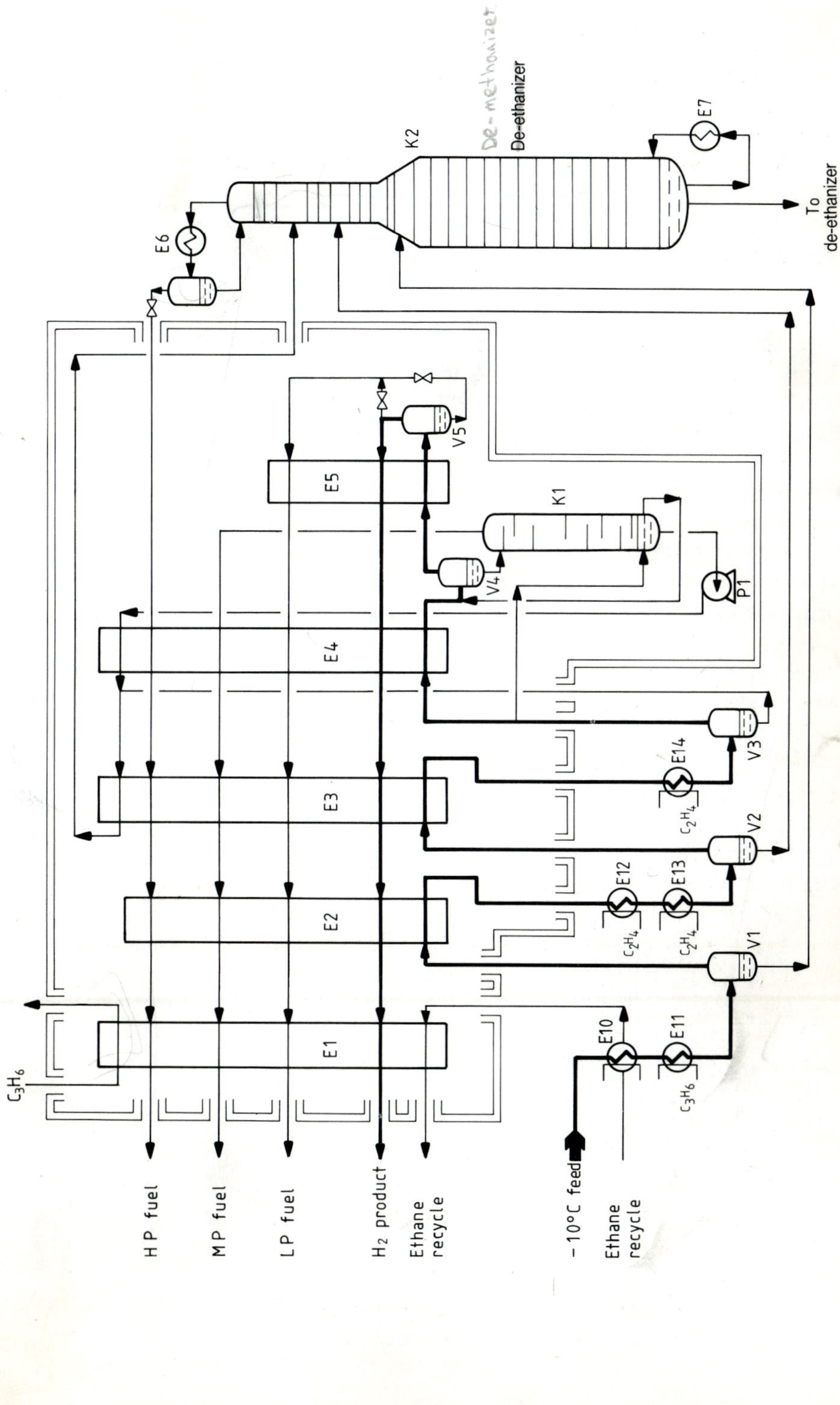

FIG. 6.17. A typical ethylene plant cooling train, hydrogen upgrading, and de-ethanizer.

$-140°C$. The ethylene-rich liquid phase is processed, after throttling, in a stripper column, which produces a relatively high purity (80 per cent) ethylene product at a high ethylene recovery. This stream is combined with the liquid phase from V3 to become the final feed to the demethanizer column K2.

Vapour-phase effluent from separator V4, consisting essentially of hydrogen and methane, is further cooled and partially condensed in heat exchanger E5 until sufficient methane has been condensed to give the remaining vapour the desired hydrogen concentration. Hydrogen product is recovered from the vapour phase at V5. Finally the hydrogen product and fuel streams, along with the tail gas stream from the demethanizer overhead, are reheated, thus completing the cycle.

The extent to which shell-and-tube heat exchangers are used in the ethylene plant cooling train depends on the size and sometimes on the designer's preference. The process description covers what is a typical example of an ethylene plant cryogenic section, and it should be noted that the de-methanizer bottoms product goes to a number of distillation stages before polymer grade ethylene is produced. Several of the distillation columns in a typical ethylene plant cooling train and separation process operate at relatively low temperatures using either ethylene or propylene as refrigerant in the overheads condensers. The column configurations are conventional in terms of equipment and therefore they will not be covered in detail here.

The ethylene plant cooling train requires the use of a significant refrigeration cycle which can produce both ethylene and propylene refrigerants to the various users on the unit. Figure 6.18 is a simple cascade refrigeration system which shows how cooling water can be a condensing medium for refrigerant which evaporates as low as $-100°C$.

A typical cascade cycle consists of separate ethylene and propylene circuits linked only by the ethylene condenser, E2. The low pressure

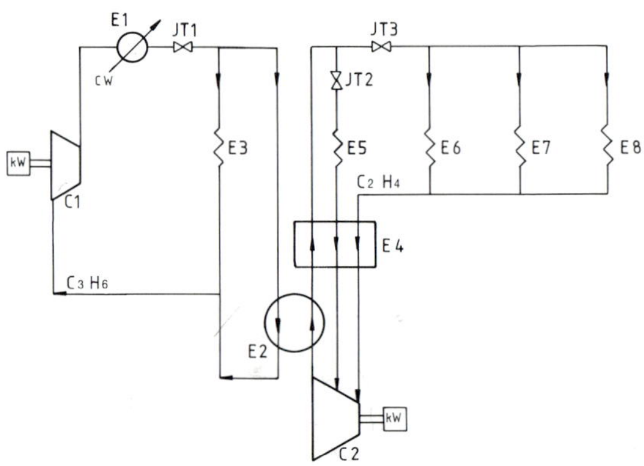

FIG. 6.18. The cascade propylene–ethylene refrigeration cycle.

propylene is compressed from about 105 kPa and −40°C to approximately 1300 kPa, whereupon it is condensed in E1 against cooling water. Flashing across JT1 results in some vaporization, but the majority of the propylene is available as liquid at −40°C to be evaporated in refrigerant consumers E3 on the process side and E2 on the cascade cycle side. The evaporated C_3H_6 leaving E2 and E3 is returned to the cycle compressor C1 to complete the circuits. Ethylene is compressed in C2 to about 1700 kPa, condensed against propylene in E2 at −40°C, and sub-cooled in E4 against returning vapour C_2H_4 streams. The sub-cooled liquid C_2H_4 is then flashed through JT2 and JT3 to provide refrigeration in various parts of the process in exchangers E5 to E8. The vapour from E5 can be at 300–400 kPa and ~ −73°C, whilst E6 to E8 vapours may be at 120 kPa and ~ −100°C. Both vapours are reheated in sub-cooler E4 before being recompressed in the cycle C_2H_4 compressor C2 to complete the circuit.

6.5.3 *Recovering ethylene from ethylene-rich purges*

A large part of ethylene is used in manufacturing polymers, of which polyethylene is a good example. Polyethylene is used for a variety of packaging applications such as bin and shopping bags, shrink wrappers, toys, and electrical and kitchen equipment. A typical low density polyethylene process (LDPE) is shown in Fig. 6.19, in which ethylene is compressed to very high pressure of ~300 000 kPa in order to polymerize it over a catalyst. Fresh ethylenc is admitted to a compressor, C1, which brings the pressure up to ~25 000–30 000 kPa. After heat exchange and mixing with initiator and moderator compounds to control the reaction, the ethylene is mixed with unconverted recycle ethylene and further compressed in a hypercompressor C2 to pressures as high as 320 000 kPa. The ultra-high pressure ethylene is then passed into

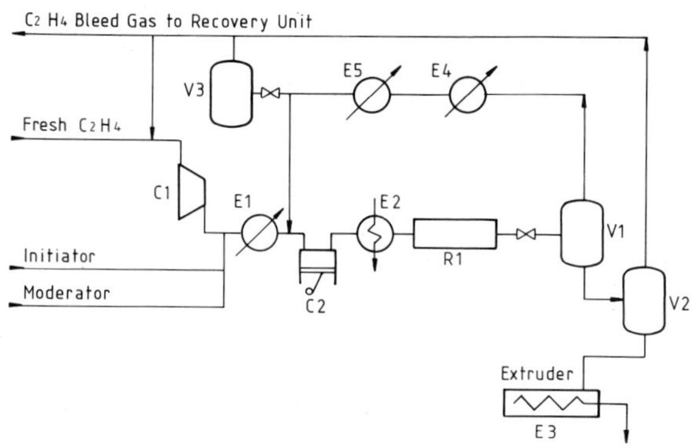

FIG. 6.19. The low-density polyethylene LDPE process.

reactors R1 where the polymerization occurs. The polymer is separated out relatively hot in V1 and V2. The product is then cooled and extruded in E3, followed by granulation or cutting into suitable particles for packaging and distribution to moulders and polyethylene film makers.

The vapour ethylene which is not converted is recycled at high and low pressure via heat exchangers E4 and E5, back to the main compressors C1 and C2. A bleed gas is taken from the low-pressure recycle and this stream can be treated to recover the ethylene for re-use in the process. There are some 150–200 plants producing polyethylene world-wide, and in the early 1980s the total polyethylene production in the UK alone was some 1500 tonnes per day. This gives considerable scope for a bleed gas recovery process.

LDPE bleed gas is a relatively small proportion of the total compressor duty, but a typical polyethylene facility can have anything from 5 to 20 tonnes per day of bleed gas, which is usually 97–99 per cent ethylene with impurities ranging from methane at the light end to oils and waxes at the heavy end. All of these impurities are present in relatively small quantities, but cannot be accepted in the reaction loop. Table 6.9 shows a typical mass balance for a cryogenic unit treating LDPE bleed gas.

TABLE 6.9. LDPE bleed gas recovery performance (Compositions in mole per cent).

	Bleed gas	C_2H_4 product	Waste
C_1 and lighter	0.3	0.1	4.0
C_2H_4	98.5	99.9	80.0
C_2H_6	0.8	traces	13.0
C_{3+}	0.4	—	3.0
Flow-rate (kg h^{-1})	500	approx. 470	approx. 30

The bleed gas recovery process is shown in Fig. 6.20 and consists of compression, oil wash, drying, and low temperature distillation. The gas is first compressed in C1, cooled, and washed with a heavy oil to remove waxes in K1, then dried in V1. The gas then enters the distillation column K2 at about 2500 kPa. The role of the distillation column is to remove light ends and heavy ends from the bleed gas, producing a side stream of pure C_2H_4 product. Because the key components in the separation are ethylene–ethane, relatively high reflux ratios are required to achieve a reasonable degree of separation within a practical sized column. The top and base of the column are relatively cold, thus requiring a refrigeration cycle to provide cold in E5. The heat-pump cycle can be based on propylene, propane, or freon, and consists of

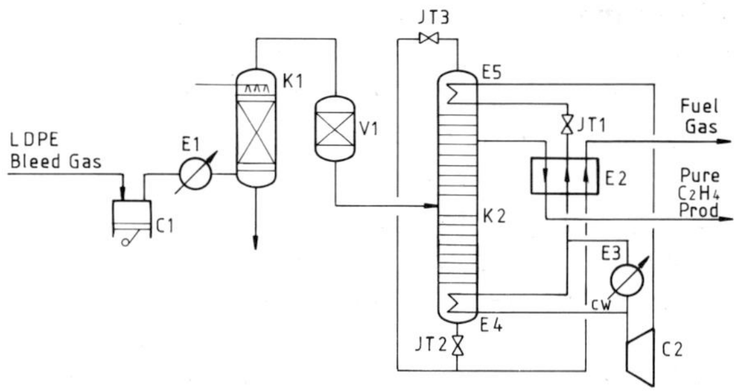

FIG. 6.20. Ethylene recovery from the LDPE process.

compression of refrigerant in C2 followed by cooling and condensation in reboiler E4 and condenser E3. The refrigerant is then sub-cooled by returning product and waste fuel gas in E2 before expansion to low pressure via JT1, which results in cold liquid being supplied to the overheads condenser E5. The evaporated refrigerant is subsequently returned to C2.

Polypropylene is made by polymerization of propylene, which is a relatively low-pressure process using hydrogen, hexane, and other components as reaction moderators and solvents. Propane is introduced into the process via impure propylene feedstock. The propane does not react in any way and behaves as a reaction inhibitor. Many of these polypropylene plants have, therefore, a bleed gas which purges light ends such as hydrogen and inerts such as propane, at the same time wasting valuable propylene. This situation is analogous to that described for LDPE bleed gas. A similar process can be used to recover the propylene and reject light ends and propane inert.

6.6 Carbon dioxide purification

Carbon dioxide is used in large quantities in the food and soft drinks industry, and is mainly obtained from the brewing and fertilizer industries, although other industrial processes such as glycol manufacture can also provide a source. Most of these processes produce a low pressure off-gas which has a high concentration of CO_2 but is unsuitable for the food industry. Therefore, many purification plants have been constructed to purify CO_2 to food grade. A typical example of a CO_2-rich source is the fertilizer plant as described in Fig. 6.3. Some of the CO_2 can be drawn off from the MEA/DEA plant and treated in a CO_2 purification and liquefaction process. A typical CO_2 purification process, shown in Fig. 6.21, resembles the LDPE bleed gas process described earlier.

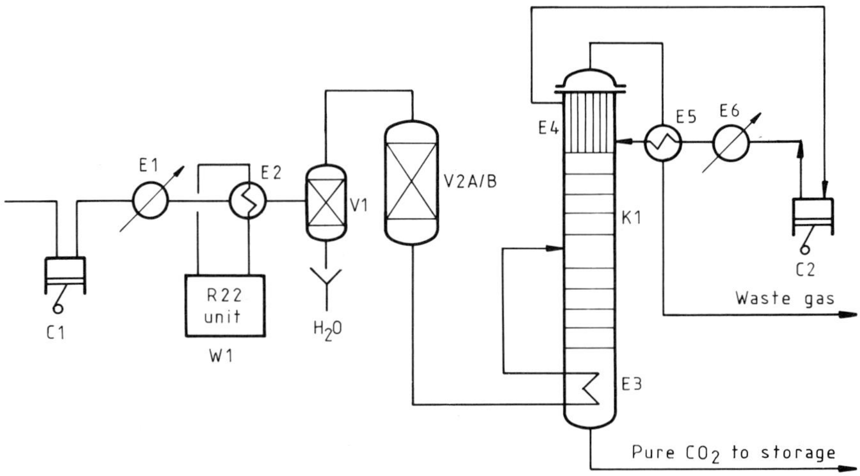

FIG. 6.21. The CO_2 purification and liquefaction process.

Impure, CO_2-rich feed gas is first compressed in C1 to approximately 1500–1600 kPa before cooling and chilling in E1 and E2 to remove the bulk of the water. Refrigeration is provided by a standard Freon unit W1. The gas is then dried in molecular sieve driers V2 A/B. The dry gas then enters the cryogenic section where it passes through the reboiler E3 before entering column K1. Reflux to the column is provided by means of a closed loop refrigeration cycle based on Freon R22. Freon is compressed in C2, condensed in E6, and sub-cooled against waste gas in E5 before being flashed to low pressure and evaporated in E4. The vapour refrigerant is returned to compressor C2. The bottoms product from K1 constitutes the pure CO_2 product which is routed to storage or dry ice formation in dry ice presses. Plants for this application range from 5 to 1200 tonnes/day of pure CO_2, which is used for food applications or enhanced oil recovery CO_2 injection. A typical mass balance is given in Table 6.10.

TABLE 6.10. The mass balance for a CO_2 purification and liquefaction unit (mole per cent).

	Feed gas	Waste gas	CO_2 product
$N_2/O_2/Ar$	0.50	2.7	<20 v.p.m.
CH_4	0.01	0.1	
C_2H_4	0.36	1.9	<5 v.p.m.
CO_2	95.21	95.3	100%
H_2O	3.92	variable	<5 v.p.m.
Flow-rate ($Nm^3\ h^{-1}$)	660	110 (dry basis)	524
Pressure (kPa)	101	200	1350
Temperature (°C)	ambient	ambient	−32

6.7 Carbon monoxide purification

6.7.1 *Carbon monoxide manufacture*

Carbon monoxide is often made by partial oxidation of hydrocarbons, fuel oils, or coal. For example a heavy fuel oil can be gasified in a 'Texaco' or 'Lurgi' type gasifier with the aid of oxygen and steam to produce a mixture of hydrogen, carbon monoxide, CO_2, and unconverted light hydrocarbons. The CO needs to be recovered from this mixture at relatively high purities so that it can be used in the production of various organic intermediates.

Carbon monoxide is used to make acetic acid, propionic acid, acrylic acid, formic acid, and formamide, and relatively high CO purities are required, typically greater than 98 per cent by volume. Therefore, very often, cryogenic processing is an effective means of achieving these purities. The integration of cryogenic units with the synthesis gas production plant is shown in Fig. 6.22. The raw material can be natural gas, naphtha, fuel oil, heavy residues oil, or even coal, and is reformed, partially oxidized, or gasified into hydrogen and carbon oxides. In the

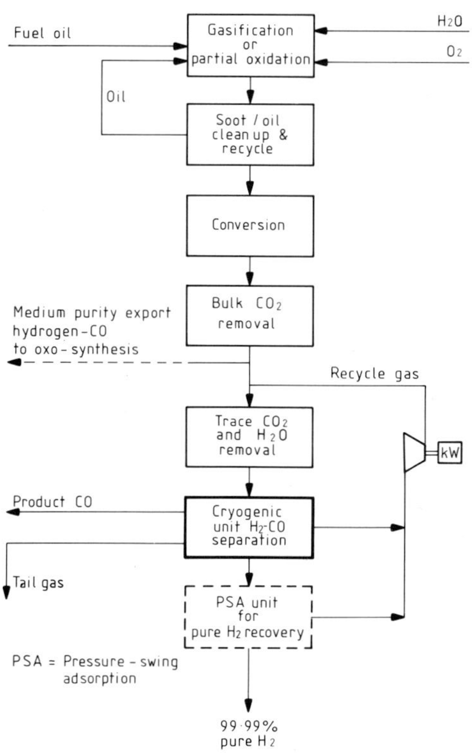

FIG. 6.22. A block flow diagram of a H_2/CO generation and purification plant.

case shown in Fig. 6.22, fuel oil is partially oxidized by O_2 and steam to produce hydrogen and carbon oxides. Following the block diagram in Fig. 6.22, the oxidation section products are cleaned up from oils and carbon before further conversion. The majority of CO_2 and any H_2S are removed from the gases by monoethanolamine ($HOC_2H_4NH_2$) and diethanolamine ($(HOC_2H_4)_2NH$) solution washing. The gas then contains significant quantities of H_2 and CO along with small amounts of CO_2, H_2O, and unconverted CH_4. Often some of this hydrogen and CO mixture can be routed to an oxo-alcohols synthesis plant. The remaining H_2 + CO mixture can be further treated to recover CO which is used to make acetic, propionic, and acrylic acids, and pure hydrogen which can be used by a variety of hydrogenation processes. The H_2–CO separation is often a cryogenic process which is preceded by a molecular sieve drying unit. High-purity hydrogen is produced by taking the medium-purity hydrogen produced by the cryogenic unit and further treating it in a pressure swing adsorption (PSA) unit to produce high-purity hydrogen. Waste gases rich in CO and H_2 can be recycled from both the cryogenic unit and the PSA unit by recompression and joining up with the feed to the cryogenic unit, thereby recovering higher proportions of hydrogen and CO.

When studying Fig. 6.22 it becomes clear that the cryogenic unit is a very important part of the whole process. The variety of products made from the H_2 and CO often determines the process route adopted by the cryogenic unit. For example if only acetic or propionic acid is the ultimate product, there is no need for large amounts of pure hydrogen. In this case, a simple cryogenic unit will suffice. However, if both hydrogen and CO products are required then a much more complex unit is needed. Also, differences in the feed gas composition to the cryogenic unit would affect the design of the unit. This kind of duty is a very good example of how differences in product requirements and feed composition affect the choice of process. Table 6.11 shows a typical range of feed gas composition for such a cryogenic unit.

Three cryogenic process alternatives are available, depending on whether:

(1) only CO product is required;
(2) CO product and H_2 product are required, and a PSA unit will follow the cryogenic unit to purify H_2 to 99.9 per cent;
(3) as for (2) but the H_2 product is of moderate purity but CO free.

6.7.2. *CO recovery in a simple cryogenic unit*

The simplest cryogenic process recovers carbon monoxide only and is shown in Fig. 6.23. In this process feed gas is admitted to the first heat

TABLE 6.11. H_2–CO recovery unit feed gas.

	Composition (mol.%)
H_2	56–76
N_2	0.1–0.4
CO	25–40
Ar	0.2–0.4
CH_4	0.5–2.0
CO_2	0.01–0.05
H_2O	Sat. at T and P
P (kPa)	3600–4000
T (°C)	40

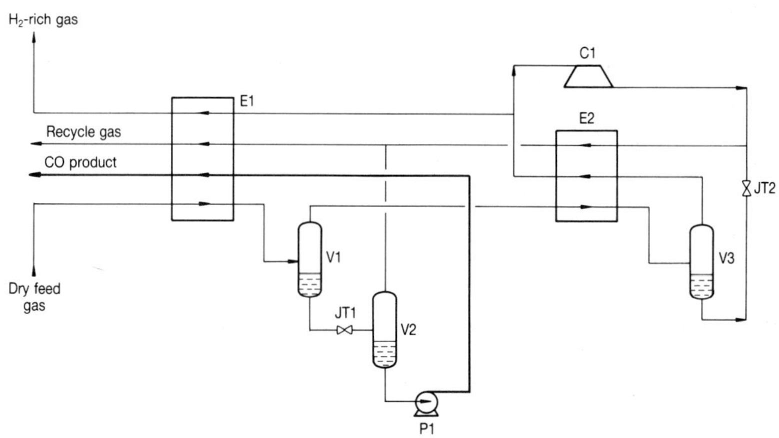

FIG. 6.23. A simple cryogenic CO recovery process.

exchanger, E1, where it is cooled and partially condensed. The liquid formed, which is relatively rich in CO, is separated in V1 and flashed across JT1 to vaporize most of the dissolved hydrogen, thereby producing virtually pure CO product. Purities of 98–99 per cent CO can be attained by this method provided the Ar, CH_4, and N_2 content in the feed is relatively low. The CO liquid product is then separated in V2 and can be pumped to the pressure required by the end user of CO before being evaporated in E1. Depending on the refrigeration balance, not all of the CO product can be pumped to high pressure. Some of the CO must be evaporated at low pressure to provide refrigeration at a colder temperature. This would then return through E1 as a separate low pressure product stream before being compressed to the required pressure. Obviously, detailed process optimization is needed to arrive at the correct split between the pumped CO product and the low pressure CO product returning streams in E1. The vapour from V1 is cooled further in E2 and partially condensed to about −203°C, thereby condensing more CO from the vapour phase. This liquid is separated in V3 and flashed in JT2 before joining low pressure hydrogen from expander C1, and evaporating and reheating through E2 and E1 providing refrigeration for the process. The vapour from V3 is reheated in E2, and whilst some of it returns through E1 as hydrogen product the remainder passes through a turbo-expander C1 to provide a significant part of the refrigeration required by the process. The expander outlet gas and the liquids from V3 mix and reheat in E2 before mixing with V2 vapours and reheating further in E1 to form a CO rich recycle gas which can be compressed back up to feed pressure to increase CO and H_2 recovery. This process produces moderate purity CO product, which can have some 0.4–0.8 per cent CH_4 present in it, and a H_2 product which will contain 2–3 per cent CO. The CO can often be used directly in further processing whilst the H_2 usually needs to be treated further to produce high purity H_2.

6.7.3 *Single-column CO recovery process*

The flow-scheme in Fig. 6.24 shows a single-column process which is often used where the methane content in the feed is relatively high at, say, 0.5 to 1.0 per cent and the CO product must have very low methane content (say 0.2 per cent CH_4 or less). The feed gas is cooled and

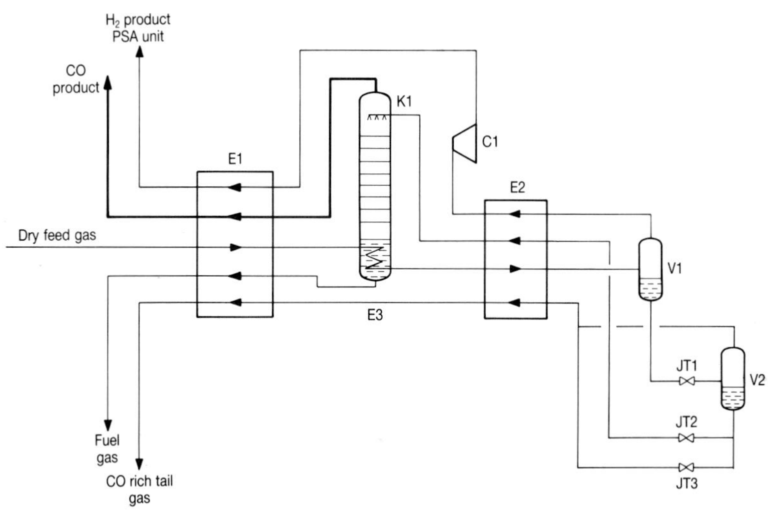

FIG. 6.24. The single-column CO recovery process.

partially condensed in E1, E3, and E2 respectively to about $-193°C$, whereupon the liquid containing the majority of the CO, Ar, N_2, and all of the CH_4 is separated out in V1. The vapour from V1, which is essentially H_2-rich gas containing some 4–5 per cent CO, is reheated to about $-183°C$ and expanded to about 2300–2500 kPa in a turbo-expander C1 before reheating in E1. This expander provides the majority of cold required to operate this process. The liquids separated out in V1 are flashed across JT1 to release a substantial amount of dissolved hydrogen, which is sent to battery limits via E2 and E1 after mixing with some CO flashed via JT3. The majority of liquid separated out in V2 is flashed to about 170 kPa in JT2 and partially re-evaporated in E2 before being fed to a separation column K1, which aims to remove some of the methane from the CO product. Reboil is provided to the column via E3 using feed gas, and a liquid is drawn off from the base of the column which contains a significant proportion of the CH_4 that enters with the feed. This stream is then reheated in E1 and becomes fuel gas. The vapour leaving K1, which represents the CO product, is relatively free from CH_4 (typically 0.1–0.15 per cent) and is heated in E1 before passing out of the cryogenic unit and being fed to the product CO compressor. The CO-rich tail gas can be recompressed and recycled with the feed gas to increase H_2 and CO yields.

6.7.4 *Methane wash process for CO and H_2 recovery*

The process shown in Fig. 6.25 depicts a unit which is not only required to produce relatively pure CO but must also produce hydrogen of 99 per cent purity or better having a CO content less than 1 v.p.m. This hydrogen can then be used in hydrogenation processes without poisoning catalysts. In this process the feed gas is cooled and partially condensed in E1, E2, and E4 before it enters the separator V1 at $-183°C$. Most of the CO is present in the liquid phase which separates out in V1, and this is then flashed to low pressure, partially evaporated in E4, and fed to the CO–CH_4 separation column. The vapour from V1 is passed into the high-pressure methane wash column where the CO is scrubbed from the hydrogen to produce a 99 per cent H_2 product which contains 1 per cent CH_4 and less than 1 v.p.m. CO in the overheads leaving the column. This hydrogen product is then reheated in E2 and E1. Part of the hydrogen product is work expanded in C2 after it leaves E2, and is then passed into E2 again and E1 to give up its cold thereby providing the necessary refrigeration for the process. This hydrogen forms the low-pressure product. The liquids from the column K1 and vessel V1 are passed via E5 and E4 respectively before entering column K2 at low pressure. K2 separates out CO and CH_4, and the liquid CH_4 is pumped from the base of K2, is partly withdrawn as fuel gas via E1, and mainly

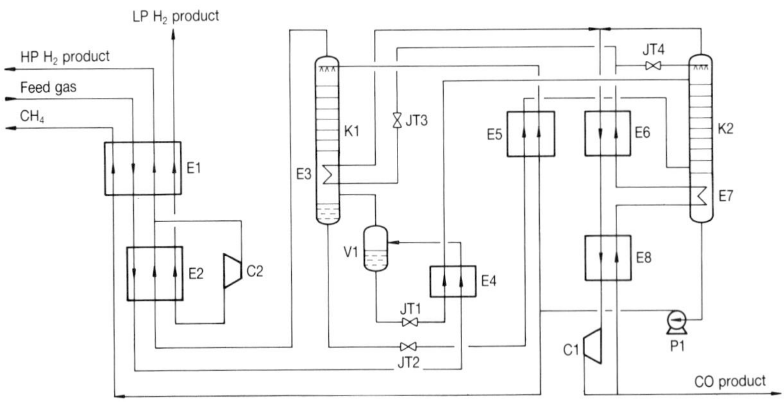

FIG. 6.25. The methane wash process for H_2/CO recovery.

sub-cooled in E5 before being used as reflux in column K1. The overheads from column K2 represent the CO product which is relatively free from CH_4. Vapour CO leaving K2 and vapour CO leaving E3 are combined and warmed in E6 and E8 before passing to the CO cycle compressor which not only compresses cycle gas but also product CO. The high-pressure cycle CO is cooled in E8, condensed in E7, and sub-cooled in E6 before being partly used as refrigerant in E3 and partly used as reflux in K2 after flashing through JT4 to complete the cycle.

It must be noted that this process requires that the CH_4/H_2 ratio in the feed gas should be greater than 1/99 in order to ensure that CH_4 does not have to be supplied from an independent source. This can normally be adjusted by the reforming/partial oxidation part of the plant to ensure that a higher CH_4 slippage occurs, thus putting the process into overall balance.

Several processes of this type were put in operation in the 1960s and 1970s, but the 1980s saw a change in the process schematic in terms of simplifying the process route to one which is more similar to that described in section 6.7.2 above. The aim of this change of emphasis was to obtain a more reliable and more efficient combination of cryogenics, PSA, and reforming technologies.

6.8 Hydrogen liquefaction

6.8.1 *Sources of feedstock for hydrogen liquefaction*

There are many sources of hydrogen which could be exploited to make, purify, and liquefy hydrogen. Some of these have been described in this chapter and chapter 5, but to summarize these and other sources the list includes the following:

- ammonia dissociation;
- H_2 recovery from ammonia purge gases;
- electrolysis of water;
- refinery off-gases;
- steam reforming of natural gas;
- partial oxidation of hydrocarbons;
- partial oxidation of coal;
- coke oven gas.

Each of these sources has its merits and disadvantages in terms of size and impurities present. For relatively large sources of hydrogen steam reforming of natural gas is probably the most economic at fuel values in the late 1980s. However it may be worth mentioning the others briefly.

Ammonia dissociation offers a source of clean hydrogen but in small quantities. This source has a relatively high cost bearing in mind that one has to make hydrogen (perhaps by steam reforming), convert it to NH_3, purify/liquefy the NH_3, and use it in a dissociation unit to produce the H_2. Ammonia plant purge gases are a potential source of hydrogen in moderate quantities (typically 10–12 tonnes per day). It would be technically feasible to recover this hydrogen and purify it to liquefaction quality. However, ammonia plants have generally not been interested in being a source of liquid H_2 to a relatively limited market. Electrolysis of water offers a good source of clean hydrogen in places where electric power is cheap, though again economics point to steam reforming despite the attraction of a pure source of H_2.

Refinery off-gases are generally not considered seriously for H_2 sources as liquefaction plant feedstock because of their varied flows and purities.

Steam reforming, partial oxidation, and coke oven gas are probably the favoured sources of H_2, particularly steam reforming. Figure 6.22 shows a typical block flow diagram of a partial oxidation unit aimed at making a CO-rich gas, but clearly H_2 is still the major constituent of such a series of stages (see Table 6.11). Steam reforming of natural gas has similar steps and the H_2 content is even higher.

6.8.2 *Conditions required for H_2 as liquefier feedstock*

Hydrogen will ultimately reside in storage at around −253°C, and hence most heavier compounds such as N_2, CH_4, Ar, O_2, etc. are below their freezing point. Therefore they will require removal. Water, NH_3, CO_2, H_2S, etc. can be removed by the traditional pretreatment systems described in Chapter 8. N_2, O_2, and other components which are much more volatile need removal by low temperature adsorption or perhaps pressure swing adsorption (PSA) at ambient temperatures. PSA is probably the most effective method for producing 99.9999 per cent pure H_2 or the kind of purity required for economic design of a H_2 liquefier. However, many of the older liquefiers use cryogenic means to achieve a relatively high hydrogen purity of around 98 per cent. The gas is then treated in low-temperature adsorbers to remove CH_4, N_2, CO, or Ar down to p.p.m. levels. Often these adsorbers will be activated carbon beds operating at −173°C. The aim here is purification to levels of around 1 p.p.m. of impurities N_2, CH_4, Ar, O_2, and CO. However, because two beds are used (one on line and the other being purged) it is difficult to maintain the H_2 with less than 1 p.p.m. impurities because of the periodic changeover activities which lead to contamination. As a result a liquid hydrogen guard filter is used to remove any solidified impurities (N_2, Ar, etc.) from the liquid product prior to its entry into storage.

6.8.3 *Conversion of ortho- to para-hydrogen*

Hydrogen molecules exist in two basic forms: ortho-hydrogen and para-hydrogen. These terms relate to different energy levels of the hydrogen molecule. Ortho-hydrogen has its two atomic nuclei spinning in the same direction while those in para-hydrogen spin in opposite directions. Hydrogen at 'normal' conditions contains about 75 per cent ortho-hydrogen and 25 per cent para-hydrogen. When this hydrogen is cooled and liquefied the liquid composition is nearly 'normal'. However the equilibrium concentration of the liquid hydrogen is nearly pure para-hydrogen, as shown in Fig. 6.26. If the fresh ortho-hydrogen liquid were left to its own devices in a perfectly insulated vessel it would eventually

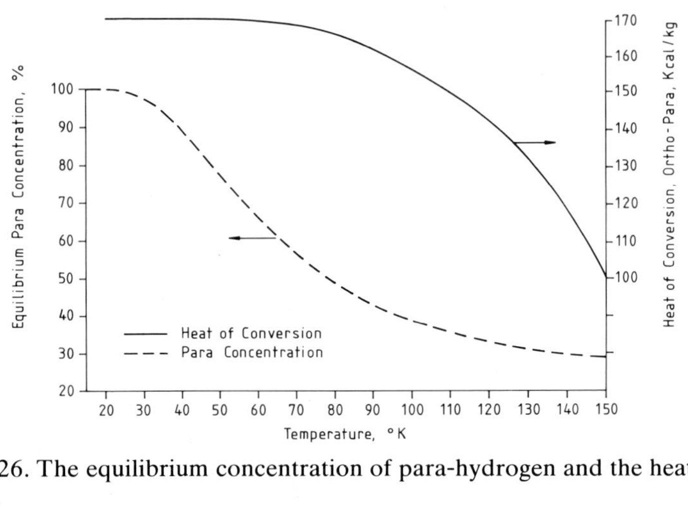

FIG. 6.26. The equilibrium concentration of para-hydrogen and the heat of conversion.

autocatalyse itself into para-hydrogen with an exothermic reaction. The heat release would cause some boil-off which would be equivalent to about 1 per cent per hour. This is clearly an unacceptable situation.

All modern hydrogen liquefiers therefore have catalyst beds designed to convert the hydrogen into para-hydrogen before it enters the storage tank as liquid. This means that the boil-off is limited to that caused by heat inleak. There are many ways in which the catalytic conversion can be effected. Designers usually adopt a staged conversion process to remove the heat of conversion at the highest practical temperature levels. The catalysts used are ferric oxide gel or a chromium oxide on an alumina base.

A third catalyst, developed by Air Products, goes under the trade name 'Apachi'. Each of these catalysts has a different bulk density and required space velocities, and all have been used successfully. In general the ortho to para conversion is done in the gaseous phase where it is easier to cope with the exothermic reaction and heat removal.

6.8.4 *Small to medium-sized hydrogen liquefiers*

Modest rocketry research programs and low-temperature physics laboratories use the small to medium-sized hydrogen liquefier. Because the prime objective in these establishments is research, little money and effort are expended on highly sophisticated and efficient process plant. Morever, the extra capital cost required for a modest power saving is not justified. Therefore, this kind of liquefier is usually inexpensive, simple, and easy to run. The capacity range is from about $5\,l\,h^{-1}$ of liquid hydrogen to about $500\,l\,h^{-1}$. This corresponds to a range of 0.0085–0.85 tonnes per day liquid H_2.

This size of liquefier usually generates the hydrogen for immediate use, and only small storage capacity (dewars) is usually provided in such establishments. Therefore it is unlikely to be economic to convert ortho-H_2 to para-H_2, and the design of small- to medium-sized H_2 liquefiers does not provide for ortho to para conversion.

Figure 6.27 shows a simplified flow-sheet for a conventional Linde liquefier. This type of unit was constructed for capacities in the range of $14\,l\,h^{-1}$ to $60\,l\,h^{-1}$. In this process, pure hydrogen is taken from bottles T2 or a gasholder T1, and compressed in C1 to a high pressure, typically 13 000 kPa, and dried in column V1 before being admitted to the cryogenic section. Here the hydrogen is cooled in E1, E2, and E3 by returning vapour hydrogen and liquid nitrogen which is admitted into V2. V2 contains E2 which provides the majority of refrigeration needed at above $-208°C$ (the temperature of the liquid N_2). The cold hydrogen emanating from E3 is flashed across JT1 to yield two phases. The liquid product H_2 is drawn off for use in the laboratory whilst the vapour is

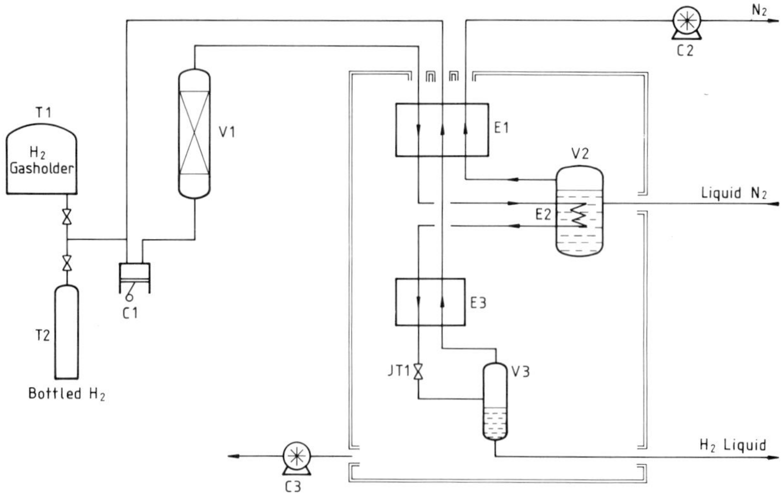

FIG. 6.27. A small hydrogen liquefier.

reheated to provide some refrigeration in E3 and E1. Vacuum pump C2 is used to ensure that the liquid N_2 boils as cold as possible. Vacuum pump C3 is used to ensure that the cold box is maintained at a good vacuum to minimize heat inleak, which is a significant factor for such small Joule–Thomson liquefiers.

6.8.5 *Large hydrogen liquefiers*

In the late 1950s the hydrogen liquefier at the National Bureau of Standards Cryogenic Engineering Laboratory, Boulder, Colorado was considered to be a large-scale liquefier. This unit had a capacity for H_2 liquefaction of a mere 0.45 tonnes per day with a process similar to that shown in Fig. 6.27. The Space Shuttle program included the Columbia and Challenger series with several launches and trials occurring annually. Each launch used about 105 tonnes of liquid hydrogen. Earlier launch vehicles such as the Titan rocket were even larger than the Shuttle launch vehicle and carried even larger amounts of liquid hydrogen. This makes it apparent that much larger H_2 liquefiers were required than the Boulder plant. The 1960s saw a massive increase in the space program activities, and several new plants were constructed with much greater capacities. Most of the information available is only for 'Western' plants, although the USSR probably has similar requirements. Table 6.12 shows the capacities of some H_2 liquefiers installed in the 1950s and 1960s in the USA. The main consumers of liquid hydrogen were the NASA space program and other research establishments. There was however a small

TABLE 6.12. Capacities of H_2 liquefiers.

Location of unit	Capacity (tonnes/day)	Approximate date of construction
NBS, Boulder, Colorado	0.45	1952
Air Products, Ohio	0.9	1956
Air Products, Florida	3.2	1957
Linde Co., California	6.4	1960
Air Products, California	30	1963
Linde Co., California	57	1965
Air Products, Louisiana	54 (approx.)	—

but growing variety of other consumers including the electronics industry, glass and gem manufacturing, hydrogenation, and on-site backup for site generated hydrogen.

The larger plants needed greater care in design to ensure an efficient process. The processes were therefore much more complex, as shown in Fig. 6.28 which is somewhat simplified for clearer understanding. Two types of cycle can be used for liquefaction—high- and low-pressure cycles—and the high-pressure cycle will be described here. Although this process may be a little more capital intensive due to the use of wound coil exchangers, it does have a power advantage. As a general principle it is often advantageous to operate liquefiers at high pressures. The larger H_2 liquefier uses a compression–expansion N_2 cycle as well as H_2 compression–expansion to supply refrigeration for the process. In this process a closed loop N_2 cycle provides both cold liquid and gaseous N_2 as refrigerants. The N_2 is compressed in C4, cooled in E1, and part of it is condensed and flashed via JT1 into V2, whilst the remainder is expanded in C3 which provides the bulk of the refrigeration. For moderate capacities, the N_2 cycle is designed for a C4 discharge pressure approaching 20 000 kPa where reciprocating machinery may be used. These pressures are of course much lower for large plants which may take advantage of more reliable centrifugal machinery.

The hydrogen feed can be compressed to a variety of pressures, but many consider 10 000 kPa to be about the optimum even for relatively large units. This, of course, requires the use of reciprocating compressors and expanders. The feed H_2 is compressed in C1, cooled by liquid and gaseous N_2 in E2 and E4, and also cooled with hydrogen from the hydrogen cycle in E4 and E5 before being expanded in JT2 cooled in a submerged exchanger E6, and delivered to storage. The H_2 cycle consists of compression in C2 and cooling of HP hydrogen in E3 against low-pressure return H_2. Further cooling is done against N_2 in E4 and

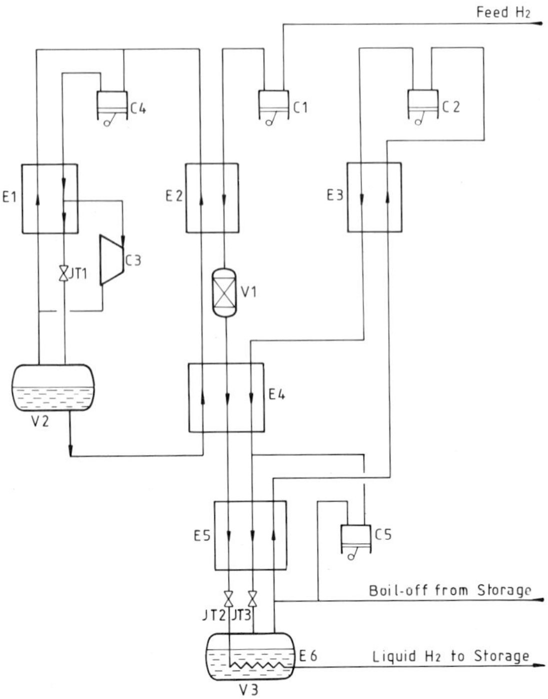

FIG. 6.28. A simplified flow-scheme for a large scale high-pressure hydrogen liquefier.

final cooling in E5 against expanded H_2 from C5, and boil-off from storage and V3. Once the HP H_2 leaves E5 it is flashed across JT3 whereupon it forms two phases, the liquid providing refrigeration for the product H_2. The adsorber V1 is used to remove trace impurities such as N_2, CH_4, etc. from the feed H_2 before final liquefaction.

There are several refinements used in such processes and these must be considered carefully depending on the size and economics of the plant. For example the H_2 refrigeration reservoir V3 can be split up into several vessels each operating at different pressures to provide a closer match for the cooling curves. This reduces the overall power consumption because it provides a more efficient process. Also this staged cooling allows the use of staged catalytic conversion from ortho- to para-H_2. A catalytic bed can be put in at each cooling stage, thereby providing the removal of heat of reaction at several temperature levels thus leading to a more efficient process. The alternative would be to remove the heat of reaction at the coldest level, requiring the same refrigeration load but *all* at the coldest level. This would clearly need a higher power consumption since power requirements rise as the refrigeration *temperature* falls.

Bibliography

L'AIR LIQUIDE (1981). Publication No. 17, L'Air Liquide, France.

BAKER, G. H. (1975). 'Corrosion-free gas sweetening'. *Hydrocarbon Processing,* (Aug.) 113.

BRITISH SULPHUR CORPORATION (1976). 'Recycling recovered hydrogen improves NH_3 plant economics'. *Nitrogen.* (July/August).

BROOK, P. H. (1983). 'Ammonia plant revamping'. *Fertiliser 1983 Intl. Conf. Proceedings,* Vol. 1, p. 159.

BUIVIDAS, L. J., FINNERAN, J. A., and QUARTULLI, O. J. (1975). Alternate ammonia feedstocks. *A.I.Ch.E. Symposium, Ammonia Plant Safety,* Vol. 17, p. 4.

COMBS, G. A. (1980). 'Improving efficiency in a 1150 STPD ammonia plant'. *A.I.Ch.E. Symposium,* 89th Meeting.

DE BLIECK, J. L., DRAAISMA, J. J. F., and MOL, A. (1977). 'Flexibility of large ethylene plants'. *Chemical Economy & Engineering Rev.,* **9**(4), 30.

DUNLOP, S. and BANKS, R. (1977). 'Cryogenics increase benzene recovery'. *Hydrocarbon Processing,* (July), 147.

ENDE, G. and SCHMIDT, K. (1979). 'H_2 recovery in the 1360 TPD NH_3 synthesis—operating experience increase in production and savings of raw material'. *XV Int. Congress of Refrigeration,* Paper A3–13.

GROTZ, B. J. and GOSNELL, J. H. (1984). 'Startup of new 1500 MTPD purifier ammonia plant'. *Ammonia Plant Safety and Related Facilities Symposium,* p. 38.

HARMON, R. (1981). 'Purge gas purification and recovery in ammonia plants'. *British Sulphur Corporation Conference,* London, Paper 6.

HASELDEN, G. G. (1971). *Cryogenic fundamentals.* Academic Press, London.

HASLAM, A. A. and ISALSKI, W. H. (1975). 'Hydrogen from ammonia plant purge gas'. *A.I.Ch.E. Symposium, Ammonia Plant Safety,* Vol. 17, p. 80.

ISALSKI, W. H. (1983). *Recovery of hydrogen and ammonia from purge gas.* UK Patent 2050 326B.

ISALSKI, W. H. (1984). '25 years of purge gas recovery'. *Nitrogen,* (Nov–Dec), British Sulphur Corporation, London.

JUNGFER, H. (1985). 'Synthesis gas from refinery residues'. *Linde Reports Sci. Tech.* **40**, 14.

LABINE, R. A. (1960). 'How the Air Force liquefies hydrogen'. *Chem. Engng.,* (Jan. 25), 86–89.

LIVINGSTONE, J. G. and PINTO, A. (1983). 'Solve ammonia plant problems with the AMV process'. *Fertiliser 1983 Intl. Conf. Proceedings,* Vol. 2, p. 421.

MORRISON, R. T. and BOYD, R. N. (1966). *Organic Chemistry* (2nd edn). Allyn & Bacon.

NEWTON, C. L. (1967). 'Hydrogen production and liquefaction'. *Chem. & Proc. Engng.* (Dec.) 51.

NICHOLSON, I. (1982). *Sputnik to Space Shuttle.* Sidgwick & Jackson.

OIL AND GAS JOURNAL (1982). 'Vast Saudi gas-gathering system moves into final phase'. April 12, p. 76.

O'NEILL, P. S., FENNER, G. W., and WILLIAMSON, E. F. (1982). 'Advanced heat transfer technology in cracked gas separation'. *A.I.Ch.E. Symposium Series,* no. 224, Vol. 79, p. 32.

PARRISH, R. W. (1984). 'Are claims for energy savings for loop hydrogen recovery justified?' *Ammonia Plant Safety and Related Facilities Symposium,* p. 162.

REDMAN, J. (1986). 'Mossmoran'. *The Chemical Engineer,* (Jan.) 10.

RUZISKA, P., SONG, C. C., WILKINSON, R. A., and UNRUH, W. (1984). 'Exxon chemical low energy ammonia process startup experience'. *Ammonia Plant Safety and Related Facilities Symposium,* p. 22.

SCHRAMM, W. and JUNGFER, H. (1985). 'Cost reduction in ammonia manufacture'. *Linde Reports Sci. Tech.* **40**, 40.

SCOTT, R. B., DENTON, W. H., and NICHOLLS, C. M. (1964). *Technology and uses of liquid hydrogen.* Pergamon Press, Oxford.

SKINNER, G. F. and KOWAL, M. W. (1983). *Synthesis gas for ammonia production.* UK patent no. 2048840B.

SMITH, W. N. and SANTANGELO, J. G. (1980). *Hydrogen production and marketing.* ACS Symposium series 116, American Chemical Society, Washington.

TRUX, J. (1985). *The space race.* New English Library.

ZDONIK, S. B., GREEN, E. J., and HALLEE, L. P. (1966–1970). 'Ethylene manufacture (reprints)'. *Oil and Gas Journal.*

7
NATURAL GAS PROCESSING

7.1 Introduction

All natural gas is extracted from below ground, and it may or may not be associated with some crude oil. In either case it is a matter of pot luck as to what composition arrives at the surface of each well or group of wells. The composition of each gas can vary from high N_2 or high CO_2 with low hydrocarbon content to a mixture composed entirely of hydrocarbons with high ethane and heavier components. In most cases the natural gas will be used as fuel or feedstock to a refinery or petrochemical operation or as a domestic fuel. In any case the composition of this natural gas must be carefully controlled so that burners cope adequately with the gas. One very effectve way of monitoring natural gas quality is the Wobbe Number:

$$W = H_v/\surd S$$

where H_v is the higher heating value in Btu/SCF and S is the specific gravity w.r.t. air.

Both of the above parameters in the Wobbe number are governed by the hydrocarbon content and the composition of the gas. For normal application N_2 content can be as high as 10 per cent and ethane content as high as 20 per cent, but generally these values are much lower with only a few percent of each and the balance methane.

Thus it soon becomes apparent that natural gas processing is about adjustment of the composition to make it suitable for combustion or some petrochemical application. Therefore one can split the processing areas into three groups:

(1) decreasing Wobbe number;
(2) increasing Wobbe number;
(3) liquefaction for buffer storage and transportation.

This chapter considers the calorific value or Wobbe number adjustment processes and liquefaction processes where shipment of liquefied natural gas (LNG) or buffer storage is needed to overcome fluctuating demands for gas. There are a number of alternative ways of recovering LPGs from natural gas. These include physical absorption solvents, adsorption beds, Joule–Thomson expansion, lean oil absorption, external refrigeration, and turbo-expander plants. This chapter is concerned mainly with the last two, but of course the Joule–Thomson effect will be mentioned as it is inevitably present in the low-temperature processes to a varying degree.

7.2 LPG extraction

In this situation the gas calorific value or Wobbe number is too high for the National Grid or the direct consumer and some high calorific value components need to be taken out. Propane and heavier components often command a higher price on the open market than fuel gas. It is often profitable to recover these from natural gas for separation into individual components—for example, commercial propane and butane—and to blend the heavier components into automotive fuel. This makes the natural gas calorific value compatible with the fuel header and can also be a profitable venture.

7.2.1 *Turbo-expander processes*

A common process used to extract propane and heavier components from natural gas is an expander plant which relies on a high pressure difference between the heavy natural gas and the fuel header. The process then converts pressure energy into rotating motion via a turbo-expander to provide the necessary refrigeration for the process. Because the natural gas composition varies from case to case, and because designers have the benefit of different experiences, few processes are exactly the same. These processes very much resemble those described for refinery applications in Chapter 5. Figures 5.5 and 5.10 in particular are excellent examples of this application. Figure 7.1 describes a simple turbo-expander process treating natural gas. In this process natural gas is first dried in molecular sieves or, in certain cases, in a glycol unit. Occasionally, for large units, both glycol and molecular sieve driers are used. The dry gas is then boosted in pressure by the compressor end of a turbo-expansion/compression assembly such that the energy dissipated by C2 is taken up in boosting the feed gas in C1. Some or all of the feed gas is then used in E2 to provide reboil in the column. On occasions, when the natural gas is particularly heavy, steam has to be used here because the bubble point is too high to use feed gas from the compander assembly.

The feed gas is then cooled and partially condensed in E1. Heavy hydrocarbon liquids are separated out in V1 and the vapour from V1 continues on to the turbo-expander end C2 of the expander/compressor assembly (C1/C2). As the gas expands across C2, the energy is dissipated across the rigid shaft connecting it to C1 and the gas gets significantly colder, thereby condensing some hydrocarbons. The total

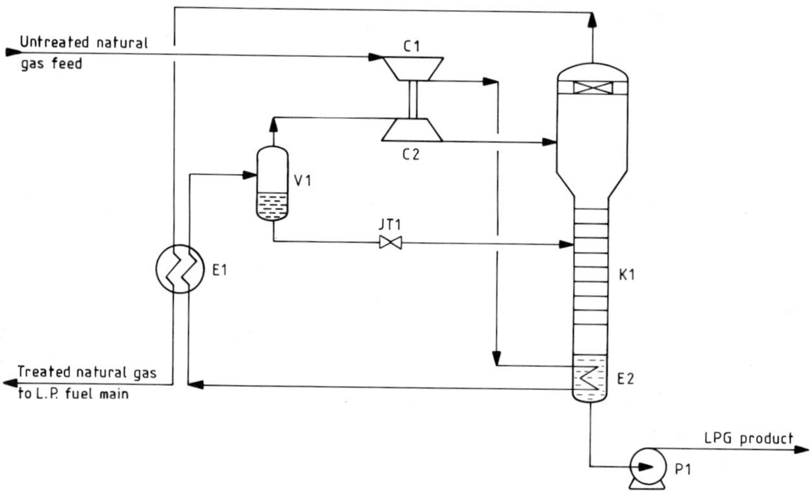

FIG. 7.1. An expander plant treating natural gas.

C_3 and heavier are contained essentially in the liquids from V1 and condensate at the turbine outlet. These two liquids form the reflux down the column K1. The vapour overheads from K1 join the uncondensed vapour from the expander C2 and pass out of the enlarged column top, are reheated in E1 and are fed to a low-pressure gas main. Alternatively, if the gas is required at relatively high pressure for the natural gas grid it needs compression. In such cases this may not be the best process, as can be seen in section 7.2.2. Notwithstanding this, the above processes themselves all consume very little energy and can be modified to suit different gas qualities. Heavy gases can be treated with these processes by adding a Freon unit to the process, probably between E1 and V1. The overall efficiency can be significantly improved by installing two turbines in series, each one making a certain amount of liquid fed into the column. To improve the process efficiency further, side reboilers can be considered for K1, but the plants need to be relatively large so that the complexity can be economically justified.

These plants can also be modified to operate in an ethane extraction mode by putting colder refrigeration at the inlet to expander C2 or V1. This pulls down more ethane into the liquid phase of V1 and the expander(s) outlet liquids. The column then takes in more ethane, but to get it out in the product the reboiler temperature must be reset to ensure no methane slippage, i.e. it must now be much colder.

These plants typically recover over 95 per cent C_{3+} in the C_3 recovery mode, with virtually total C_4 and heavier recovery. When operating to recover C_2 these plants can also produce high proportions of ethane, but this depends on the level of assisting refrigeration supplied. Typically figures of over 80 per cent are achievable. Each plant must be considered on its economic merits taking account of the pressure drop available, the cost of utilities, and the value of LPGs and ethane before a decision is made concerning the nature and complexity of the process.

7.2.2 *Evaporating refrigerant processes*

Sometimes natural gas has too high a calorific value, making it desirable to remove some C_3 and heavier compounds, but the waste or fuel gas pressure must be close to that of the feed. In such situations, a turbo-expander plant is often not the best solution. To resolve this problem a process commercially available called PetroFlux can be used with negligible pressure loss from feed gas to fuel gas. This process relies on mechanical refrigeration to provide cold and a novel design of heat-exchange system which utilizes simultaneous heat and mass transfer to improve overall process efficiency.

The first natural gas liquids (NGL) recovery plant using the new PetroFlux process was started up in 1985, and has since demonstrated its ability to achieve 97 per cent propane recovery with a power consumption markedly lower than competing processes. Exciting new applications for the same basic concept include hydrocarbon dewpoint control and flexible plants which can respond easily to market fluctuations. Figure 7.2 shows a simplified flowsheet for a typical PetroFlux unit.

The feed to all cryogenic processes must be dehydrated to prevent formation of ice or hydrates which would foul or plug heat exchangers and valves. Water vapour is reduced to the required level by adsorption on a molecular sieve or similar desiccant V1A/B. At least two towers are provided so that one bed can undergo thermal regeneration while the other is on stream, drying the feed gas. Part of the residue gas from the

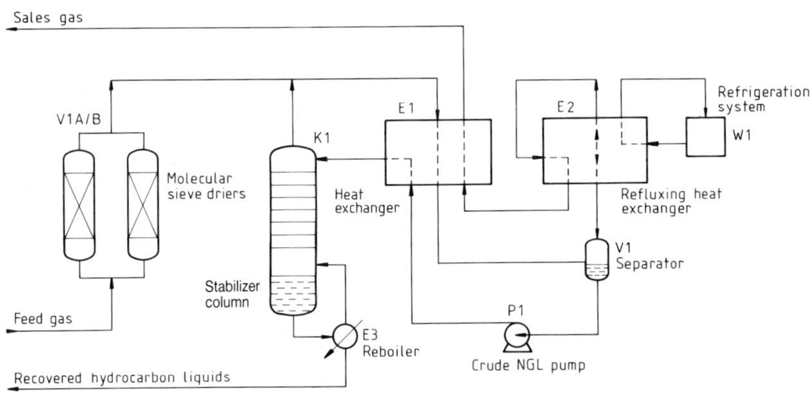

FIG. 7.2. The Petroflux process for NGL recovery.

plant is heated and used for the molecular sieve regeneration. The molecular sieve unit may sometimes be preceded by a glycol unit for bulk water removal.

Dry feed gas is cooled, usually in an aluminium plate-fin heat exchanger E1 by returning cold product streams, i.e. residue gas and recovered NGL, plus a small amount of propane refrigeration, if required. Further feed cooling and partial condensation are achieved in the refluxing exchanger E2. A refluxing exchanger, sometimes termed a 'dephlegmator' is a vertical plate-fin heat exchanger, designed so that the condensed liquid runs back down in the same passages counter-current to the rising vapour. Flowing in adjacent passages is either evaporating ethylene or ethane refrigerant or cold residue gas from the top of the exchanger. These two streams supply the 'cold' to condense the natural gas liquid components from the feed within the refluxing passages.

The feed-cooling process corresponds to a large number of partial condensation steps in series in which liquid formed at each step becomes partially stripped or stabilized as it flows back down the refluxing exchanger. The exchanger may be designed and operated with very close temperature differences throughout its length. These factors result in high thermodynamic efficiency and, in turn, low energy consumption by the refrigeration unit.

In the PetroFlux process much sharper separation occurs between the condensed NGL components and the lighter undesired components, methane and (perhaps) ethane, than could be achieved in a single condensation step. In practice, this produces almost total propane recovery in a fairly concentrated NGL stream that needs only a relatively small amount of further stabilization in an external column. Potential loss of propane in the overheads stream from the stabilizer column may be recovered by recycling this stream into the feed, or by an auxilliary overhead condenser.

To provide a low enough temperature to achieve high propane recovery, an ethylene (or ethane)/propane cascade (W1) is typically used. Ethylene (or ethane) vaporized in the refluxing exchanger is compressed and condensed at high pressure against evaporating low-pressure propane, before being expanded and returned to the refluxing exchanger. The low-pressure propane vapour is compressed and condensed against air or cooling water. Instead of the ethylene/propane a mixed refrigerant could be used. Alternatively, non-flammable halocarbon (Freon) refrigerants may be preferred in some locations.

The new refluxing exchanger process has excellent turndown capabilities compared with expander plants. Also, for a given recovery, the new process can tolerate fairly high carbon dioxide levels without risk of freeze-up, because the pressures and temperatures are kept at higher values throughout the process. For a propane recovery of over 95 per cent, the process has a power consumption only 40 per cent of the turbo-expander NGL process.

Thermodynamic efficiency is one of the reasons behind the impressive energy efficiency of the new process. Since products are delivered at ambient temperature and a pressure similar to the feed, there is no real requirement for net refrigeration, except for what is needed internally in the process to provide driving forces for the separation. For high efficiency, excessive driving forces are to be avoided. In the expander plant recovering propane, large driving forces can occur at the top of the column where cold liquid from the expander is fed as reflux; liquid and vapour leaving the top tray may be significantly warmer than the expander exhaust. Also a wide temperature difference usually exists at the cold end of the gas/gas exchanger. Thus even with a 100 per cent efficient expander this process has some inherent irreversibilities. In practice, the turbo-expander will have an isentropic efficiency of 80–85 per cent at its design point.

Refrigeration provided for the PetroFlux process is by a mechanical vapour compression cycle and is acknowledged to be more efficient than the turbo-expander at these temperature levels. The high efficiency may also be explained in terms of the separation achieved in the refluxing exchanger. This is more akin to a multi-stage rectification (or at least multi-stage partial condensation) as opposed to the one, two, or perhaps three stages occuring in an expander process. This multi-stage rectification leads to production of a more stable liquid, thus reducing the load on the de-ethanizer column. Furthermore, the rectification is non-adiabatic—that is, the reflux cold is provided over a range of temperature. This is known to improve the thermodynamic efficiency of rectification, particularly at low temperature.

Table 7.1 summarizes the results of a thermodynamic loss analysis (energy analysis) carried out on a turbo-expander and PetroFlux process for propane plus recovery.

7.3 De-nitrogenation of natural gas

7.3.1 *Fixed composition gases*

In the preceding section we have considered natural gases that are too rich to be admitted to a fuel gas system. At the other end of the spectrum we have natural gases which are too lean to be fed into any fuel network or national grid. For example, Dutch natural gas contains 14 per cent N_2 and Polish natural gas contains 42.75 per cent N_2. Two methods exist for making these gases reach the correct calorific value and Wobbe number:

(1) blend with large amount of rich gas;
(2) reduce the nitrogen content to an acceptable level.

TABLE 7.1. Comparative loss analysis expressed as actual shaft power. The basis is 28 000 $Nm^3\,h^{-1}$ feed and 95 per cent propane recovery.

	Turboexpander	PetroFlux
Minimum theoretical work on separation (kW)	67	67
Plant losses (kW)		
Compression	194	134
Warm exchangers	351	134
Cryogenic exchangers	246	149
Expander/Compressor	328	—
Distillation	157	31
Valves, mixing pumps, etc.	—	37
Total	1276	485

In the first option, often very large amounts of rich gas are required to dilute the lean gas so that the mixture represents a gas of saleable quality. This is in fact done by British Gas with Morecambe Bay gas, which contains somewhat less than 10 per cent N_2, making dilution a reasonable proposition.

The second option is often used for natural gases with more than 10 per cent N_2 and in both the Dutch and Polish Natural gas examples de-nitrogenation is carried out. Most of the processes used to de-nitrogenate natural gas are based on the classical double-column system, and to illustrate the process we shall use the Polish natural gas upgrading process as an example. Table 7.2 gives an overall mass balance around one of the two trains treating Polish natural gas.

The basic process consists of a significant pretreatment section first of mono-ethanolamine (MEA) washing to remove the bulk of the CO_2. The next phase is to remove the remaining CO_2 and to dry the gas, which can be done on molecular sieves. The final pretreatment phase removes heavier hydrocarbons and traces of aromatics which could freeze up in the cryogenic section. These are removed using an activated carbon bed which is thermally regenerated, along with the molecular sieve beds, using hot waste gas. The clean dry feed gas is then admitted into the cryogenic unit, which is depicted in Fig. 7.3.

The dry feed gas enters the main exchanger E1 where it is cooled and partially condensed. At the outlet, it is flashed to about 2700 kPa before entering the lower column K1 of a double-column system K1 and K2. A nitrogen-rich liquid waste gas is withdrawn from the top of K1, subcooled in E3 against waste N_2 product, flashed in JT2, and passed to K2 as reflux. Feed to K2 is the methane-rich bottoms from K1, which are in turn subcooled against CH_4 and waste gas products in E2 before being flashed to 200 kPa across JT3 into K2. Methane product liquid is withdrawn from the base of K2, pumped up to about 2200 kPa in P1,

TABLE 7.2. The overall mass balance for a de-nitrogenation unit (compositions in mole per cent).

	Lean feed	Methane product	Waste N_2	Helium product
He	0.40	—	0.09	100.00
N_2	42.75	4.00	98.95	—
CH_4	56.01	95.09	0.96	—
C_2H_6	0.51	0.87	—	—
C_3	0.02	0.03	—	—
C_{4+}	0.01	0.01	—	—
CO_2	0.30	—	*	—
Sulphur	6 v.p.m	—	*	—
Water	sat. at 0°C	—	*	—
Flow-rate ($Nm^3\ h^{-1}$)	137 400	80 340	56 110	480
Pressure (kPa)	5620	2100	105	110
Temperature (°C)	10–15	20	20	−269 (liquid)

*The trace impurities are adsorbed and absorbed before entry into the cryogenic unit and are regenerated with waste gas, which will contain varying concentrations of sulphur, water, and CO_2.

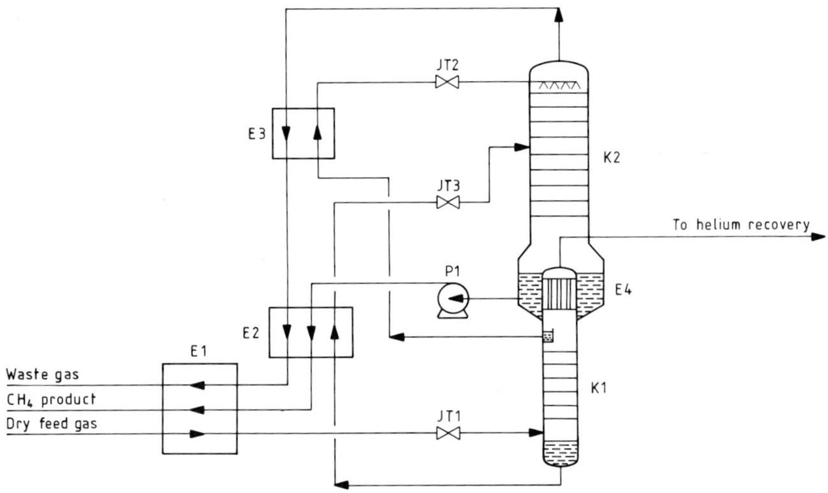

FIG. 7.3. The cryogenic section of a typical denitrogenation plant.

and evaporated and reheated in E2 and E1 respectively before being supplied to battery limits. This process works entirely on the Joule–Thomson principle which takes effect over JT1, 2, and 3. We can see that there is a significant amount of Joule–Thomson effect available when we consider that the feed is available at 5620 kPa and the products are needed at 2100 kPa and 105 kPa. It is clear that if turbo-expanders were introduced into parts of the process they would produce more cold and give more potential to make higher quality products or better recoveries. However, higher product pressures were not required and the CH_4 quality was acceptable. Therefore, the designers decided to minimize the extent of machinery in order to keep capital and maintenance costs down and process reliability as high as possible. Thus each case should be considered on its own merits with high regard for capital cost, maintenance costs, and process reliability. For a different application with the same gas the only feature in the process that is likely to remain the same may be the double column—everything else could be different.

The above process and its variants is generally suitable for cases where enhanced oil recovery (EOR) using nitrogen injection is not desired. This is because the nitrogen invariably emanates from the plant at near atmospheric pressure. For EOR applications single-column processes with perhaps expansion turbines and/or heat pumps have found more favour, because the nitrogen can be generated at pressure. This reduces N_2 injection compression costs making the EOR scheme more attractive.

7.3.2 *Enhanced oil recovery*

7.3.2.1 Background

The preceding section covered de-nitrogenation of a constant composition feedstock which, until the 1970s, was the usual situation. Since then, enhanced oil recovery (EOR) has become fashionable because of regular scares of shortages and political unrest in countries with abundant oil supply. Primary oil recovery withdraws somewhat less than 50 per cent of the oil and associated gas present underground, and in many cases this may be as low as 5–10 per cent of the original amount. If the remaining 90–95 per cent could be economically recovered the reserves would be beyond imagination. Therefore the terms 'secondary oil recovery' and 'tertiary oil recovery' have become part of the oil-man's vocabulary in defining degrees of EOR.

A number of techniques have been developed to carry out EOR, including the following:

(1) Pressure maintenance/gas cycling. This means re-injection of associated gas into the well when gas prices are low. This enhances the amount of recoverable oil.
(2) Driving of miscible gas slugs. This gas can be CO_2 or a hydrocarbon injected at high pressure into the well at above 7000 kPa. The gas becomes miscible with the oil at that pressure and oil recovery can be enhanced. Nitrogen can be a substitute for CO_2 in certain cases.
(3) Steam injection of polymerizing compounds into the well.
(4) Injection of polymerizing compounds into the well.

This section will primarily be concerned with method 2, where N_2 is used as the injected gas for EOR. Nitrogen can be generated either by a combustion inert gas generator or, preferably, a cryogenic unit, which gives N_2 of higher purity. The nitrogen is then compressed to very high pressures (typically 20 000 kPa or even more) and injected into the depleted oil well. At first the associated gas contains no N_2 and the gas is of the same approximate composition as that encountered when the well was first put into production. Then as more nitrogen is injected over the months of EOR practice, the N_2 begins to show in the associated gas. This may occur a few months or perhaps one or two years after starting the injection procedure. This concentration continues to rise gradually to very high values over, say, 10 to 20 years. At some point the nitrogen breakthrough is so high that it becomes uneconomical to process the associated gas and the oil rate has decreased so that production is terminated.

7.3.2.2 Enhanced oil recovery processes

Therefore we arrive at a scenario for de-nitrogenation where over a period of, say, 20 years associated gas starts essentially as all hydrocarbons with perhaps a trace of inerts and CO_2, and after 20 years we may have 70–80 per cent N_2 and the remainder useful hydrocarbons. The gas is probably acceptable as sales gas with up to 4 or 5 per cent N_2 but beyond that it is almost certain that adjustment of calorific value will be required. Therefore a de-nitrogenation unit is required to reduce nitrogen content down to below 4 per cent for treating a feed gas with a typical composition as shown in Fig. 7.4.

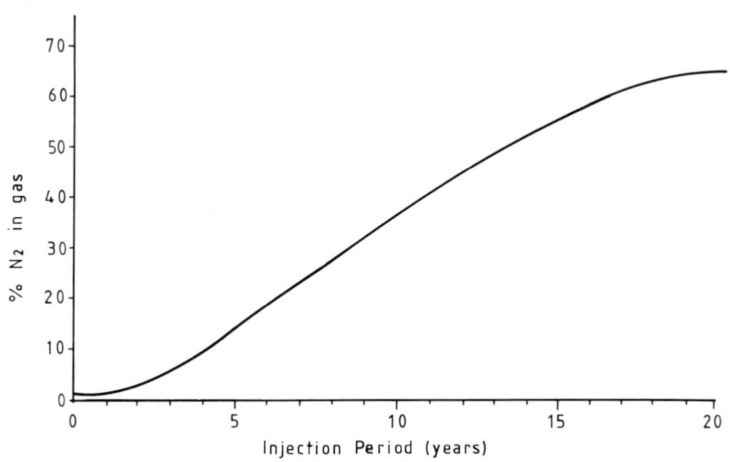

FIG. 7.4. The variation of associated gas composition with time in EOR cases.

Therefore, one can envisage three cases for nitrogen rejection:

(1) low nitrogen content— ~ 10 per cent N_2;
(2) medium nitrogen content— ~ 40 per cent N_2;
(3) high nitrogen content— ~ 70 per cent N_2.

In each of the cases the optimum process is likely to be very different. For example, in the first case with low nitrogen content, the extent of Joule–Thomson expansion is likely to be low because most of the gas will end up as sales gas which is usually required at moderately high pressure. Therefore one could envisage a single-column distillation process with a heat pump cycle to drive the reboiler and condenser. Because the amount of nitrogen is low, the vapour and liquid traffic are likely to be much lower near the top of the column than the base of the column.

Consequently there may be some advantage in having intermediate condensers, as shown in Fig. 7.5.

In this kind of process methane can be used as a heat pump cycle fluid, compressed by C1, cooled in E5 and E4, and then routed to the three condensers E1, E2, and E3 via Joule–Thomson valves JT1, JT2, and JT3. The role of the intermediate condensers E2 and E1 is to reduce energy consumption in C1, because the return pressures for E2 and E1 vapour are progressively higher than E3. This is because the condensing temperature rises as we go down column K1. An alternative way of processing low N_2 content could be to use a process similar to that shown in Fig. 7.3—to add a third 'pre-separation' column upstream of

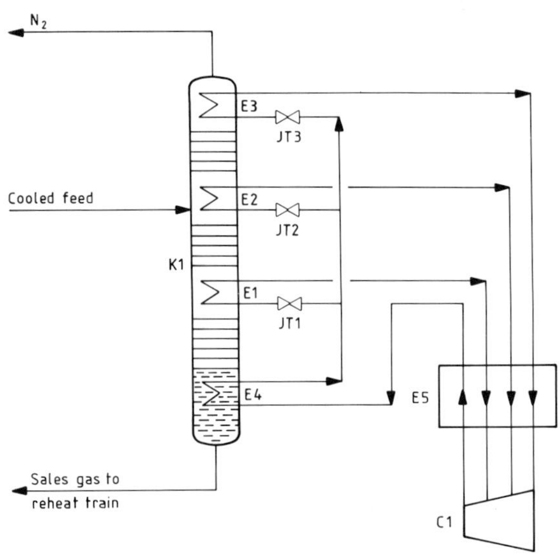

FIG. 7.5. The low nitrogen case de-nitrogenation unit—single column.

column K1 in order to increase the nitrogen content of the gas entering the double-column system. This process has successfully been applied to Dutch natural gas (14 per cent N_2) in Alfortville, France.

For medium N_2 content gas, the classical double-column process can most likely be used, as shown in Fig. 7.3. However, with case (3)—high N_2 content gas—this process is no longer suitable and one might suggest a classical turbo-expander plant. This process is quite a logical choice (Fig. 7.1) because the N_2 content is high and thus there should be plenty of gas for expanding through a turbine to drive the process.

Of course one cannot have three different processes which operate only for five or six years and then shut down. Therefore, designers approached the problem of flexibility and the result is a successfully modified double-column process which has features built in to cater for varying composition. The following process has been designed to accommodate a range of gas compositions from 5 per cent N_2, 92.2 per cent C_1, and 2.8 per cent C_2 to 80 per cent N_2, 19.4 per cent C_1, and 0.6 per cent C_2. The feed gas is treated in molecular sieves to remove water and CO_2 before entering the double-column cryogenic unit. The basic process is shown in Fig. 7.6. for a plant operating for Dome Petroleum, Willesden Green, Canada.

The dry, CO_2-free feed gas enters the cryogenic unit E1 at about 2700 kPa and 40°C where it is cooled and partially condensed against sales and vent gases. The small amount of heavy hydrocarbons that condense in E1 are separated in V1, while the vapour continues cooling with

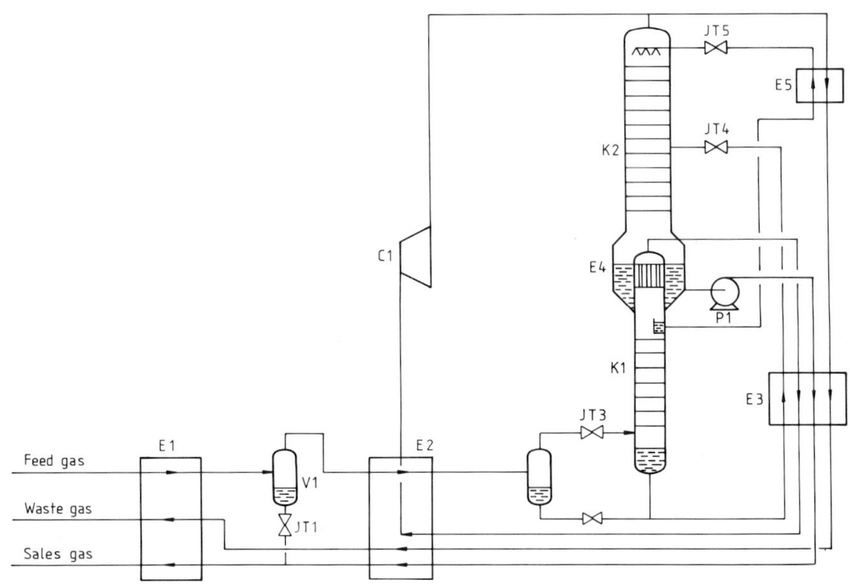

FIG. 7.6. A double-column de-nitrogenation unit for EOR applications.

varying degrees of condensation in E2 depending on N_2 content in the feed. For the early period with low N_2 content, the feed is slightly subcooled as it enters V2 and, therefore, the separator operates completely flooded. The feed then enters the double-column system K1 and K2 near the base of the column. Rich liquid from the base of K1 is subcooled in E3 and expanded across JT4 into K2. Column K2 is driven by reboil from E4 which acts as the condenser for K1, and reflux for K2 is provided by excess nitrogen condensed in E4, subcooled in E5, and expanded across JT5 into the top of column K2. Methane product (sales gas quality) is withdrawn from the base of K2, pumped in P1, and reheated in E3, E2, and E1 before being passed to battery limit at the required pressure. For feed gases with less than 33 per cent N_2 the turbine C1 does not need to operate, but as the N_2 content rises above this, a gaseous bleed from the top of K1 is reheated in E3 and E2 and fed into C1 to provide extra refrigeration. What becomes apparent in this process concept is that in order to cater for the varying feed N_2 content, the operating conditions change very significantly as time goes on. Table 7.3 shows the extent to which the parameters vary.

It is clear from the Table 7.3 that the designer must consider the plant sizing for such a wide variety of conditions that he must take care not to create any problems by oversizing for certain cases. For example the cold exchanger E2 must be carefully designed to ensure good flow distribution for both the subcooled outlet liquid and the two-phase cases. The warm exchanger E1 must be designed to cool the feed gas in later years to −85°C, as well as the initial cooling down to about −40°C. The designer must also take great care in his column design to ensure that the appropriate approach to flood is maintained for early and later years. Moreover, because of the wide variation in fluid traffic over the years in each of the columns, the designer is forced to use valve and bubblecap trays instead of traditional sieve trays for cryogenic service. This puts the

TABLE 7.3. Variation of operating parameters for a de-nitrogenation unit in EOR applications.

Parameter	Early years (low N_2)	Later years (high N_2)
N_2 content (%)	10–22	65–75
V1 guard separator temperature (°C)	−40	−85
V2 separator temperature (°C)	−135	−146
K1 feed quality	subcooled liquid	two phase
K1 pressure (kPa)	1000	1700
K2 pressure (kPa)	165	255
Expander C1 flow	nil	design
Sales gas outlet pressure (kPa)	1100	950

tray spacing up somewhat, and hence adds to the cost of the columns. Traditional air separation columns can accept a tray spacing of around 100 mm but valve and bubblecap trays usually require spacing of 200 mm or more, depending on the physical properties of the fluids. It is clear that designers in the 1980s and beyond must pay a great deal of attention to varying conditions and flexibility to ensure the highest versatility of the plant in a tough economic climate.

7.4 Liquefaction of natural gas

7.4.1 *Requirement for liquefaction*

The trouble with sources of energy and high population densities is that the two are seldom in the same place. Natural gas as a domestic and an industrial fuel and feedstock has been taken for granted by all but the suppliers of it. Some of the largest gas fields in world are in low population or poorly developed areas. For example the North Sea, Algerian, Lybian, Brunei, and Indonesian gas fields are all vast but they cannot be effectively utilized locally. Consequently, the gas is sold to developed nations to fuel their industry. At times it is practical to bring the gas to the users via a large pipeline. Notable examples of this are the Trans-Siberian pipeline supplying gas to Europe and the Trans-Alaskan pipeline supplying gas to Northern parts of the USA. North Sea pipelines containing both oil and gas are routed under the sea onto the shore before treatment and distribution.

In some cases, however, it is not economic to build a pipeline because of large distances across land and deep sea. It becomes more advantageous to liquefy the natural gas and transport the liquefied natural gas (LNG) by ship to a distribution point where it can be vaporized and fed into a fuel grid. For example Algerian natural gas is liquefied on the north coast of Africa and transported by tanker to various parts of Europe (France, Spain, and England). It is then stored, vaporized and distributed to users by pipeline. The capacity of the liquefaction facilities on these sites is very large and hence called the 'base load'. Typical capacities range from 10 000 and 25 000 tonnes per day liquefaction capacity, with one or two approaching 70 000 tonnes per day LNG. This means that each site may have several liquefaction trains of up to 5000 tonnes per day capacity. The process required to achieve such large-scale liquefaction will be considered in this section.

Once the gas has been supplied in the correct quantity to a given country it still requires distribution to each of the users. Usually this will be by using a gas pipeline which will accept vaporized LNG or treated sales gas. However, in some parts of the world there is a high population density which leads to high fluctuations in gas consumption both daily and seasonally. Within reason such fluctuations can often be covered by buffer capacity in a long distribution gas main or some buffer gas holder storage. This is not always possible, and alternative methods of buffer storage are required. Taking the UK as an example, the London and South East area has a population in excess of 10 million, the majority of whom utilize natural gas as a fuel. Apart from daily variation which is coped with by the gas 'buffer' within the pipeline, the seasonal variation can be very large with winter consumption rising to four times the summer consumption. To alleviate any problems of short supply in the winter, British Gas operate small LNG plants during the summer months. This allows them to store LNG over low demand periods and vaporize LNG during high winter demands. This is called 'peak shaving' in the gas industry, and is a highly acceptable means of coping with gas consumption variation in relatively populated areas. Such facilities also exist in other parts of the UK, i.e. near Glasgow, Manchester, South Wales, and Bristol. Some of the process options available for 'peak-shaving' plants resemble the 'base load' processes and are considered here.

7.4.2 *Base load LNG plants*

7.4.2.1 Cascade processes

The first base load LNG plants were constructed in the early 1960s to liquefy North African natural gas for shipment into Europe. Since then the LNG industry has developed world-wide with large facilities in North Africa, the Middle East, the Far East, and the USA. Recipients of the LNG include industrially developed parts of the world, namely Japan, Western Europe, and the USA. Since the 1960s, the capacity of a single LNG train has risen from around 1000 tonnes per day to over 5000 tonnes per day in the 1980s. Various processes have been used to carry out the liquefaction, each aiming to improve the efficiency of a process which uses a large amount of power. The first base load LNG processes used a cascade comprising three closed-loop refrigeration cycles supplying cold at different levels, often referred to as the Phillips Cascade process. Figure 7.7 shows a simplified process flow-sheet for this cascade process as applied to an LNG facility in Alaska. Only the cryogenic section is shown but extensive pretreatment was required to prevent freezing of CO_2 and water in the cryogenic equipment. The natural gas was first scrubbed with di-glycol amine (DGA) solution to remove CO_2 and then dried on molecular sieves to reduce water to less than 1 v.p.m.

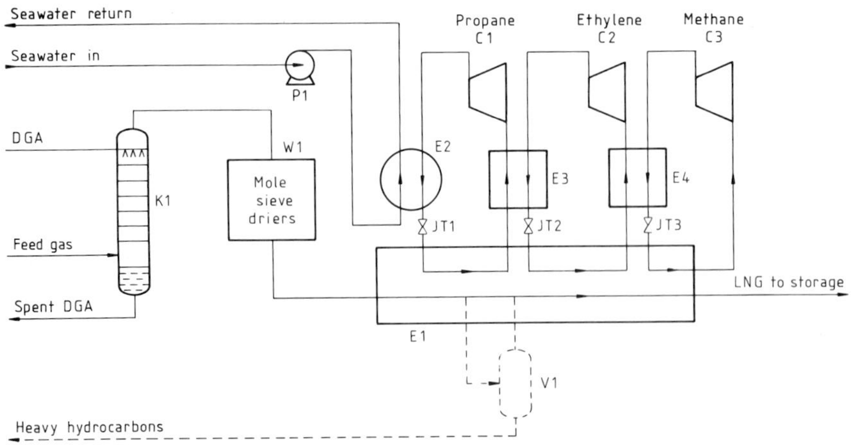

FIG. 7.7. The cascade process for LNG plant.

The dry, CO_2-free feed gas is passed into the cryogenic unit where it is cooled and condensed in E1. If the natural gas contains heavier hydrocarbons than C_5 or C_6, or perhaps traces of aromatics, they would have to be removed using a separator or separators strategically located along the cooling train in E1. V1 is shown in the Fig. 7.7 as the hydrocarbon guard pot, although it is not always designed into such processes. The remainder of the plant is a large refrigeration cycle or cycles. We have learnt from Chapter 3 that condensing/evaporating refrigeration cycles show the greatest promise. Because of this LNG designers sought this kind of cycle for base load units. Firstly, we must find a cheap cooling medium to start our condensing process—seawater in Alaska is fairly cold all year round and is therefore a sound choice. The first stage of the process is to compress commercial grade propane in C1 and condense it at elevated pressure (1200–1400 kPa) against seawater in E2. The liquid propane is then flashed across JT1 to yield a two-phase mixture at about −40°C. Some of the liquid is evaporated in E1 to provide the necessary cooling and partial condensation of LNG, whilst the remainder is used to condense the ethane or ethylene in E3. The gaseous spent propane is then fed back to C1 to complete the propane cycle. The ethylene and methane cycles operate in exactly the same way, except that they use propane and ethylene respectively to provide the necessary condensing loads in E3 and E4. They in turn are expanded across JT2 and JT3 and provide the necessary refrigeration for E1 and E4 in the case of ethylene, and for E1 in the case of methane.

7.4.2.2 Higher efficiency processes

The above process is generally suitable for small to medium-sized base load plants, but for larger units a more efficient process is required.

Before we discuss the more efficient process we shall cover the factors which make a given liquefaction process more efficient. This is where the designer looks at the fundamental thermodynamics. One of the major factors affecting process efficiencies is the irreversible losses in heat exchangers. Over 40 per cent of the irreversible losses are due to pressure drops and temperature differences in the heat exchangers in a process of this kind. To put things in simple terms, the greater the pressure drop the greater the irreversible loss. Also if we exchange heat or cold between two fluids with a small temperature difference between them we are carrying out the process more effectively than if the difference was very large. The bigger the driving force, the greater the irreversible losses. Therefore, it is paramount to reduce pressure and temperature differences if we are to improve process efficiency. Obviously, this can only be done up to an economic point because if we, say, reduce pressure losses to extremely low values, our heat transfer coefficient will be very low and the area required very large. A similar argument can be drawn for temperature differences.

Nevertheless, since refrigeraton at the colder levels costs a great deal of energy to produce there is considerable incentive to improve efficiency and save power, as demonstrated in Chapter 3, Table 3.6. This can readily be done with the cascade process shown in Fig. 7.7 by returning refrigerant into each of the compressors C1, C2, and C3 at different points in the machine. This means that evaporation of the same refrigerant is divided into two or three temperature, and hence pressure, levels. The result is that the overall temperature difference in the heat exchanger reduces and the power of recompression of each refrigerant reduces. This can be illustrated more cleary by studying Fig. 7.8 and 7.9. Figure 7.8 shows the methane compressor C3 operating with three return streams at different pressure levels, but with the same total heat load being supplied to the LNG condensing train. Figure 7.9 shows the heat load curves for a single-level and a three-level methane refrigeration loop. The methane cycle begins with the evaporated methane leaving the main exchanger E1 at about −161°C, being reheated in E4/C, and compressed in the first stage of C3 from about 110 kPa to about 350 kPa. A second vapour stream at about −145°C and 350 kPa leaves the main exchanger E1, reheats in E4/B, joins the gas leaving stage 1 of C3, and is further compressed to about 950 kPa. The same procedure is repeated with CH_4 returning at 950 kPa through E4/A until all of the CH_4 is compressed to about 2800 kPa. If we study the head load curves in Fig. 7.9 we can see that for the cold end of E1 methane is primarily used as refrigerant at below −97°C. If we assume the three-stage cycle for methane the respective heat loads for each stage are: Stage 1 entry—$A_1B_1 = 9$ per cent; Stage 2 entry—$A_2B_2 = 8$ per cent; Stage 3 entry—$A_3B_3 = 10$ per cent, giving a total of 27 per cent or the same as A_1C.

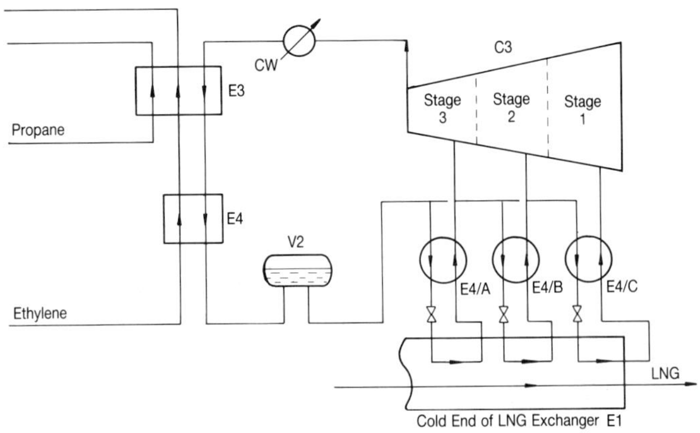

FIG. 7.8. The cold end of the cascade cycle showing three levels of refrigeration.

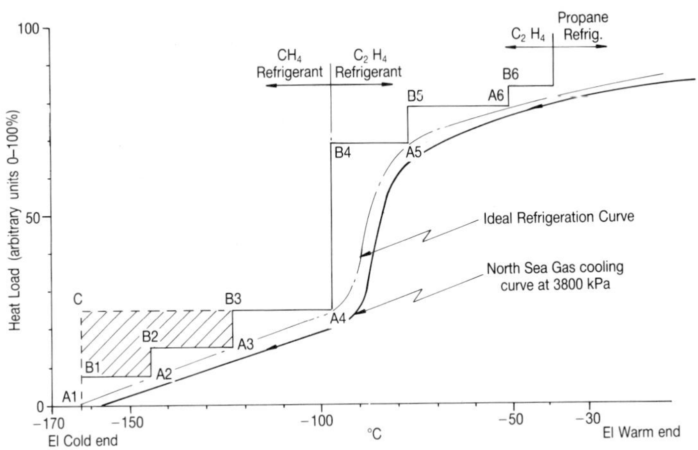

FIG. 7.9. A comparison of the heat load curves for one-stage and three stage evaporation of CH_4.

Thus it is clear that if we adopt the three stages, considerable refrigeration is supplied at warmer levels with compression from 350 and 950 kPa, and less than one-third is supplied through compression from 110 kPa. If only one pressure level is used to supply refrigeration at −162°C then *all* of the methane needs to be compressed from 110 kPa. Thus the shaded area could be termed as process inefficiency or a large irreversible loss, because heat transfer occurs in regions where the temperature difference is large (over 60°C in some parts). One can extend this to the whole process covering a three-stage propane and a three-stage ethylene cycle, so that ultimately we have *nine* temperature levels at which refrigerant is evaporated giving up cold to the natural gas.

These are depicted on the heat load diagram, Fig. 7.9, by the steps A_1B_1, A_2B_2 and A_3B_3 for methane; A_4B_4, A_5B_5, and A_6B_6 for the ethylene, and so on for the propane refrigerant at −40°C and warmer. Ethane has been used instead of ethylene in cases where ethylene is difficult to obtain to make-up cycle losses. This represents a commercially proven cycle from several base load LNG plants.

The ideal reheat curve is a line parallel to the natural gas cooling curve (shown dotted), following the curve at some 3–4°C colder all the way from the cold and to the warm end of E1. Theoretically we could have an infinite number of evaporation pressures over which each component evaporates, so that the steps A_1B_1, A_2B_2, etc. would be very small. This would give a power approaching the minimum possible to achieve the liquefaction. Of course this is totally impractical but it does give a very good clue as to how a mixed refrigerant cycle (MRC) works.

Pure components evaporate at a constant temperature for a given pressure, hence the vertical lines A_1B_1, A_2B_2, etc. Multicomponent mixtures evaporate over a range of temperatures for a given pressure. The reason for this is that the content of the less volatile components change into the vapour phase as evaporation takes place. Thus during the initial phases of boiling there is a low vapour content of heavier components, giving rise to a low partial pressure of the heavier components. As boiling continues the content of heavier components in the vapour increases, raising the partial pressure of the heavier components in the vapour. This means that the temperature must be raised to promote further boiling of heavier components which further raises the partial pressure of the heavier components, and so on. Thus we can see that the pressure (partial pressure) of the heavier components changes, and to overcome the change the temperature must change. This provides us with a situation close to that postulated above with an infinite number of pressures. The difference is that this process does not physically see many refrigerant pressure levels, but only the partial pressures of the various components change. In fact the main cycle compressor for the MRC plant operates with one feed pressure and one discharge pressure.

7.4.2.3 The mixed refrigerant process

The MRC process dominates the modern base level LNG plants around the world, with the largest being operated using a modified MRC cycle. Figure 7.10 shows a classical MRC process which uses N_2, CH_4, C_2H_4, C_3H_8, and C_4H_{10} as components in the mixed refrigerant fluid. The pretreatment for the LNG is the same as for any LNG plant, so we shall concern ourselves only with the cryogenic unit. The feed natural gas enters the cryogenic section at the first of a series of heat exchangers, which are likely to be wound coil type E2 to E6, where the gas is cooled,

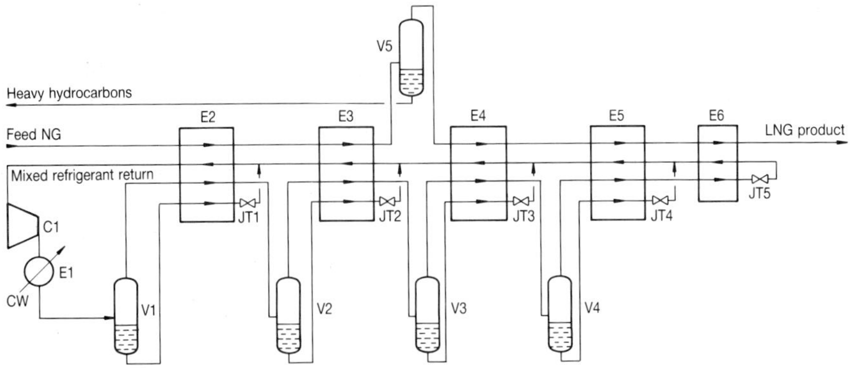

FIG. 7.10. The basic mixed refrigerant cycle (MRC) liquefaction unit.

condensed, and subcooled. After this it is expanded into LNG storage tanks. Refrigeration is provided by a closed cycle of five components (N_2, CH_4, C_2H_4, C_3H_8 and C_4H_{10}), the composition of which varies depending on the composition of the natural gas to be liquefied.

The mixed refrigerant is compressed from about 300 kPa to about 3000 kPa in C1. The gas is then cooled and substantially condensed in the water cooler/condenser E1. The liquid, essentially propane and butane, is separated out in V1. The liquid is then subcooled and the vapour is cooled and partially condensed in E2 against returning low-pressure mixed refrigerant. The subcooled liquid is then flashed to about 330 kPa across JT1, where the temperature falls dramatically causing some evaporation and leaving a substantial amount of liquid but at a much colder temperature. Upon mixing with the mixed refrigerant (MR) return from E3 the temperature falls a little more which provides the necessary cooling for the feed LNG and both the liquid and vapour leaving V1. This process repeats itself in sets of exchangers, separators, and Joule–Thomson valves V2 + E3 + JT2, V3 + E4 + JT3, and V4 + E5 + JT4. Finally at the cold end the cooled and condensed vapour from V4 is expanded across JT5 to give the necessary subcooling load for the LNG product. Understandably, the flow of MR in the cycle is considerably greater than the natural gas in order to give the necessary cooling and condensation load to the cycle streams, as well as to the gas which is to be converted to LNG.

Many natural gases contain heavier hydrocarbons in relatively low concentrations, and to avoid freezing in the cold parts of the process or the LNG tank they must be removed. This can be done by adding a separator, V5, part way down the cooling train, between say E3 and E4. If the temperature of this separator is strategically chosen (−40 to −80°C) then it removes the majority of heavier hydrocarbons, thereby alleviating any freezing problems. For large base load plants this separator achieves a second aim—to provide the MRC with hydrocarbons. In remote locations (often where the large base load plants operate) shipment of make-up hydrocarbons for the MRC is uneconomic. Therefore on-site generation is successfully done, but instead of ethylene ethane is used because natural gases contain only saturated hydrocarbons. In order to control the composition of the MRC gas, pure components must be added and thus on-site generation of MRC needs distillation of the heavy hydrocarbon fraction to obtain pure CH_4, C_2H_6, C_3H_8, and C_4H_{10}. Figure 7.11 shows a simplified flow-scheme of a purification system for MRC make-up hydrocarbons. This process consists simply of a series of distillation columns K1, K2, K3, and K4. Each column recovers one of the MRC components CH_4, C_2H_6, C_3H_8, and C_4H_{10} respectively. They are then blended with N_2 from a cryogenic unit in the correct proportions to give the best match of cooling and reheat curves.

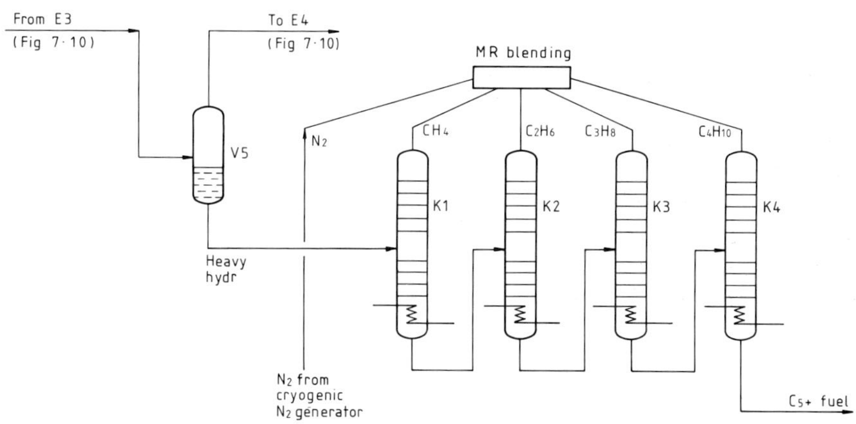

FIG. 7.11. MRC make-up hydrocarbon component purification.

The start-up of one of these units is a very involved affair—in order to mix the correct proportions because as the plant cools down the composition changes due to build-up of liquid in various parts of the cryogenic section, i.e. in the exchanger headers and the separators. However, the large power saving justifies the complexity when comparing simpler less efficient processes for base load application.

7.4.3 *Peak-shaving LNG processes*

Several developed countries operate peak-shaving LNG plants to overcome seasonal demand variation. Such countries include the UK, USA, Holland, West Germany, and France. As already mentioned, many peak-shaving processes resemble the MRC and cascade cycles described in section 7.4.2. There are two other alternatives which are suitable for peak-shaving LNG plants which merit some consideration. This section shall consider both the use of a closed-loop nitrogen cycle as a means of peak-shaving LNG production, and a turbo-expander process which utilizes the pressure energy in the natural gas to produce the LNG.

In certain parts of the world electricity costs are relatively low, e.g. in Norway and British Columbia where hydroelectric power is abundant. In other areas the operating requirements for a peak-shaving plant are relatively short, making the power cost a less significant factor in the capital cost analysis. In such cases a process that is less efficient but cheaper to install and simpler to operate merits consideration. The closed-loop nitrogen cycle is one such process, and a simplified process flow-sheet is shown in Fig. 7.12. In this process the feed gas is dried and rendered free of CO_2 before it enters the cryogenic unit. The pre-treatment is done in the traditional manner using molecular sieves. The clean gas is first cooled in the warm end of the plate-fin heat exchanger battery E1 (usually made of several blocks in series and parallel, depending on the designer's philosophy). If the natural gas contains heavy hydrocarbons some intermediate separation is required in V1. Sometimes, more than one separator is used at various temperatures. For a typical North Sea gas at about 4000 kPa, heavy hydrocarbon removal might be designed to operate at temperatures of −40°C in V1 and in additional separators at temperatures down to −85°C. The vapour from V1 is further cooled and condensed in E1 and leaves E1 at around −145 to −150°C before it is flashed in JT2. The LNG is separated from the flash gas in V2 and fed to an LNG storage tank. The flash gas returns through the E1 battery, joins up with flashed liquids from V1 and any

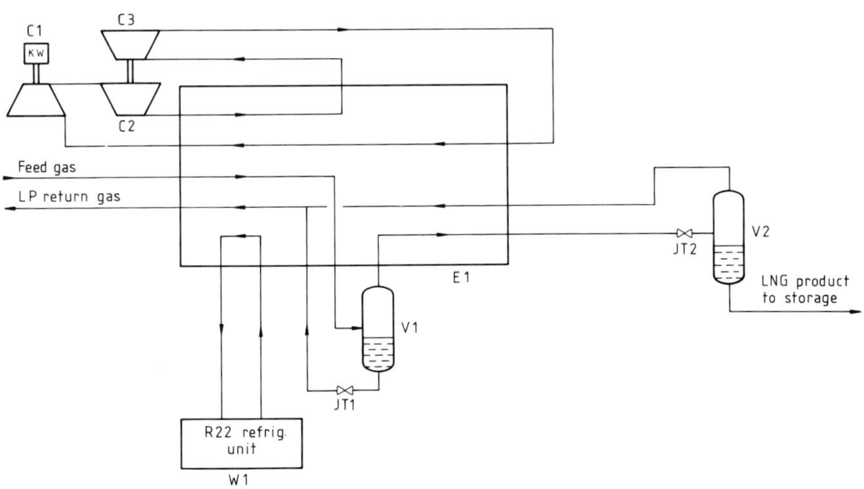

FIG. 7.12. A closed-loop N_2 cycle peak shaving LNG plant.

other pots, and leaves the cryogenic unit from E1 as ambient low pressure gas. Refrigeration is supplied mainly by a closed loop nitrogen cycle assisted by a refrigeration unit using Freon (R22) or propane. The nitrogen cycle can operate over a variety of pressures, but for this kind of cycle and medium-sized peak-shaving units an operating range of 600 to 5200 kPa is considered beneficial to the selection of machinery and their efficiency. The main cycle compressor C1 compresses N_2 from about 600 kPa to about 3200 kPa. This N_2 is passed through the expander brake end (C2) whereupon it is boosted to about 5200 kPa. Of course, inter- and after-coolers are used in both C1 and C2 before the N_2 enters the cryogenic unit. The N_2 is first cooled to about −35°C against Freon and then further cooled against return N_2 to about −70°C before being expanded in C3. The cold produced in the expander C3 provides the bulk of refrigeration needed to liquefy the natural gas. After the nitrogen gives up its cold at 600 kPa in E1 it returns to the cycle compressor C1. The beauty of this process is that it uses inert single fluids (R22 and N_2) as refrigerants, making the process easy to start up, shut down, and turn down. This kind of flexibility is ideal for a peak-shaving situation which may require a rapid response to climate changes.

An alternative to the above process is a turbo-expander plant, which generates all of its refrigeration from a large volume of high pressure gas which is let down to a medium- or low-pressure grid. In such a case the pretreatment for the process is approximately the same, depending on the CO_2 content. The cryogenic section of this process is shown in Fig. 7.13.

In Fig. 7.13 the liquefied gas takes a similar path to that described for the cycle process, i.e. cooling, partial condensation, condensation in

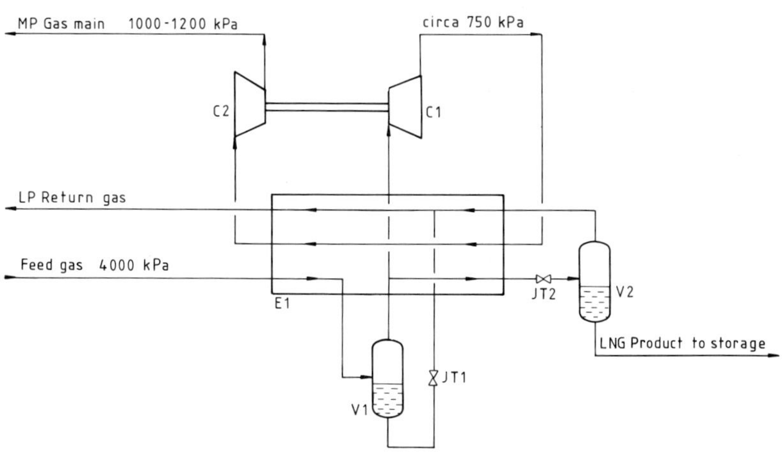

FIG. 7.13. The expander LNG peak-shaving process.

total, flash, and separation before routing the LNG to storage. The difference in this process lies in the amount of feed gas admitted to the plant. For the N_2 cycle process, only that portion of gas requiring liquefaction is fed to the plant through E1, V1, JT2, and V2. In this process a much larger flow of natural gas is supplied to E1, and after partial condensation in E1 and separation in V1, the vapour is split. The minor part goes on to be liquefied whilst the major part goes to the expander, C1. This provides all of the refrigeration required to make LNG, which after reheating in E1 is re-compressed to a level suitable for entry into a gas main. If the gas flow at high pressure is large, then this process has the advantages of being simple and consuming no power. The three basic types of processes for peak shaving have all been used successfully in industry. Selection of process route is made on economic grounds having in mind capital cost, power cost, amount of gas available for let down from high pressure, number of days per year operation required, etc. Table 7.4 gives some of the more salient points which affect the choice of cycle for LNG peak shaving.

TABLE 7.4. Comparison of peak shaving processes to make 200 tonnes/day LNG.

	Process route		
Parameter	Mixed refrigeration	N_2 cycle	Expander
Power consumption (kW)	5400	6700	25
Approximate capital cost ratio	1.4	1.0	0.8
Gas used for expansion ($Nm^3\ h^{-1}$) (assumed $P_1 = 4000$ kPa and $P_2 = 1000$ kPa)	nil	nil	100 000
Start-up time (h)	36	8	8

Bibliography

ALVAREZ, M. R., *et al.* (1984). 'Dome's NRU is successfully treating gas from EOR project'. *Oil and Gas Journal (Technology),* (Aug. 20), 95.

ALVAREZ, M. R. and VINES, H. L. (1985). 'Nitrogen rejection/NGL recovery for EOR projects'. *Energy Progress,* **5**(2), 67.

BAGGIO, J. L. and SAINTHERANT, P. (1980). 'NGL refrigerant improves recovery'. *Hydrocarbon Processing,* (Apr.), 136.

CRYOGENICS AND INDUSTRIAL GASES (1970). 'LNG liquefaction cycles'. (Oct.), 33.

DI NAPOLI, N. R. (1975). 'Estimating costs of base load LNG plants'. *Oil and Gas Journal,* **17**, 53.

DI NAPOLI, R. N. (1975). 'Trends in base load LNG plants'. *Hydrocarbon Processing,* (Dec.) 87.

FINDLAY, A. J. and SPICER, G. W. (1983). 'The Isle of Grain installation'. *Gas Engineering and Management,* (Feb.), 47.

FORG, W. and ETZBACK, V. (1970). *Fractional condensation of natural gas.* US Patent 3721099.

GOETHE, A. and MAWER, D. J. (1987). 'A new integrated nitrogen rejection process with NGL recovery'. *Proc. of the 66th GP Annual Convention,* p. 89, Tulsa, Oklahoma, USA.

HARRIS, R. A. (1980) 'Consider nitrogen rejection design'. *Hydrocarbon Processing,* (Aug.), 111.

HASELDEN, G. G. (1979). 'LNG and world energy needs pt. 1: gas liquefaction and world resources'. *Intl. J. Refrigeration,* **2**(5), 139.

HÖRNER, B., *et al.* (1983). 'Offshore liquefaction of natural gas'. *Linde Reports Sci. Tech.* **36**, 3.

JONES, J. K. (1973). 'Upgrade low-BTU gas'. *Hydrocarbon Processing,* (Sept.), 193.

JONES, J. K. (1974). *Cryogenics upgrading of low-Btu gases containing air, CO_2 and N_2.* GPA Conference, Denver, Tulsa, Oklahoma, USA.

JUNCKER, F., *et al.* (1970). *Single-stage fractionation of natural gas containing nitrogen.* US Patent 3797261.

KENNETT, A. J., *et al.* (1981). *Offshore liquefaction of associated gas—a suitable process for the North Sea. 13th O.T.C. Conference,* p. 31. Houston, Texas.

KHAN, S. A. (1985). 'Expander-gas processing plant converted to boost C_3 recovery at Canada's Judy Creek'. *Oil and Gas Journal,* (June 3), 78.

KINARD, G. E. and GAUMER, L. S. (1973). 'Mixed refrigerant cycles for LNG'. *Chem. Eng. Progress,* **69**(1), 56.

LIMB, D. I. and CZARNECKI, B. A. (1987). 'Reflux-exchange process lifts propane recovery at Aussie site'. *Oil and Gas Journal (Technology),* (Dec. 14), 35.

LOONEY, S. K., *et al.* (1984). 'Integrated N_2 rejection facility produces fuel and recovers NGLs'. *Energy Progress,* **4**(4), 214.

PARNAG, J., *et al.* (1968). *Cryogenic process for separation of a natural gas with a high nitrogen content.* US Patent 3531943.

PRICE, B. C., *et al.* (1981). 'N_2 rejection unit will improve NGL recovery and save fuel'. *Oil and Gas Journal (Technology),* (Apr. 13), 99.

RATHMANN, U., *et al.* (1984). 'Comparison of various natural gas liquefaction processes and driver systems for peakshaving plants'. *Linde Reports Sci. Tech.* **37**, 3.

RATHMANN, P. and GRIMM, P. (1982). 'Separation of NGL, N_2 and helium from natural gases and associated gases'. *Linde Reports Sci. Tech.* **34**, 14.

RUHEMANN, M. (1984). 'Cryogenic techniques in enhanced recovery of oil and gas'. *Indian J. Cryogenics,* **9**(4), 256.

SEAY, J. G., *et al.* (1978). 'LNG base load projects continue to grow'. *Oil and Gas Journal,* (Dec. 18), 70.

STREICH, M. (1971). 'Removal of nitrogen and recovery of helium from Dutch natural gas'. *International Cryogenics Conference,* Brighton, England, British Cryogenics Council.

TEMPLE, R. W. and DI NAPOLI, N. R. (1970). 'Ethane and LPG recovery in LNG Plants'. *Hydrocarbon Processing,* (Apr.) 89.

THOROGOOD, R. M. (1986). 'Gas separation and liquefaction plants'. *Cryogenic Engineering,* (ed. B. A. Hands), p. 391. Academic Press, London.

TOMLINSON, T. R., and BANKS, R. (1985). 'LPG extraction process cuts energy needs'. *Oil and Gas Journal,* July

VIELVOYE, R. (1978). 'Gas supply/demand gap forecast for W. Europe'. *Oil and Gas Journal,* Aug. p. 31.

WALL, J. D. (1975). 'LNG goes world-wide'. *Hydrocarbon Processing,* (Apr.), 141.

ZOLLNER, H. R. (1981). 'Liquefaction of natural gas with the aid of refrigerant mixtures'. *Linde Reports Sci. Tech.* **32**, 19.

8

PRE-PURIFICATION AND SYNERGISTIC PROCESSES

8.1 Introduction

The cryogenic section of a gas separation plant, though crucial, is often only a relatively minor part of the process. Before one can admit a gas into a cryogenic unit it may need feed compression and significant pretreatment. Most feed gases will contain constituents that may freeze inside or even corrode the cryogenic equipment and will therefore require removal. Wet hydrocarbons can form hydrates at temperatures above 0°C (typically 5–15°C) and in such cases water removal may be necessary. Freezable components include H_2O, CO_2, and NH_3. In dry conditions Hg and CH_3OH can cause corrosion problems. These are usually removed from the feed gas by absorption, adsorption, reversing exchangers, or combinations of these in a pretreatment section. These shall be covered in a little more detail in this chapter.

Product treatment can also be an extensive part of a cryogenic plant, depending on whether large storage facilities are required or whether any special bottling or send out facilities are envisaged. Disposal of cryogenic liquid is also done in a manner specific to this technology. These are also considered in more detail here.

Mature technologies often look for synergies with other technologies, and cryogenics is a good example. Synergies with membrane and adsorption technology are considered in this chapter as a means of enhancing the performance of cryogenic technology.

Table 8.1 gives a list of typical gases treated by cryogenic units. Typical impurities are listed for each of these gases. In general, these impurities are removed by absorption, adsorption, chilling, permeation, or even distillation prior to cryogenic processing to p.p.m. levels (typically less than 1 p.p.m.). However, in some cases the liquids formed in the process dissolve the freezable components to a significant extent, making it possible to allow several hundred p.p.m. to enter the cryogenic unit. Each case has to be considered carefully by the cryogenic unit designer, who uses reliable solubility and equilibrium data to determine the levels acceptable in the cryogenic unit.

The impurity concentration in the feed gas determines the nature of the pretreatment process used. Table 8.2 shows a typical breakdown of pretreatment processes used for various impurity levels. This table is a very general guide and each case must be carefully analysed to select the proper pretreatment.

Other impurities, such as HCl, NH_4Cl, Hg, and arsenic compounds, are generally found in much lower concentrations and, therefore, if removal is required, catalytic methods are often used.

TABLE 8.1. Freezable impurities present in cryogenic plant feed gas.

Feed gas	Typical impurities present
Air	H_2O, CO_2, traces of hydrocarbons
Natural gas	H_2O, CO_2, COS, Hg, heavy hydrocarbons, aromatics, CH_3OH
NH_3 purge gas	H_2O, NH_3
NH_3 plant synthesis gas	H_2O, CO_2
Petrochemical plant off-gases	H_2O, CO_2, H_2S, aromatics, CH_3OH
Refinery off-gases	H_2O, CO_2, H_2S, COS, NH_3, NH_4Cl, HCl, HCN, arsenic compounds, aromatics, heavy hydrocarbons

TABLE 8.2. Pretreatment methods used to treat freezable components.

Component	Typical concentrations	Pretreatment used
H_2O	0–2000 v.p.m.	Adsorption, freezing*
	> 2000 v.p.m.	Chilling + adsorption, freezing*
CO_2	0–5000 v.p.m.	Adsorption, freezing*
	> 5000 v.p.m. to 5%	Absorption + adsorption, freezing*
	> 5%	Absorption, permeation
H_2S	0–5000 v.p.m.	Adsorption, reaction
	> 5000 v.p.m.	Absorption, reaction, permeation
HCl	0–100 v.p.m.	Adsorption, absorption
COS, CS_2	0–50 v.p.m.	Reaction, adsorption
	> 50 v.p.m.	Absorption, adsorption
CH_3OH	0–500 v.p.m.	Adsorption
	> 500 v.p.m.	Adsorption, chilling
NH_3	0–2000 v.p.m.	Adsorption, absorption
	> 2000 v.p.m.	Absorption, chilling
Heavy hydrocarbons and aromatics	> 5000 v.p.m.	Adsorption, absorption, chilling

*By using reversing heat exchangers.

8.2 Adsorption processes

8.2.1 *Non-regenerable adsorption*

Mercury (Hg) is an interesting example of an impurity which is sometimes present at very low levels in natural gas (typically 0.1 v.p.m. or less). The natural gas treatment cryogenic plant is often made up of aluminium equipment which is very sensitive to the presence of Hg, since aluminium dissolves in liquid Hg to form an amalgam. This can lead to significant weakening of the aluminium equipment after a period of years or even months, and can cause breakdown or failure. Hg must therefore be removed for such plants and this is effected by the use of a catalytic guard bed. A convenient method is to use a bed of coal-based carbon granules impregnated with at least 10 per cent sulphur (by weight). The sulphur readily reacts with Hg thereby removing it. This is a non-regenerable process but popular because of the relatively high capacity of sulphur for Hg on a weight-for-weight basis. Other, less attractive, methods are available, including iodine-impregnated carbon beds, adsorption onto carbon, and molecular sieve beds.

Another example where a small amount of potentially corrosive constituents is removed is the use of molecular sieve or activated alumina to remove traces of HCl from a refinery off-gas prior to cryogenic treatment in aluminium equipment. Typically, refinery off-gases contain 1–4 v.p.m. HCl, and this can be adsorbed onto molecular sieve or alumina. Because HCl is very reactive it breaks down some sieves making them less effective after regeneration. Therefore it is fortunate that the HCl levels are low and the adsorbent can be used on a 'throwaway' basis. Typically a vessel may be filled with 'catalyst' beads or pellets, as shown in Fig. 8.1, and the gas is passed downwards through the bed. This design can apply to the above examples as well as to, say,

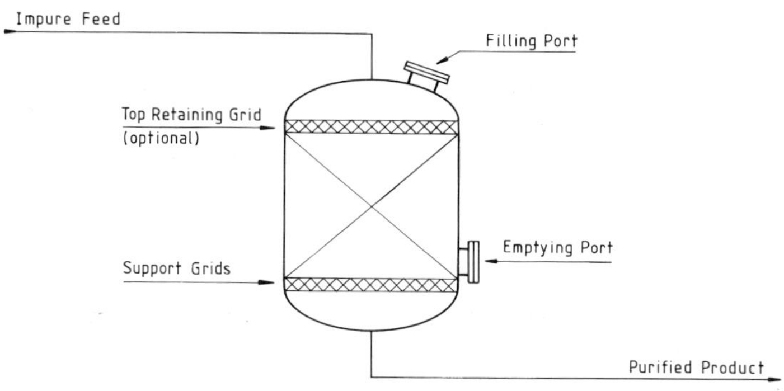

FIG. 8.1. The fixed bed non-regenerable system for the removal of trace impurities.

removal of H_2S traces using ZnO pellets or hydrated iron oxide. In each case the on-stream life must be relatively long to minimize the labour intensity of the plant.

8.2.2 *Regenerable adsorption systems*

When impurities are present in quantities significantly greater than a few v.p.m. it is economic to provide a regenerable adsorption system. A typical example may be the removal of 500 v.p.m. H_2O and 700 v.p.m. NH_3 from ammonia purge gas. Typical loadings on molecular sieve at elevated pressure for these components are shown in Fig. 8.2. It is clear from these that a one-year 'throwaway' bed for 1000 kg mol h^{-1} of feed gas flow in impractical and a regenerable system is required. In such cases two beds are usually employed, one on-stream while the other is being regenerated. A convenient cycle time is chosen to give an economic bed size and to fit in with shift changes on the plant, typically 8, 16, or 24 hours. Both beds are made the same size and the regeneration of the off-stream bed is carried out by heating a clean and dry waste gas at low pressure to about 200 to 300°C and passing it through the 'spent' bed. Figure 8.3 shows a typical arrangement of a two-bed regenerable adsorption system. In this diagram, V1 is the on-line vessel while V2 is being regenerated with a dry waste gas. We shall assume that, for this example, water and ammonia are being removed from the impure feed gas. The waste gas, heated in E1, is passed through the spent bed V2, cooled in E2, and some regenerated waste is knocked out in V3.

The design of the adsorption bed depends on the amount of impurities requiring removal. In this example both NH_3 and H_2O are removed. Each has its own saturation capacity on a given molecular sieve and each

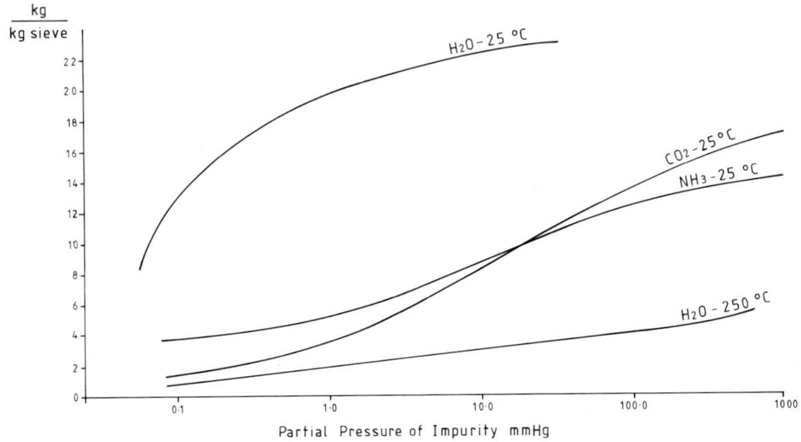

FIG. 8.2. Adsorption isotherms for ammonia, water, and carbon dioxide.

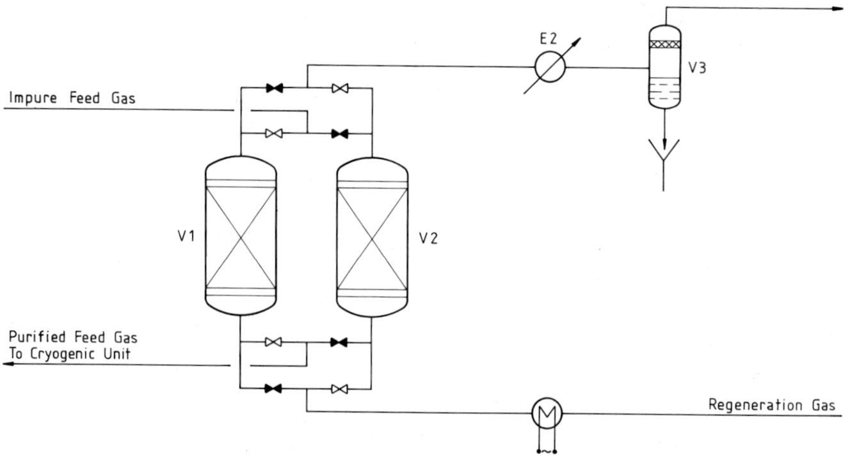

FIG. 8.3. The two-bed thermally regenerable adsorption system.

has its own mass transfer zone (length of bed over which adsorption occurs). The designer calculates the former based on the on-stream time and the latter on the fluid properties and velocity. He also allows for deterioration of the molecular sieves and adds a factor to cover design condition uncertainties. In multicomponent adsorption, such as H_2O and NH_3 mentioned above, a significant amount of co-adsorption occurs, i.e. two or more components adsorb in the same molecular sieve pores onto neighbouring sites. This makes for a complex design technique which is outside the scope of this chapter. Modern design techniques use computer programs for accurate design of multicomponent adsorption systems.

Some compounds, for example H_2O, NH_3, and CH_3OH, are strongly adsorbed onto desiccant leading to a large concentration deposited onto the sieve—up to 23 per cent by weight. Other adsorbates are less strongly adsorbed with much lower saturated loadings, for example H_2S, CO_2, and aromatics. In each case adsorption heat is released, the amount of which varies a great deal for each absorbate with H_2O being the highest at about 1000 kcal kg^{-1} H_2O. This often manifests itself as a gas temperature rise of a few degrees across the bed. If very low concentrations of adsorbable impurities are present this rise is negligible.

As can be seen from Fig. 8.2, saturation concentration for the desiccant is partial-pressure dependent, although this dependency is less pronounced for strongly adsorbed components such as water and ammonia. Figure 8.2 shows this very well with H_2O and CO_2. One can therefore conclude that by releasing pressure from a spent (or loaded) bed, some desorption occurs. This phenomenon is utilized in pressure swing adsorption (PSA) which is described in more detail in section

8.6.2. For strongly adsorbing components such as H_2O and NH_3 thermal swing regeneration is usually employed to render a bed clean.

To calculate regeneration requirements the designer takes into account heat requirements to the metal, sieve, insulation, desorption, and exit gas losses. The generalized equations often used to determine the cooling and heating times for a given regeneration gas flow are as follows:

$$\theta = t_h + t_c + t_x$$

and

$$t_c = \frac{W_B C_{PB}}{G C_{PG}} \ln \left\{ \frac{(T_1 - T_3)}{(T_2 - T_3)} \right\}$$

where θ = total on-stream time for one bed ($\frac{1}{2}$ cycle time)
t_h = heating time
t_c = cooling time
t_x = changeover time (depressurization, equalization, pressurization—often fixed values)
W_B = weight of bed
C_{PB} = specific heat of bed
W_G = gas flow
C_{PG} = specific heat of gas
T_1 = initial bed temperature
T_2 = required bed temperature
T_3 = coolant temperature

No account needs to be taken of water or any other contaminants on the bed in the above equation because it is for the cooling period after H_2O has been driven off.

The designer is required to consider various heating times to ensure that he has adequate cooling time to achieve a given outlet temperature for the regeneration gas at the end of the cycle. He also has to consider parameters to prevent bed lift or attrition during adsorption and regeneration. In adsorbers where significant amounts of water are removed the heat desorption can be as high as 35 per cent, but is usually less for most other components and conditions. Where vessel dimensions are large with a high design pressure it may be prudent to consider internal lining so that the bulk of the metal is not heated, thus reducing energy consumption. The choice of system depends on economics and costs of utilities.

8.3 Absorption processes

As the concentration of a given impurity rises, adsorption or chemisorption onto a solid bed becomes less attractive. In such instances absorption into a suitable solvent is often preferable. Typical examples of solutes and solvents to deal with them are given in Table 8.3. This table is meant only as a guide to what is available in the market place, and does not indicate any optimum solution.

TABLE 8.3. A summary of solutes and solvents in absorption used for pretreatment.

Solute	Physical solvents	Chemical solvents
NH_3	H_2O	—
H_2O	Ethylene glycol, di- and tri-ethylene glycol, methanol, 'Drizo'	—
HCl	H_2O	—
Aromatics + heavy hydrocarbons	Toluene, kerosene, naphtha wash	—
CO_2	H_2O, 'Selexol', 'Sulfinol', methanol ('Rectisol')	Monoethanolamine (MEA), diethanolamine (DEA), hot carbonate ('Benfield', 'Girdler', 'Catacarb', 'Vetrocoke')
H_2S	H_2O, 'Selexol', 'Sulfinol', methanol ('Rectisol')	MEA, DEA, hot carbonate, quinone ('Stretford'), 'LO-CAT'.
COS, CS_2	'Sulfinol', 'Stretford'	DGA, MEA, DEA

Bulk ammonia removal is usually done with water because it is a cheap solvent. This may be done on a once-through basis with pure water or a recirculating system as depicted in Fig. 6.5. In such cases a lean solution of, say, 1 per cent NH_3 in H_2O is used to wash gas containing ammonia. The 'clean' gas may still contain some 500 v.p.m. NH_3, though it may have started with 1–5 per cent. NH_3 has a very high heat of absorption (in the region of 7500–8500 kcal kg^{-1} mol^{-1} of NH_3 absorbed in water). This means that the heat is released into the 'clean' gas as well as the NH_3-rich liquid. The designer must take account of this in calculation of heat balances. The rich solution is then distilled ito pure ammonia and water or 'lean' solution, which is either cooled and discarded or recycled into the absorber depending on the system preferred. On occasions designers have used condensation by cooling the impure gas to $\sim -70°C$, thereby condensing out most of the NH_3. This process leaves the equilibrium concentration of NH_3 in the vapour phase at a level of several thousand p.p.m. depending on pressure. For this reason and controllability the process is not often used. Bulk removal by adsorption is expensive because of the high heats of regeneration involved. However, for low NH_3 content it can be attractive in some cases.

Semi-permeable membranes established themselves in the early 1980s and show some promise for ammonia removal. Ammonia readily permeates such membranes in preference to most other components, although its chemical effect on the membranes requires careful consideration. At the time of writing this method of NH_3 removal was in the developmental phase.

Water removal is effected by the use of standard glycol contactor packages for many natural gases but only a few refinery and petrochemical applications. This is a very effective way of removing relatively large quantities of water from the vapour phase down to dewpoint levels of −40 to −50°C (or dewpoint depression by 30–60°C). A typical process schematic for a glycol drier/regenerator system is shown in Fig. 8.4.

In this process sour gas is admitted into K1 and scrubbed with dry glycol solution to yield a product gas of −40 to −50°C dewpoint, typically. Spent glycol (containing water) is preheated in E1 against cooling regenerated glycol. The hot spent glycol is then heated in the stripping still V2 using steam or fuel gas in E2. The water is thus driven off and the clean (water-free) glycol is pumped via P1 and returned to K1 via E1.

Using this above process, the lean glycol leaving V2 contains residual water which, when supplied to K1 as reflux, prevents the attainment of very low dewpoints. A relatively new process came on the scene in the 1980s which alleviated this problem. The 'Drizo' process uses hydrocarbon solvents (pentanes and heavier) as azeotropic stripping agents to remove trace quantities of H_2O in glycol (TEG). These agents, once passed through the 'rich' glycol stripping column, are condensed along with stripped-out water and the hydrocarbon solvent is recovered in a

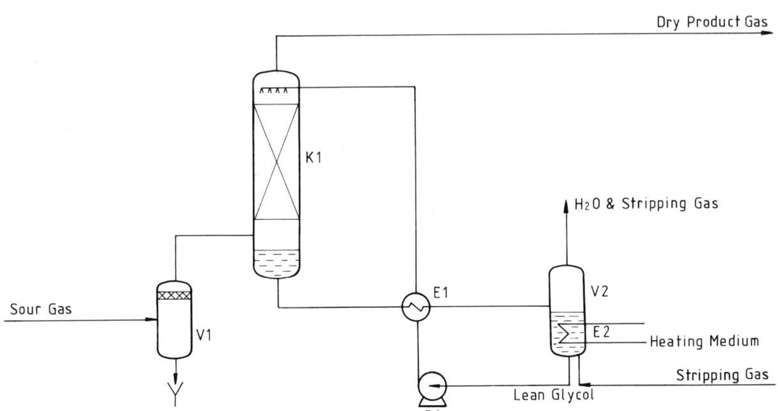

FIG. 8.4. The glycol drying package.

three-phase separator, pumped into a heater/evaporator, and then used as stripping gas again. In this way the glycol can be dried to a concentration of 99.999 per cent or better, thus making it highly suitable for use in achieving a lower water content. Using a 'Drizo' process to attain a better dryness for the glycol drying liquid allows the achievement of dewpoints of − 70 to − 100°C in the dry process gas.

HCl is usually absorbed in a simple, once-through water-wash tower because concentrations are usually very low in the impure gas. Other impurities (i.e. aromatics, sour gases, etc.) are generally removed by washing in a lean solution which is then regenerated in some way. The method of regeneration differs somewhat depending on whether the absorption is physical or chemical. Figure 8.5 shows a typical chemical absorption/regeneration process such as MEA or DEA.

Figure 8.5 has some similiarities to the NH_3 wash and recovery system in Fig. 6.5. Much of the same principles apply in that lean MEA, DEA, or other suitable solvent is pumped in P1, cooled in E4, and fed into the top of the absorber K1 where it scrubs the incoming gas. 'Rich' solvent is flashed in V1 to remove dissolved gases, and heated in E1 against fresh 'lean' solution from K2. The rich solution fed to K2 is regenerated, producing acid gas overheads and fresh lean solution. This is done by applying steam in reboiler E2. Cooling E3 reduces losses of water and useful chemicals from the system. Depending on duty various additives can be added to prevent degradation, foaming, and losses of useful chemicals, but this is not within the scope of this chapter.

Physical solvents are often easier to regenerate, but can result in more complex processes or higher impurity levels in the clean gas. A typical physical solvent absorption/regeneration scheme is depicted in Fig. 8.6. The sour gas is admitted into K1, possibly via an inlet guard pot and perhaps a preheater which cools sweet gas. Semi-lean and lean solvents

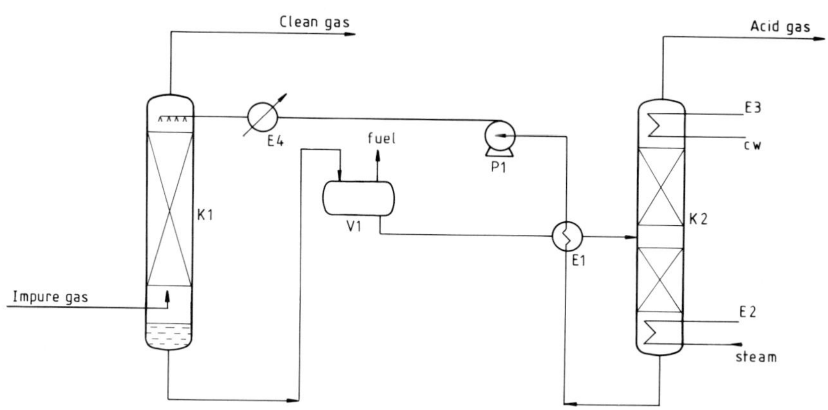

FIG. 8.5. A typical chemical absorption/regeneraton process.

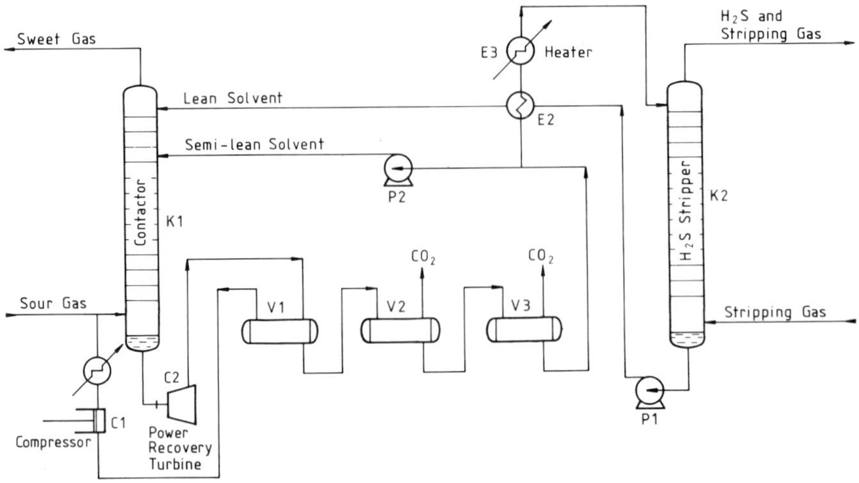

FIG. 8.6. Typical gas-sweetening by physical absorption.

are used to scrub the gas in K1 to sweeten it. The rich liquor at the base of the column is flashed (sometimes expanded in C2 on large plants) to degas the liquor. Vapour recovered from the degassing in V1 is recompressed in C1 and cooled in E1 before re-mixing with sour feed gas. V2 and V3 are used to flash the liquid further to release CO_2 to decreasing levels. The resultant liquor is 'semi-lean', part of which can be recycled to the tower K1 via P2. The remaining semi-lean solvent is preheated in E2 and E3 before entering the stripper K2, where it is stripped with gas free from sour constituents. The H_2S is thus released in the stripper and the truly lean liquor is pumped from the base of K2 in pump P2, cooled in E2, and fed to the top of K1 as lean liquor.

It is clear from this description that the bulk of CO_2 is removed by flashing and not regeneration heat, and this constitutes a major energy saving which more than offsets the extra capital cost and complexity.

8.4 Regenerators and reversing exchanger technology

The air-separation industry has used regenerators to remove water and CO_2 for many decades, and it is the USSR which has had the longest experience in this area, dating back to the early twentieth century. Regenerators are in essence vessel containing pebbles onto which H_2O and CO_2 deposit as solid from the passing pressurized airstream. Two such vessels are employed with one being regenerated by low-pressure waste nitrogen. The nitrogen picks up the H_2O and CO_2 in vapour phase and regenerates the spent vessel. This cycle is repeated in a matter of minutes with changeover being effected by a series of valves. The design of a regenerator is determined by a need to vaporize H_2O and CO_2 from the pebbles into the regeneration gas. A typical calculation procedure looks at a clean-up factor F which is governed by the following equation:

$$F = f\frac{M_R(P'_R - P^\circ_R)P_a}{M_a(P'_a - P^\circ_a)P_R}$$

where f = approach to equilibrium assumed to be ~ 0.8 for low pressure air-separation plant

M_R, M_a = molar flow for regeneration and feed gases, respectively

P_R, P_a = total pressure for regeneration and feed gases, respectively

P°_R, P°_a = vapour pressures for CO_2 for regeneration and feed gas as at regenerator cold end

P'_R, P'_a vapour pressures for CO_2 for regeneration and feed gases at any point in the regenerator.

From this kind of relationship a practical design needs a regeneration flow of greater than 60 per cent of the incoming equivalent air flow to remove the water and CO_2.

The early 1960s saw the advent of reversing heat exchangers, which work in a similar way except they are not a large heat sink like regenerators. The main application area for this technology was and still is in air-separation pretreatment. These also work on the principle of freezing a layer of H_2O and CO_2 onto the surface of the heat exchanger layers through which the feed gas passes. These also work on a short cycle basis where feed gas is being 'cleaned' in one section and the other is being regenerated. The difference is that regenerators are generally two separate vessels with purging facilities and cooling (Trumpler) coils, whereas the reversing heat exchange duty is often done in one multi-stream plate-fin exchanger. Figure 8.7 shows a typical arrangement of such an exchanger.

Feed gas passages are streams A and B, except that whilst feed passes through A at ~ 600 kPa, waste gas purges stream B at about 105 kPa. After some 5–10 minutes the valves X1, X2, X3, and X4 change over, pressurizing B and depressurizing A. Then the feed gas passes through B depositing H_2O and CO_2 while A is being regenerated with dry waste gas at low pressure. Streams C and D are the other process streams which continue to pass through unchanged.

8.5 Relief and disposal of cryogenic liquids

In general process plant of a non-cryogenic nature, equipment is protected with relief valves. When a process upset occurs a relief valve may be called upon to protect the equipment and, more importantly, life

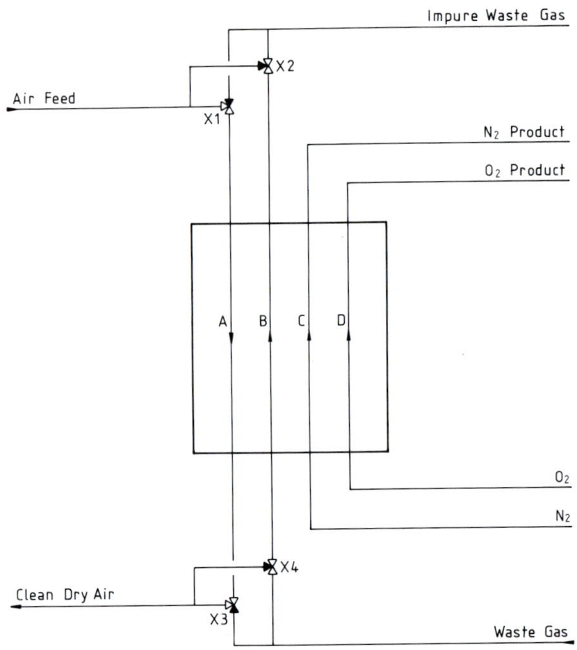

FIG. 8.7. Arrangement of a reversing plate-fin heat exchanger.

and the environment. The relieved fluid is usually routed in a relief or flare header to a safe location for disposal or combustion. After a process unit has shut down, gases are vented and liquids drained into dump tanks or storage as appropriate. Cryogenic liquids are disposed of in a similar manner. However, to avoid the hazards of cold and brittle fracture of carbon steel, the liquids are evaporated and reheated to at least ambient conditions. The gases are then released into normal carbon-steel headers and flare. A typical cryogenic relief and drains system is shown in Fig. 8.8

In the scheme depicted in Fig. 8.8 we assume that a cryogenic process has both gaseous and liquid reliefs and drains. During normal operating conditions non-cryogenic parts outside the cryogenic unit are protected by large heat sinks (plate-fin heat exchangers), and temperature alarms and trips to prevent cold gases emanating from the cryogenic unit and causing embrittlement problems to carbon-steel equipment. The cryogenic unit itself is usually constructed from aluminium and/or stainless steel to prevent such problems. During abnormal conditions relief valves RV1, 2, and 3 may relieve gas and/or liquid at low temperatures. These are usually catered for by collecting liquids in V1 and heating vapours in E1 using a steam/water bath or ambient air

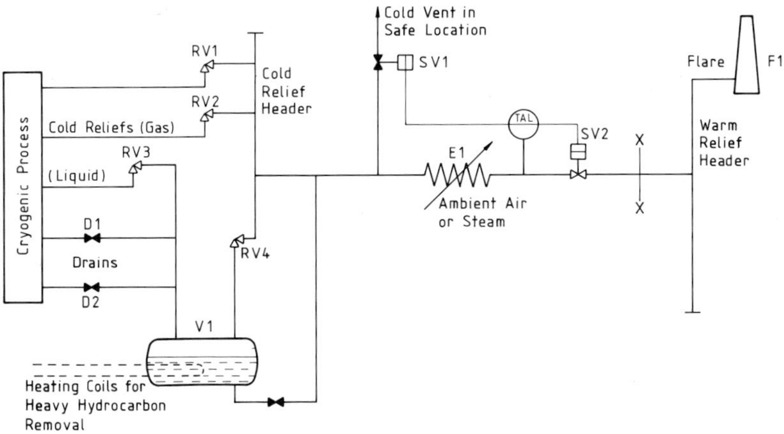

FIG. 8.8. A typical cryogenic fluid drain and relief system.

heater to heat the vapours to such temperature as to render them compatible with carbon steel equipment. E1 is always in the standby mode to ensure that heat can be applied to the maximum possible flow of cryogenic fluid from V1 or the plant. XX shows the point at which there may be a change of material from stainless steel/aluminium to carbon steel. One may consider a cold vent in the event of failure of E1 in which case SV1 and SV2 should be mechanically linked to prevent both valves being closed at once. If cryogenic liquid dump rates are small or more easily quantified then the alarm or switchover system with SV1 and SV2 can be deleted, with perhaps the addition of a second E1 using an alternative heating medium for added security.

Liquid drains are catered for very easily by draining D1 and D2 manually, either directly to a vaporizer E1 or via V1 if part of the process is operating. The exact mode of operation depends on the philosophy developed by individual sites. In all cases cryogenic valves use extended spindles to prevent handles getting cold and guards are fitted on the surfaces of dump equipment to prevent manual contact. Cold 'burns' are just as serious as hot burns and result in skin peeling off in a similar way and damage to subcutaneous tissue.

8.6 Synergies with non-cryogenic processes

8.6.1 *General aspects*

Technologies currently used in conjunction with cryogenics include absorption, adsorption, refrigeration, and chemisorption. Much of this is described in earlier parts of this chapter as pretreatment methods to render the gas free from freezable constituents. Therefore, they can be described as necessary, rather than synergistic, steps. However, to achieve synergy one must produce a better result using two technologies together compared with what could be achieved using each of the technologies separately. It is therefore appropriate to include other technologies in an assessment of synergistic processes. To see how synergies might develop it is best to tabulate the advantages and disadvantages of each route, as shown in Table 8.4. This generally applies to wide boiling mixtures such as hydrogen/helium and hydrocarbons or inerts.

One can conclude from Table 8.4 that cryogenics can be used to achieve most of the gas-separation duties encountered. In air separation, cryogenics is at the forefront because high recoveries *and* high purities are required for most products. Moreover many plants are required to produce liquid products, reinforcing the need for cryogenics. However, even with this typical scenario many other options are evolving. For example, N_2 generation using pressure swing adsorption is quite common in places where the product is required at 99.5 per cent purity as gas in relatively small amounts. Membranes and PSA are being considered as a method of producing enriched air for effluent treatment. Because required N_2, O_2, and Ar purities are likely to be high and they will often be needed in both liquid and gaseous form, cryogenics is likely to remain the preferred processing technology for merchant plants.

However, new uses are rapidly increasing the demand for medium purity products, which will open up a technology niche for permeation and PSA. Synergies between the technologies are likely to be less prevalent in air separation because of the relatively fixed product characteristics.

The picture is far less clear in terms of gas separation in the hydrogen and hydrocarbon fields. There appears to be a myriad of gas mixtures, conditions, and product qualities to consider. There is far greater scope for process synergies in this area in order to improve overall performance or reduce potentially expensive cryogenic routes. The following sections highlight this fact.

8.6.2 *Adsorption and synergies*

Thermal swing adsorption has already been described in section 8.2.2. This is an operation often used to remove final traces of impurities such as water and CO_2 prior to cryogenic processing. Usually, only strongly adsorbable components are removed in this way and these usually require heat to regenerate them. Adsorption is often at ambient temperatures and regeneration is at elevated temperatures.

TABLE 8.4. Comparison of technologies.

Technology	Advantages	Disadvantages
1. Cryogenics (without distillation)	(a) High recovery of products. (b) Moderate purity light product (e.g. H_2 up to 98%). (c) Can operate at high pressures. (d) Low cost. (e) Requires low pressure loss for light product.	Cannot achieve very high purity of light product.
2. Cryogenics (with distillation)	(a) High recovery of products. (b) Moderate purity of light product (e.g. H_2 up to 98%). (c) Can operate at high pressures. (d) Good purity of heavy products. (e) Low pressure loss for light product. (f) High H_2 purity (99.5%+) when using hydrocarbon wash processes.	(a) Cannot achieve very high purity light product. (b) High cost. (c) High energy consumption.
3. Membranes	(a) Good means for bulk separation. (b) Low capital cost. (c) Simple process.	(a) Lower purity or recovery than other routes. (b) High pressure loss for light product (e.g. H_2).
4. Adsorption (PSA)	(a) Very high purity for light product (e.g. 99.99% H_2) (b) Simple process.	(a) Lower recovery. (b) Operates most favourably at lower pressures (2000–3000 kPa).
5. Hydrides	(a) High purity product. (b) Simple process.	(a) Limited commercial scale. (b) Low capacity.
6. Absorption	(a) Simple process. (b) Low pressure drop for light product.	Poor separation characteristics, i.e. low purity light ends or low recovery of heavy ends.
7. Adsorption (thermal swing)	Can remove minor components to virtually zero.	Expensive for bulk removal of impurities.

In some instances adsorption is carried out at subambient temperatures. Examples of this include ammonia adsorption at −70°C and hydrocarbon (acetylene mainly) adsorption at about −170 to 180°C. In the first example, NH_3 purge gas can be treated by chilling to −70°C and the resultant vapour passed through low temperature adsorbers at high pressure (7000–14 000 kPa). Regeneration is carried out at elevated temperatures with a low pressure fuel gas.

The second example is in the treatment of enriched liquid air to remove acetylene as a safety measure. The adsorption takes place at elevated pressures and in the liquid phase. Regeneration is carried out by passing low pressure gas (waste nitrogen) at elevated temperatures to render the bed clean.

In both the above instances the result is efficient and safe plant operation, but they do not represent synergistic operations.

Pressure swing adsorption (PSA) is a means of removing compounds from a light product without applying any heat. This is particularly useful in the case of H_2 or purification where contaminants are the less volatile hydrocarbons or inerts. The simplest form of PSA unit can be a two-bed system where one vessel is on line and the other is being depressurized, purged, and repressurized. The process flow-sheet for this unit looks very similar to that depicted in Fig. 8.3 for a thermally regenerable system. There is one fundamental difference though, in that the regeneration gas is passed through at low pressure to purge the bed V2 without prior heating in E1. The majority of desorption occurs upon depressurization of the spent bed. The purging phase with product gas renders it free of desorbed impurities, and repressurization with product gas prepares the bed for changeover to a fresh on-stream cycle. Because desiccant capacities are relatively low and often impurity levels are high, cycle times are short—typically 4–10 minutes on line for each bed. This means that venting, purging, and pressurizing must be done over a very short time. Moreover, the frequent venting, purging, and pressurizing means significant losses in product gas, thereby limiting PSA unit recovery to 75–85 per cent. This, of course, depends on the impurity levels, but in many cases H_2 recovery via PSA is required from sources with only 75–80 per cent H_2 in the feed with the balance being hydrocarbons.

In such cases it is often economic to use more than two beds for the PSA system. Four or more beds have successfully been used to improve H_2 recovery. The design of such systems would typically allow for the following events to occur simultaneously in each of the four beds:

Bed 1	Adsorption	Bed 3	Purging
Bed 2	Depressurization	Bed 4	Pressurization

Figure 8.9 shows a typical arrangement for a four-bed system.

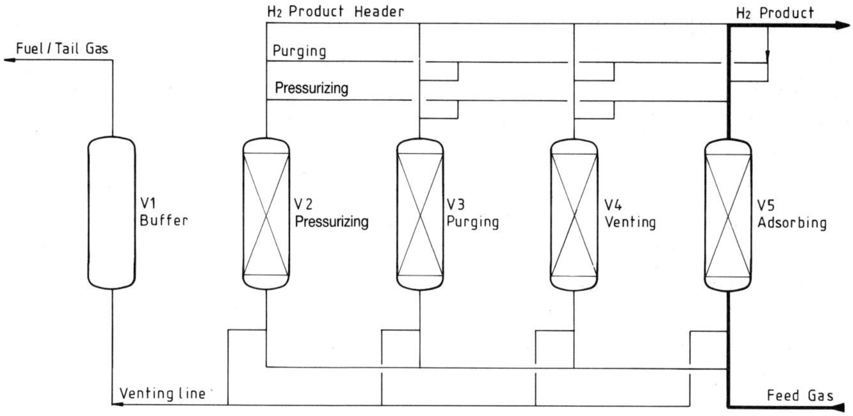

FIG. 8.9. The four-bed PSA system.

In this process V5 is adsorbing hydrocarbons at elevated pressure from a hydrogen-rich gas, and the resultant outlet is a high-purity hydrogen product. V4 is venting down to fuel gas pressure, V3 is being purged with product, and V2 is being pressurized with H_2 in preparation for coming on-line. V1 is used as a buffer vessel to balance out the varying fuel gas compositions as one proceeds through the complete regeneration cycle in V2, 3, and 4. To improve recovery various designers have designed improved systems which partially repressurize with feed gas, and other variants with more vessels. One quickly sees that the valve arrangement to carry out the sequences becomes rather complex. Thus programmable logic controllers (PLCs) are used to ensure the correct sequence of operation of the complex systems. The result is a plant capable to provide moderate recoveries of a light product at very high purity (e.g. H_2 at 99.99 per cent purity from 85 per cent H_2 in hydrocarbons). These processes operate favourably at pressures in the region of 1500–3000 kpa, although there are some which operate as high as 6000 kPa but with reduced recoveries.

PSA technology has widened its application into air separation to produce pure N_2 and enriched O_2 as well as hydrocarbon separation (ethylene from methane). However, its main application is in the field of hydrogen recovery. For air separation, both zeolite-based molecular sieves and activated carbon (as manufactured by Bergbau Forschung of Essen) have been successfully applied.

In cases where there is a high incentive to recover high-purity hydrogen at high levels one can select a synergistic scheme combining cryogenics and PSA. This scheme offers the best of both worlds by producing 99.9 per cent purity hydrogen with recoveries in excess of 99 per cent. This scheme operates very simply by recycling the PSA unit vent gas to a cryogenic unit and allowing the cryogenic unit to recover moderately pure hydrogen for subsequent processing by the PSA unit. Figure 8.10 illustrates this mechanism. Taking the example of a PSA unit giving only 84 per cent hydrogen recovery and the cryogenic unit recovering 95 per cent of the hydrogen present in the PSA unit vent gas, the overall recovery of hydrogen at 99.9 per cent purity is 99.2 per cent.

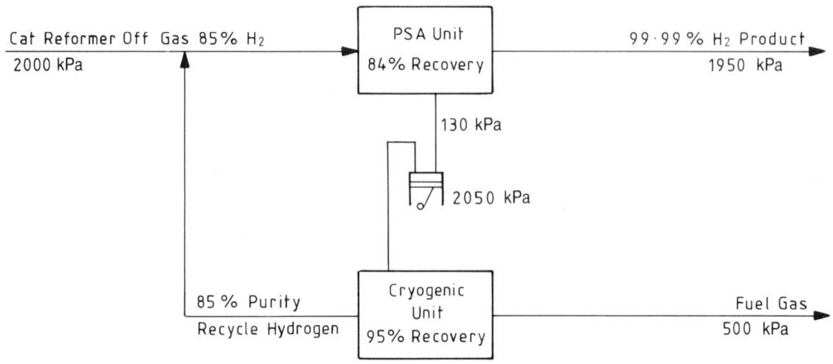

FIG. 8.10. A combination of PSA and cryogenics.

8.6.3 *Permeation and synergies*

8.6.3.1 Principles of permeation

In 1866 Thomas Graham published an article recognizing that gas separation can be carried out by means of different permeation velocities through a membrane. Tentative attempts at commercial plant occurred about 20–25 years ago. The serious advent of permeation plant occurred in about 1980 with the Monsanto hollow fibre membranes for gas separation. The mid 1980s saw the arrival of many players in the field of gas separation using membranes.

The mechanisms of permeation basically fall into two categories, each governed by the type of membrane involved. The first is through non-porous or dense-film membranes. At temperatures exceeding the glass transition, segments of amorphous polymer chains in the membrane undergo motion similar to that observed in liquids. This transport of gas through these membranes is by a solution diffusion mechanism. Temperature has a significant effect on molecular movement in the membranes and hence on membrane performance, with transport rates often following the Arrhenius relationship. The membranes are highly selective but have low capacity, and their effectiveness depends on manufacture of ultra-thin films without sacrificing selectivity.

The second mechanism is through porous membranes, which generally contain larger voids than dense-film membranes. These have inter-connected pores which are significantly larger than the molecular diameters of gases passing through them. Transfer of gases occurs through these pores at varying rates, depending on the pore structure and size distribution. Selectivity is determined primarily by the relative molecular sizes of the gases being separated. Unlike the dense-film membranes, porous membranes exhibit no molecular interaction between the membrane polymer and the diffusing species.

Although dense-phase membranes have the ability to be highly selective, they need to be very thin to achieve reasonable capacities per unit area. This limits their mechanical strength. Porous membranes can be made much thicker and retain high capacities with good mechanical strength. However, as they rely on diffusion of small molecules through larger pores, their selectivity is much lower. Therefore permeation through porous membranes has become an effective means of bulk separation because of their high capacity and good mechanical strength, at the expense of selectivity.

Synthetic membranes are made from a variety of polymers including polyethylene, polyamides, polyimides, cellulose acetate, polysulphone, and polydimethylsiloxan. Generally the membrane is manufactured as either a flat film or a hollow fibre. There are three basic types of modules in which membranes are contained:

(1) hollow fibre bundles in a metal module;
(2) spiral wound bundles in a metal module;
(3) plate and frame module constructed like a plate and frame heat exchanger.

The first two are of interest in gas separation, with the majority being made using hollow fibre but with increasing use of spiral wound-types. Hollow-fibre membranes, though less robust, have the largest area/volume ratio. Typically, spiral-wound modules have a packing density of 1500 $m^2\ m^{-3}$ surface, compared with about 9000 $m^2\ m^{-3}$ for hollow fibre membranes. Figure 8.11 shows the arrangement of spiral wound and hollow fibre membranes. Usually, membranes are made up of composite materials to give them favourable mechanical and selectivity properties. In such cases the mechanism of permeation becomes more complex, relying on solution diffusion and diffusion through pores due to a pressure gradient. To calculate the performance of the membrane an equation can be drawn up assuming Fickian diffusion of a dissolved gas in a dense membrane. For each component:

$$Q = \frac{PA(P_f - P_p)}{L}$$

where Q = flux rate
P = permeability coefficient
A = membrane area

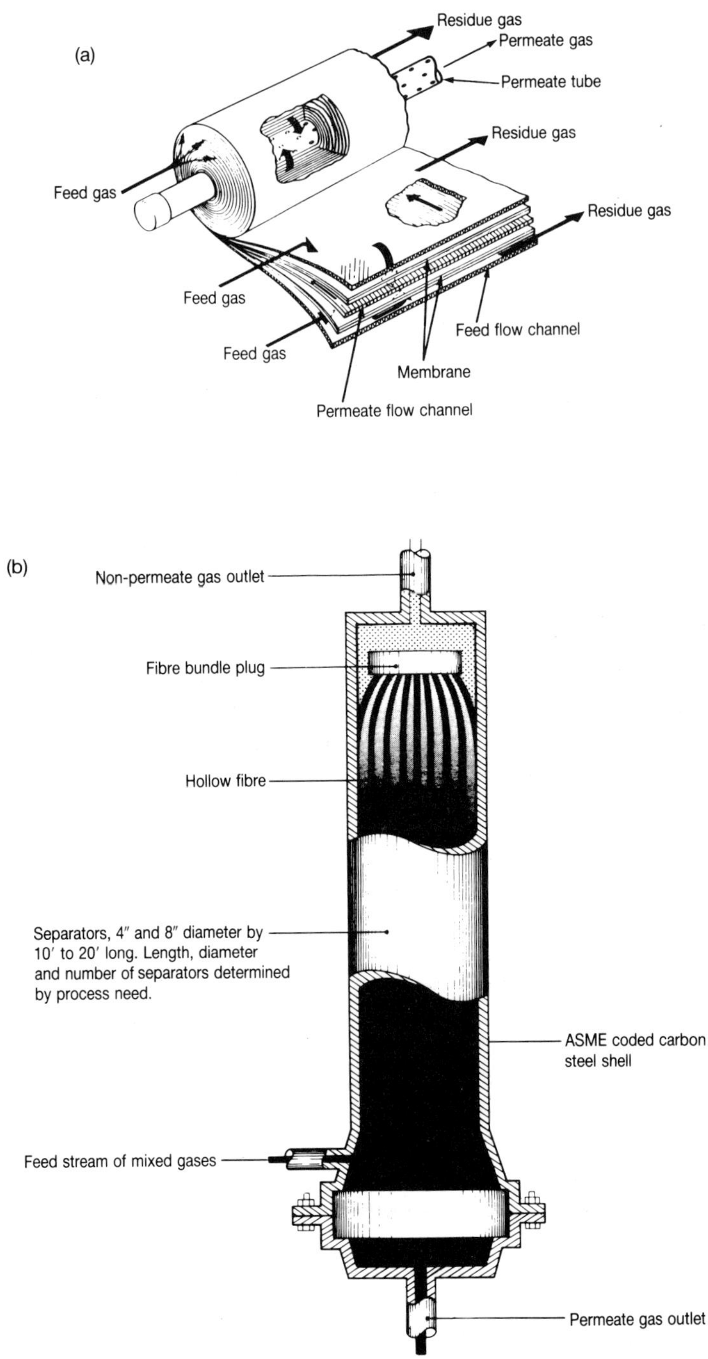

FIG. 8.11. (a) Spiral-wound and (b) hollow-fibre membranes.

P_f = feed partial pressure
P_p = permeate partial pressure
L = thickness of membrane

Transport across the membrane can be assumed to consist of:

(1) solution of molecule at the high partial pressure surface;
(2) diffusion across the membrane due to pressure gradient;
(3) desorption at the low pressure surface.

Thus the permeability for each component is the product of solubility and diffusion coefficients. It is important to pack as much area as possible into each bundle of hollow fibres. The fibres themselves are, therefore, very narrow in bore, typically about 0.2–0.4 mm diameter. The thinner the membrane the higher the flux of gases, but the weaker the membrane itself. Therefore designers of such membranes seek a happy medium for a given application.

Different gases have widely differing permeabilities, and different membrane construction/formulation can yield very different flux rates. Table 8.5 shows the relative permeabilities of various gases through different types of membranes. The relative basis in this table is unity for O_2 on a typical polysulphone membrane. One can see that often low molecular weight gases and strongly polar gases have high relative permeabilities. These are often called 'fast gases'. In contrast, slow gases are of larger molecular weight and are non-polar.

TABLE 8.5. The relative permeabilities of different components through different membranes.

	Component permeability					
Typical membrane	H_2	N_2	O_2	CH_4	CO_2	H_2O
Polysulphone	13	0.2	1	0.22	6	—
Cellulose acetate	12	0.17	1	0.20	6	120
Polyimide	9	0.05	0.5	0.05	—	—
'Dow' product	136	8	32	—	93	—
'Permea' product	22	0.4	2.3	0.4	9	—
'PDMS'	649	281	604	3230	1070	—

8.6.3.2 Use of membranes with other technologies

Membrane technology competes successfully in its own right in a number of areas, covering:

- hydrogen recovery from ammonia purge gas;
- hydrogen recovery from refinery and petrochemical plant off-gases;
- enhanced oil recovery, bio-gas, and other CO_2/C_nH_m separations;

- helium recovery from diving gas or balloon gas;
- inert gas generation from N_2/O_2.

A total installation list approaching 200 plants has been secured by a number of competing companies over a period from 1979 to 1987. With more efficient 'second' and 'third generation' membranes, this technology will undoubtedly become very widespread. Whilst new generations of improved performance membranes are sought and tested, there is considerable interest in synergies with other technologies. Since membrane separation is essentially a bulk separation process yielding moderate recovery it may be sensible to link it with PSA or cryogenics.

This gives the following beneficial synergies for the recovery.

1. *PSA + membranes.* A high-purity product with moderate recoveries. By recycling waste gases from PSA unit, recoveries can be increased.
2. *Cryogenics + membranes.* A moderately high-purity product with a high recovery. Judicious use of membrane units can improve the performance of a cryogenic unit.

Because of the resultant poor efficiency there are a few PSA+ membrane applications that show any merit. Cryogenic + membrane combinations are more attractive and can be used in many synergistic instances. A typical example might be a hydrogen-rich gas at 2000 kPa which contains 80–85 per cent H_2 with hydrocarbons as the balance in the feed, and the product required is of 96 per cent purity with a 99 per cent recovery. Cryogenics alone can carry out this duty, but with a large refrigeration cycle to achieve the performance. A typical process for this is described in Chapter 5, Fig. 5.4. However, the inclusion of a membrane step as shown in Fig. 8.12 can allow the use of a simple cryogenic unit

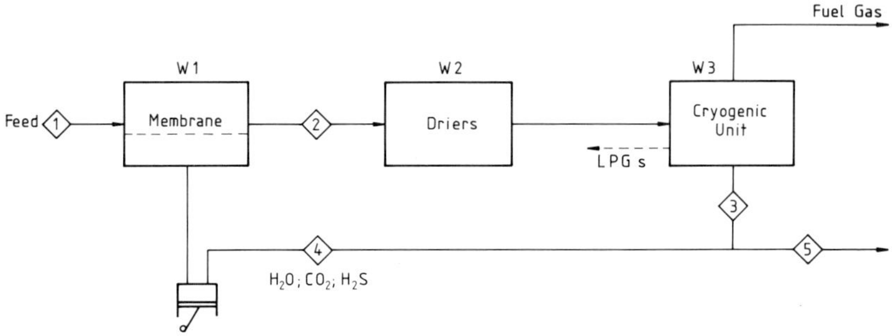

Purities of H_2 at various points:

Point No.	1	2	3	4	5
H_2 Content mol %	85	60–70	90–94	99	96

FIG. 8.12. A combination of membrane and cryogenic units.

using Joule–Thompson refrigeration and no power input. This process reduces the hydrogen content of the non-permeate to a level that will allow cryogenic processing with a simple, low-cost, no-power unit. Because H_2O and acid gases readily permeate through the membrane, the driers W2 are smaller. The permeate can often be of relatively high purity since it is only a small portion of the hydrogen in the feed (10–30 per cent). This means that compression using C1 is generally less than the power usage on a cryogenic-only process. However, higher recoveries can be achieved than with any of the routes operating separately.

Membrane technology can be applied in a variety of such applications where improved recovery or purity of light ends or, alternatively, improved recovery of heavier ends is required, e.g. LPG recovery from refinery off-gases. One can see that by reducing H_2 content in the feed to driers (2), recovery of LPGs by cryogenic means becomes easier. A similar philosophy can be applied to ammonia purge gases for argon recovery.

Air separation using membranes alone has been successfully used in N_2 generation. Combination with cryogenics may be advantageous when one sees that O_2 permeates faster than N_2. The non-permeate constitutes a better feed for a cryogenic unit recovering nitrogen. Therefore, one can envisage a situation where air is compressed into a process similar to that described in Fig. 8.12: enriched oxygen is recovered as permeate and used in sewage treatment to satisfy BOD (Biological Oxygen Demand) and the N_2-enriched non-permeate is passed to driers and a cryogenic unit to recover nitrogen in a single-column process, as described in Chapter 3. The cryogenic process would have to be somewhat different with a larger expansion turbine to satisfy the overall cold balance and reflux rates to the distillation column. Thus one could achieve a higher N_2 recovery than hitherto possible using cryogenics alone. Membranes alone would not produce adequate purities of O_2 or N_2 economically, even with several stages.

Although membrane technology has currently only a small foothold in the gas separation business, this is likely to change over the coming years. New generations of membranes with improved permeabilities will greatly increase applications for cryogenics.

In broad terms cryogenics will be more limited to pure products and membrane techology will be increasingly used for bulk separation.

8.6.4 *Absorption, distillation, and synergies*

Refinery and natural gases can be treated by washing with a heavy hydrocarbon liquid (e.g. naphtha, kerosene, toluene) to recover LPGs. This has often been applied to refinery off-gases yielding moderate to low recoveries of LPGs. Cryogenic processes compete favourably in this area because they potentially provide much higher recoveries of LPGs. Depending on circumstances, there are opportunities for synergies between absorption and cryogenics. We can take as an example a situation where a refinery off-gas is relatively rich in LPGs and a naphtha wash process is adopted. Often the naphtha is regenerated in existing regeneration columns which may be limited in capacity. Moreover, this process does not recover high proportions of LPGs. It would seem worthwhile to combine this route with a cryogenic route, as shown in Fig. 8.13, to increase significantly the overall LPG recovery without overloading existing equipment.

This process uses naphtha in K1 to absorb some 40–60 per cent of LPGs present in the off-gases at ambient temperatures. Chilling the column to about 5–10°C raises LPG recovery, but only to 70–80 per cent at best. By using a turbo-expander process (described in Chapters 5 and 7) one can raise C_{3+} recovery to over 95 per cent. This often represents an attractive solution. A stabilized C_{3+} product can be provided via P1 and K2 without contamination with naphtha.

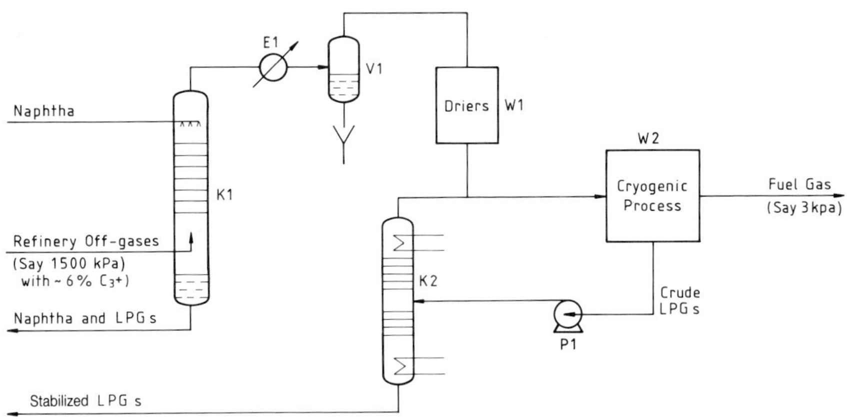

FIG. 8.13. A combination of absorption and cryogenics for recovering LPG from refinery off-gas.

Another area where membranes and distillation are used in synergy is the area of natural gas treatment. It can be advantageous to remove bulk quantities of CO_2 using membranes. An example of this is a process developed by Fluor for bulk removal of CO_2 from natural gas and subsequent C_2 and C_{3+} recovery. Figure 8.14 shows this process where bulk CO_2 removal is achieved in M1 and sales gas is produced in column K1. Subsequent distillation in K2 yields an azeotrope of CO_2/C_2H_4 in the overheads which if left as waste gas reduces hydrocarbon yields. By

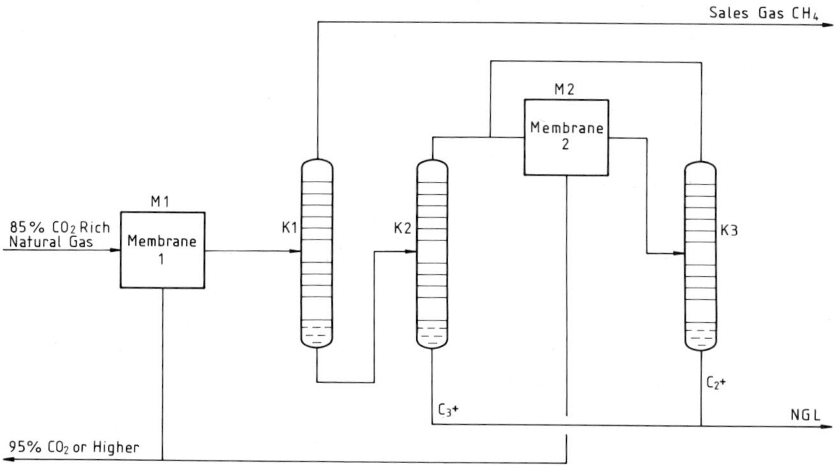

FIG. 8.14. The removal of CO_2 from natural gas using membranes.

using a membrane unit M2 the azeotrope can be separated, and, by subsequent distillation in K3, more C_{2+} can be recovered. This process is particularly useful for gases containing high levels of CO_2 where conventional distillation and absorption techniques are not economical.

8.6.5 *Hydrides*

In the late 1970s and early 1980s developments in hydride technology have encouraged a number of applications in industry. Among these are:

- non-evaporable getters for vacuum maintenance
- hydride-powered fuel cells;
- thermal hydride/hydrogen compressors;
- hydrogen storage;
- gas separation to recover hydrogen.

Hydrogen storage is described in more detail in Chapter 9. The chemisorption of H_2 onto metals has also found a use in all of the above, and gas separation is of particular interest here.

Most elemental metals will form hydrides of one form or another given appropriate conditions of pressure, temperature, and large metal surface area. The reaction is reversible and can be represented thus:

$$M + H_2 \rightleftharpoons MH_2$$

where M is a metal.

Alloys can be compounded and pulverized to produce a large surface area and used as chemisorptive media to 'store' H_2 (see Fig. 8.15). A large number of metals and alloys are known to form hydride complexes,

FIG. 8.15. Metal alloy pellets for hydride applications (courtesy of SAES Getters (SpA).

each of which has a different 'storage' capacity and requires different conditions to form the hydride. Table 8.6 shows a typical list of hydride complexes with their potential storage capacity on a weight-for-weight basis. Each requires heating to a different temperature to ensure desorption/decomposition. By appropriate selection of hydrides for a given duty a system similar to the molecular sieve adsorbers described in Fig. 8.3 can be designed, but with hydrides in the vessels. Other design considerations must be taken into account, including:

- the metals/alloys 'swell' by up to 25 per cent when forming hydrides;
- the weight of the vessel contents is very high;
- the decomposition (desorption) temperature may be higher than for molecular sieve regeneration;
- high temperature rises are encountered when reaction occurs.

Taking all of these obstacles into account, commercial-scale process plant applications have been limited in numbers to date. Ammonia purge gas has always been a favoured gas for trying out new processes, and hydrides are no exception. Air Products has installed a pilot plant for this duty with reasonable success, but has as yet only one unit which has been operating since 1983.

It is likely that ammonia plant economics will dictate that cryogenic processes will be more attractive in the forseeable future if both H_2 and

TABLE 8.6. Hydride complexes and their capacity for H_2.

Material	Capacity (wt.% hydrogen)
Liquid H_2	100
$FeTiH_{1.95}$	1.75
$LaNi_{4.7}Al_{0.3}$	1.4
MgH_2	7
UH_3	1.3
TiH_2	4.0
VH_2	2.1
$Mg_2NiH_{4.2}$	3.2
$LaNi_5H_{6.7}$	1.4

N_2 are to be recovered at substantial pressures. Hydrides will thus remain an option for lower capacity, higher H_2 purity units for which there is little market at the time of writing.

Bibliography

ARNOLD, M. (1987). 'Handling hydrogen'. *Process Engineering,* (Aug.) 24.

BEAVER, E. R., *et al.* (1987). 'Integration of membranes with other air separation technologies'. *A.I.Ch.E. Summer National Meeting,* Minneapolis, USA. paper no. 22c. AIChE, New York.

BODLE, W. W., *et al.* (1980). 'Considerations for Mercury in LNG operations'. *LNG 6 Conference,* Session II, paper 1.

BRITISH CRYOGENICS COUNCIL (1970). *Cryogenics safety manual.* 2nd Ed. Mechanical Engineering Publications Ltd., London.

GAS PROCESSORS SUPPLIERS ASSOCIATION (1987). *Engineering Data Book* (10th edn), Vol, II. GPSA, Tulsa, Oklahoma.

GOLLAN, A. and KLEPER, M. H. (1986). 'Membrane based air separation'. *A.I.Ch.E. Symposium Series,* **82** (250), 35.

GREGORY, E. J. (1987). 'Heat Exchangers'. In *Cryogenic Engineering,* (ed. B. A. Hands), Chap. 8, p. 193. Academic Press, London.

HARDISON, L. C. (1984). 'Catalytic gas-sweetening process selectively converts H_2S to sulphur, treats acid gas'. *Oil and Gas Journal,* (June 4), 60.

HUSTON, E. L. and SANDROCK, G. D. (1980). 'Engineering properties of metal hydrides'. *Intl. Sym. Prop. and Appl. of Metal Hydrides,* Colorado, Elsevier Sequoia, The Netherlands.

JOHNSTON, I. W. and KING, M. (1987). 'An introduction to membrane technology and the role of membranes in gas processing'. *I. Chem. E. (SONG) Symposium.* I.Chem.E., Rugby, England.

KIRKBY, N. F. (1987). 'Pressure swing adsorption methods'. *4th BOC Priestley Conference,* p. 218. BOC, England.

KNOTT, T. (1987). 'Gas separation: a question of selectivity'. *Offshore Engineer,* (July), 44.

LAVERTY, B. W. and O'HAIR, J. G. (1987). 'Application of membrane technology in the gas industry'. *4th BOC Priestly Conference,* p. 291.

PETTY, L. and NAEGER, R. J. (1987). 'NGL treating: a review of today's practices'. *Proc. 66th GPA Annual Convention,* p. 29.

REILLY, J. J. and SANDROCK, G. D. (1980). 'Hydrogen storage in metal hydrides'. *Scientific American,* **242** (2), 118–29.

RIQUARTS, H. P. and LEITGEB, P. (1985). 'Gas separation using pressure swing adsorption plants'. *Linde Reports Sci. Tech.* **40**, 26.

ROHDE, W. (1985). '20 years of molecular sieves in air separation'. *Linde Reports Sci. Tech.* **40**, 48.

SANDROCK, G. D. and HUSTON, E. L. (1981). 'How metals store hydrogen'. *Chemtech,* **11**, 754–62.

SCHENDEL, R. L. (1987). 'Separation of acid gases and hydrocarbons'. *4th BOC Priestly Conference,* p. 311, BOC, England.

SMITH, R. S. and TSAI, M. J. (1987). 'Application of DRIZO dehydration in North Sea gas processing'. *I. Chem E. Unit Ops in Offshore and Onshore Prodn. Conference,* I.Chem.E., Rugby, England.

THOROGOOD, R. M. (1987). 'Gas separation and liquefaction plants'. In *Cryogenic Engineering,* (ed. B. A. Hands), Chap. 16, p. 391. Academic Press, London.

9

PLANT HARDWARE CONSIDERATIONS

9.1 Introduction

The preceding chapters cover the various process routes involved in recovery, purification, and liquefaction of some 16 industrial gases. In all of the process schemes depicted in earlier pages, one can identify many of the normal unit operations encountered by a chemical engineer. One could therefore shorten this chapter greatly by referring to the myriad of textbooks available which describe these in detail. However, low-temperature gas-separation technology has a number of attributes which warrant highlighting. Many of these are closely linked with the high thermodynamic efficiencies attained by cryogenic processes. These in turn are linked to the equipment used.

Throughout the history of cryogenics there has always been a relatively high emphasis on process efficiency compared with traditional 'hot' processes in refineries and petrochemical complexes, largely because it is so difficult to attain low temperatures. Moreover, once the valuable component has been removed from a cooled gas mixture, one is left with a waste gas at low temperature which, if discarded cold, wastes energy and can cause metallurgical as well as environmental problems. It is therefore highly desirable that it is reheated to its original condition. Hence, effective heat exchange plays a large part in any cryogenic process. This is achieved in the main by plate-fin heat exchangers, but wound-coil and shell-and-tube exchangers also play a part in some areas.

Compression equipment also warrants a mention here because of its importance in air separation and in many gas processing areas. The major difference between compressor equipment for cryogenic plant compared with non-cryogenic plant is in the lubrication system. Many gas compressors preceding cryogenic processes are oil-free on the process side. If oil-lubricated compressors are used then sophisticated and expensive oil-removal systems are included after the machines. For these types of reciprocating compressors, piston rings are made of composite materials which include graphite, thereby promoting lubrication. Packing glands are purged with dry gases to form a seal between the oil-lubricated crankcase side and the process end. In centrifugal machinery a dry-seal system is provided to prevent process gas being contaminated with oil. Therefore the metal bearings are allowed thorough lubrication while the process gas is kept dry.

Expansion turbines are used extensively to provide refrigeration to a process in an efficient manner. These require dry operation for the process gas in a similar manner to 'dry' centrifugal compressors, and therefore employ similar lubrication and seal systems. Reciprocating expanders were the forerunners of the expansion turbine and have now largely been made obsolete. There are however a few exceptions where reciprocating expanders are used today, including many East European air-separation and gas-separation processes. Another case where reciprocating expanders are successfully used today is in helium liquefiers where volumes are relatively small, because design of rotary expanders for low molecular weight gases to a high efficiency is difficult without using very high speeds.

Pumps and distillation equipment are fundamental parts of many cryogenic processes, but will only be covered briefly here because of their similarity to conventional process unit operations. Storage and send-out warrants some discussion because of the high degree of insulation applied to cryogenic storage and the complexity this introduces.

In considering the above hardware one soon finds that the term 'mature technology' should not really apply to cryogenics. Many developments have continued in all of these areas allowing greater efficiencies and reliabilities, and making cryogenics a suitable learning forum for designers of 'hot' processes.

9.2 Compressors

Air and gas separation relies on compression of the feed gas to an acceptable pressure. Because of the importance of process efficiency, the designer is looking for the best efficiency for the given duty. This may present him with a problem of choice of type of compressor. Over the years, the machinery industry has developed a range of different types of compressors, i.e. reciprocating, centrifugal, and screw. Different duties warrant different approaches. Figure 9.1 shows a typical 'map' of compressor applications which the designer can use to select types of machine for his duty. This will generally give him a starting point to continue the search for the most efficient and reliable machine.

The power (Fig. 9.2) for a given duty can be calculated from the classical equation given below. The theoretical adiabatic power (assuming 100 per cent energy conservation) to compress a volume of ideal gas is given by:

$$\text{Power} = KP_1V_1\left\{\left(\frac{P_2}{P_1}\right)^{(k-1)/k} - 1\right\}\frac{k}{k-1} \tag{1}$$

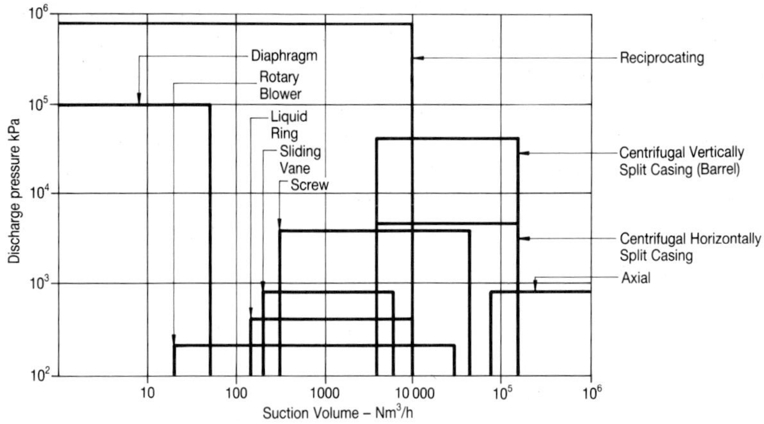

FIG. 9.1. A selection 'map' for compressors.

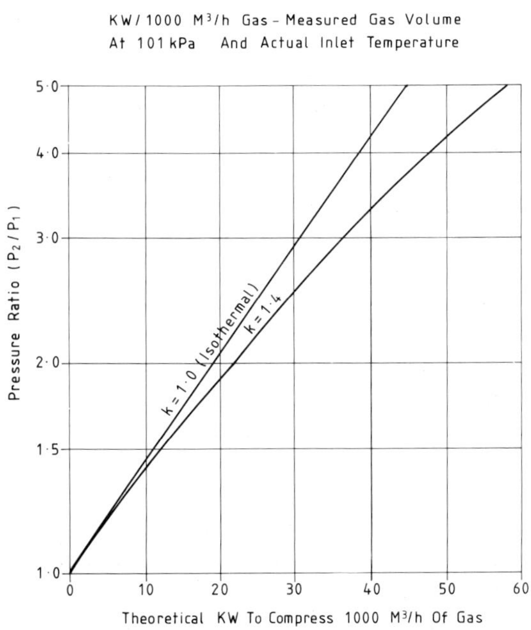

FIG. 9.2. The theoretical power to compress a gas.

For non-ideal gases, this power should be multiplied by the average compressibility, Z. The discharge temperature is calculated from:

$$T_2 = T_1 \left(\frac{P_2}{P_1}\right)^{(k-1)/k} \tag{2}$$

The usual situation is that compression is not fully adiabatic and extra energy is needed to overcome losses in the machine which result in a temperature higher than T_2.

If compression is isothermal then the theoretical compression power is governed by:

$$\text{Power} = K_2 P_1 V_1 \ln\left(\frac{P_2}{P_1}\right) \tag{3}$$

Here P_1 = inlet pressure
P_2 = discharge pressure
V_1 = inlet volume
k = ratio of specific heats
Z = compressibility
T_1 = inlet temperature
T_2 = discharge temperature
K, K_2 = constants

The major cause of inefficiency in compression by reciprocating compressors is valve losses, which cover friction, inertia effects, and damping losses. The first is due to the throttling of the gas as it passes through the valve, the second due to the delay in the valves' plates lifting because of their mass, and the third due to the imperfect damping characteristic of the valve plates. The result of these three effects is to enlarge the area of the pressure–volume (PV) diagram.

The losses occur at the inlet and discharge of the gas into the cylinders and are constant for any particular gas valve configuration, regardless of the pressure ratio. If the pressure ratio is small then these effects are a large percentage of the total work done, i.e. the area of the PV diagram. As the pressure ratio rises the valve losses become a smaller percentage of the total work input, and hence the compressor efficiency increases.

Figure 9.3 shows this effect clearly and is based on practical experience of compressors. The valve effect is shown clearly up to a pressure ratio of about 4, where secondary effects due to the high compression ratios produce a fall-off in efficiency. For high compression ratios (greater than 4) it might be necessary to use a multistage compressor, depending on the gas handled, cylinder lubrication and piston sizes, and cooling. For multistage compressors the power can be calculated by taking the nth root of the pressure ratio (assuming n stages) and calculating the power using this new stage pressure ratio. The power of the compressor is then given by the sum of the stage powers.

The efficiency of centrifugal compressors, like other head-developing thermodynamic machines, depends principally on throughput. Just as the efficiency of centrifugal pumps increases with capacity, the efficiency of centrifugal compressors increases with the actual volumes of gas within

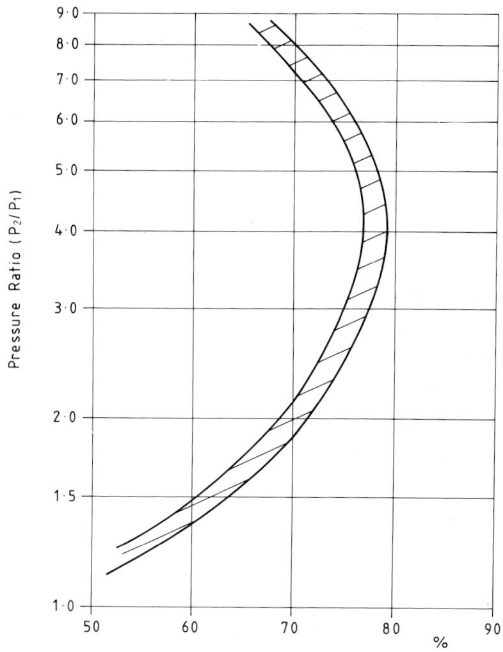

FIG. 9.3. The stage efficiency of reciprocating compressors.

the compressor impellers. For high actual throughput, large wide impellers give a good streamlined flow with little losses, and polytropic indices near to the adiabatic ones. For small capacities small narrow impellers result with high friction losses, high recycle losses through interstage labyrinths, and low polytropic efficiencies.

With centrifugal compressors there are many other restraints which prevent efficiency optimization, such as the number of barrels, number of coolers, and avoidance of gearboxes, both interbarrel and between compressor and driver. These problems generally become more acute with reduction in the actual volume throughput of the compressor. Figure 9.4 shows the efficiency variation with capacity for centrifugal compressors.

The criterion used, i.e. the flow/discharge pressure ratio, approximates to the actual volume in the last stage of the compressor, the last stage being the most critical one for efficiency considerations.

Other compressors are generally of the rotary positive-displacement type, i.e. Roots blowers, screw compressors, liquid ring compressors, sliding-vane compressors, etc. They generally have a high constant efficiency of about 75 per cent, variations between different manufacturers being greater than between different duties.

Figure 9.1 shows the limits of reciprocating compressors with

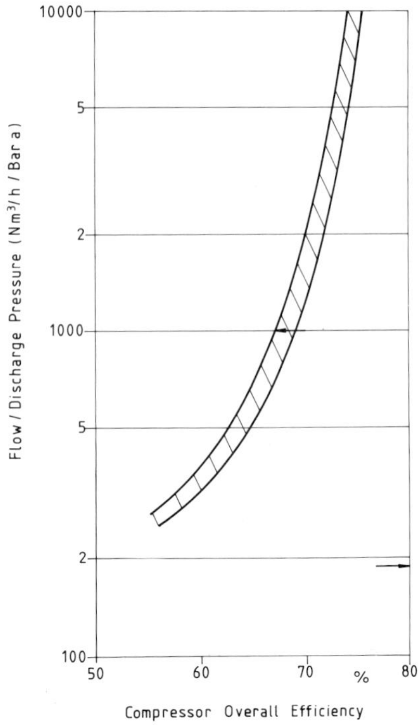

FIG. 9.4. The efficiency of centrifugal compressors.

centrifugal machines generally being used for larger capacities. Some of the largest machines proposed in the 1980s were in the air separation industry with air compressors handling 160–250 m^{-3} s^{-1} of actual inlet air and compressing to ~600–700 kPa. This kind of machine would require drivers of 40–60 MW size. The compressors themselves reach relatively high efficiencies of ~75 per cent isothermal, and a number of manufacturers are capable of making them. Beyond this size transport and manufacturing limitations begin to play a significant role. Moreover, the demand for single sources of such large oxygen producers is limited even though the economics of scale favour large plants. The other aspect of such large compressors is the driver end. Whilst smaller machines are usually electric or gas-engine driven, the larger ones are usually steam- or gas-turbine driven. Happily, because the demand for large drivers is significant in the power industry, there is adequate experience in drivers to meet the large compression duties, even above 75 MW driver power.

The above paragraphs cover large volume flows and modest discharge pressures. At the other end of the pressure scale there is a reasonable demand for 'hypercompressors' in gas compression of ethylene for polyethylene manufacture. These machines are generally reciprocating with discharge pressures up to 300 000 kPa, and generally low-speed and electrically driven. Nevertheless manufacturers feel capable of carrying out such duties in centrifugal hypercompressors and have proposed machines to compress 125 tonnes per hour of ethylene from 24 500 kPa to about 245 000 kPa. To achieve reasonable efficiencies relatively high speeds are proposed for these duties of ~ 30 000 r.p.m. requiring step-up gearing from steam turbines which tend to run at about half this speed.

9.3 Expanders

Providing refrigeration to a cryogenic process is fundamentally different to the provision of heat to a hot process. Heating is achieved by combustion either directly or indirectly connected with the process and at some point or other involves chemical reaction. The provision of cooling is generally via a physical change. Some examples of refrigeration cycles have been described in Chapters 3 and 7. A very effective way of providing refrigeration is to expand a gas from high pressure to low pressure by means of an expansion turbine, as already explained in Chapter 3. The importance of efficient refrigeration is also illustrated in Chapter 3.

For the refrigeration process the theoretical minimum reversible work, W_R, is:

$$W_R = \Delta H - T_0 \Delta S$$

where ΔH = the enthalpy difference between the final and initial states
T_0 = reference temperature of the 'sink' into which the heat is pumped
ΔS = entropy difference between final and initial states (negative)

Of course no process is ideal, and an analysis of process losses or irreversibilities shows them to occur in three main areas:

(1) the result of finite temperature differences in heat exchange;
(2) heat inleak from surroundings;
(3) compression and expansion losses.

Compression and expansion equipment losses are usually the largest and thus require a great deal of attention. To make a process work the 'irreversible element' of the work, W_{IRR}, needs to be put in:

$$W_{IRR} = T_0 \sum \Delta S_{IRR}$$

where $\sum \Delta S_{IRR}$ is the sum of the irreversible entropy increases in the process. It is this which must be minimized.

The actual work, W, for a process is therefore the sum of W_R and W_{IRR}.

One can arrive at 'process efficiency' by the ratio $W_R/(W_R + W_{IRR})$ and use this as a yardstick for process optimization.

The use of efficient expansion methods can greatly reduce W_{IRR} in a liquefaction process. Expansion can be carried out using valves or reciprocating or centrifugal expanders. Valves are not efficient because they are isenthalpic. Reciprocating expansion machines have been used since 1902 when G. Glaude constructed an air-separation plant which included one. This expander operated at an efficiency of around 60–70 per cent. Over the decades improvements in design by Heylandt in 1912 and Collins in 1947 led to subsequent efficiencies of around 80 per cent for relatively small machines. Reciprocating expanders are still used, though in relatively limited applications in helium and hydrogen liquefiers. The hardware itself looks very like a reciprocating compressor but of course works in reverse. The gas at high pressure is admitted to the cylinder via an inlet valve where it pushes the piston giving circular motion to a crankshaft via a connecting rod. The crankshaft has a pulley which may be connected to an electricity generator (the reverse of a motor) which absorbs the energy from the process gas. Because the gas converts its energy into motion the process is not isenthalpic and so gives rise to a large temperature drop, thus providing refrigeration.

Expansion turbines have a far wider application and range of capacities than reciprocating expanders. They have a potentially higher efficiency for many applications are more reliable and simpler in construction. The first commercial expansion turbines were probably constructed in the mid 1930s by Linde in Germany. These achieved relatively low efficiencies of only about 53 per cent. Improvements in turbine design were made by Kapitza in 1939 approaching 80 per cent efficiency on a turbine rotating at 42 000 r.p.m.

In 1942 the Elliott and Sharples Corporation designed and built a turbine running at 22 000 r.p.m. with an efficiency of 82.5 per cent. During the 1950s a gas bearing turbine manufactured by Lucas for Petrocarbon Developments Ltd became very popular on a range of nitrogen-generation plants. This machine runs at between 25 000 and 90 000 r.p.m. using N_2 gas as a bearing and seal gas, and achieving efficiencies of between 70 and 80 per cent depending on application. In the 1960s Jekat of the Worthington Corporation produced a novel design which with subsequent modification achieved 92 per cent efficiency. This unit was not commercialized in any significant manner and had a relatively narrow range of application. However, it did prove that efficiencies of over 90 per cent were achievable in practice.

As applications varied, designs for low molecular weight gas expansion led to greater and greater speeds. Very small diameter rotor (8 mm) helium expanders were constructed and operated at 500 000 r.p.m. and about 50 per cent efficiency in the 1960s. Speeds of 900 000 r.p.m. are known in helium units designed in the 1980s. Happily, most applications operate a long way from those incredible speeds. Typical applications for natural gas service range from 15 000 to 40 000 r.p.m. with efficiencies in the region of 80–86 per cent. Some manufacturers claim 90 per cent or higher efficiencies on moderately sized commercial units. Typical ranges of power developed in such expansion turbines are from 50 to 3000 kW, although units of only a few kW to above 10 MW have been constructed for cryogenic service. Turbines of 70 MW are available for geothermal power applications, so if the demand was evident for such machines in cryogenics, the technology is available. Figure 9.5 shows a typical expander–compressor used for natural gas service.

FIG. 9.5. An expansion turbine for natural gas service (courtesy of Rotoflow Inc.).

Often the expander releases enough energy to justify its recovery and conversion to electricity or compression of another gas. These aspects are shown in Fig. 9.6 (a) and (b) respectively. In the first case (a) the turbo-expander is connected to the process taking the appropriate high-pressure gas radially and expanding it to produce refrigeration in part A. The gas evolves axially and is routed to the process to provide refrigeration. The energy released (converted into rotating motion) is used to drive a generator, C, via a rigid shaft, through gearing B. Given an expander efficiency of 85 per cent, shaft and gear losses of 10 per cent, and a motor efficiency of 94 per cent the overall conversion of pressure

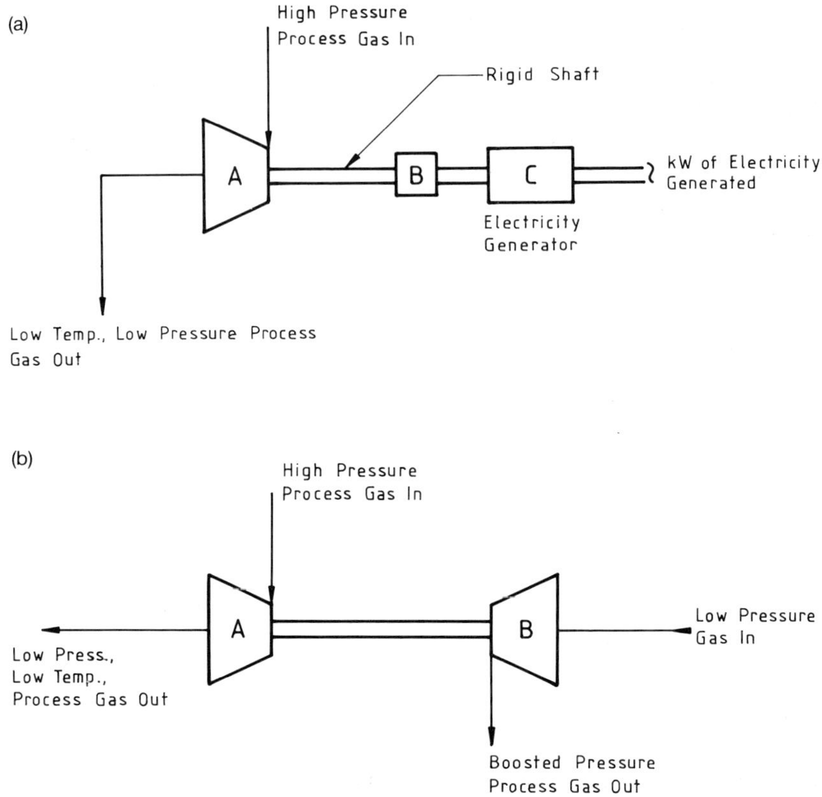

FIG. 9.6. The utilization of expansion energy: (a) conversion to electricity; (b) conversion to compression power.

energy to electricity is around 70 per cent. Thus in addition to providing refrigeration, the machine generates electricity.

The second case (b) shows the expansion turbine A coupled to a compressor, B, via a rigid shaft. This is analogous to a steam or gas turbine driving a compressor. However, it is usual to have the compressor end running at the same speed, i.e. without gearing. This limits the efficiency of the 'brake' or compressor, B, because it cannot be operated at the optimum design speed for the compressed gas conditions. It is therefore usual for the brake to have an efficiency some 8–15 per cent less than the expander. If it is important to increase the efficiency of compression gearing can be used, but the losses must be taken into account when calculating the merits of such a scheme.

For a given set of design conditions, an expansion turbine can be designed for relatively high efficiency. However, it is most important to know what turndown characteristics are required by the process, as this can make a marked difference to the performance of the expansion turbine treating the process. Figure 9.7 shows the variation of efficiency with capacity for a single-stage turbine. To maintain high efficiency at turndown several units in parallel are often constructed. This of course leads to higher capital cost which must be justified by the improved process efficiency. As with most things economics tends to dictate the nature of equipment used in process plant, and hence the extent to which new technology is developed.

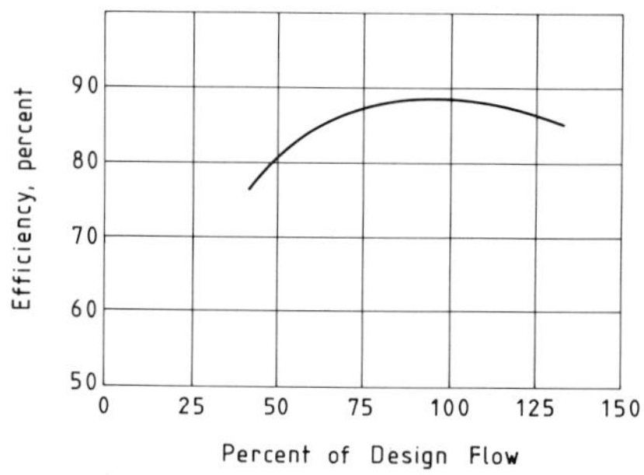

FIG. 9.7. Turbo-expander efficiency versus flow.

9.4 Pumps

Once liquid has been produced in a cryogenic process it is transferred into storage tanks for holding until it needs to be transported to the consumer. Where processes operate at relatively high pressure, liquids can be conveyed into storage by virtue of the pressure generated in the process. This is done wherever possible to avoid any machinery, i.e. pumps, thus making the plant more reliable. This process can also be achieved in certain cases in transfer of bulk liquids from storage tanks using pressure raising coils. However, there are many instances where this cannot be done and some form of pumping is required. These instances can include the following:

(1) the process operates at low pressure;
(2) the storage is at high pressure to reduce boil-off/heat inleak;
(3) the liquid flow is high;
(4) the product draw-off from storage is intermittently transferred to road/rail tankers and rapid transfer is desirable;
(5) the product pressure required is high, i.e. it can be pumped, vaporized and bottled at high pressure.

These cases demand the use of pumps suitable for cryogenic service. In general, the principles of pumping cryogenic liquids are the same as for normal liquids. The same criteria of preventing cavitation and process liquid leakage, selection of construction materials, and type of pump must be considered in each case. Therefore, the calculation of net positive suction head (NPSH) requires the same careful consideration as for any liquid. The typical equation is given below for symbols based on Fig. 9.8.

$$\text{Required NPSH} = P_T + H_L - P_v - h_f - h_a$$

where

H_L = head of liquid to be pumped (m)
h_f = frictional loss (m)
h_a = acceleration loss for reciprocating pumps (m)
P_T = tank pressure (metres of liquid)
P_v = liquid vapour pressure at pumping temperature (metres of liquid)

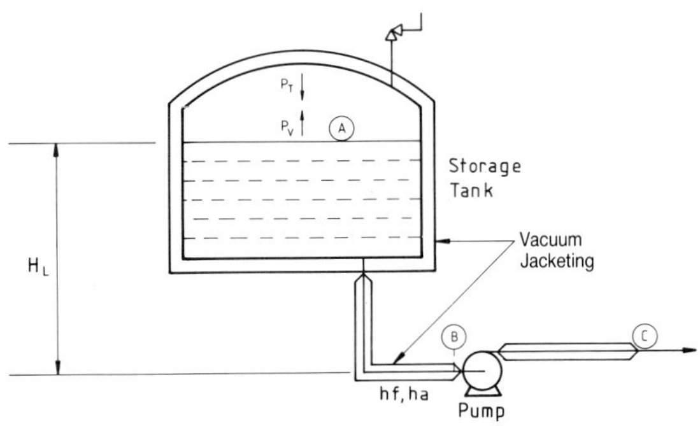

FIG. 9.8. NPSH in pumping a cryogenic liquid.

Vacuum insulated lines are used to prevent excessive heat inleak so that the liquid vapour pressure at pump inlet B does not rise significantly above P_v at point A.

Each pump manufacturer has his own safe limit for the margin *above* the *required* NPSH, and this varies from about 1–6 m of liquid.

In the cryogenic industry there is a variety of pumping needs, as shown in Table 9.1. Clearly, there is a wide variety of flows and delivery heads requiring different types of pumps. Reciprocating and centrifugal pumps are used in cryogenic service, with the latter being grouped into 'normal' and canned or submerged pumps.

Reciprocating pumps are generally used when a high discharge head is required, or a relatively low capacity, for example, cylinder gas filling. This type of pump requires a pulsation damper to minimize the pulsations in the flow-rate which contribute to the acceleration head losses. Because of the need to insulate, the complexity of this system

TABLE 9.1. The range of duties for cryogenic pumping systems.

Application	Demand pattern	Typical head (m)	Typical flow ($m^3\ h^{-1}$ of liquid)
1. Bulk liquid transfer from storage to tanker or vice versa	2–6 h once or twice daily	6–12	15–40
2. Plant output to storage	Continuous	6–12	1–20
3. Delivery to customer from tanker	0.5–1 h several times per day	50–60	30–60
4. Bulk gas supply to grid via vaporizer, e.g. O_2 to steelworks or LNG to gas grid	Continous operation for several weeks or months per year	100–400	20–40
5. Gas cylinder filling	Intermittent, several times per day	400–600	0.5–2

increases and hence costs can be moderately high. Sometimes the pumps are encased totally in a vacuum jacket to prevent heat ingress. Start-up also requires great care to prevent running the pump while the liquid is boiling-off in the casing during the 'cool-down' phase. Appropriate 'priming' equipment must be provided and operated correctly to ensure that all parts of the suction side are properly cooled down and flooded with liquid prior to start-up. This ensures that wear due to 'two-phase' fluid pumping is kept to a minimum. Figure 9.9 shows a typical reciprocating cryogenic pump.

Centrifugal pumps are used generally for low heads and larger capacities, and would probably be appropriate for applications 1, 2, and 3 in Table 9.1. The lower delivery head applications of 4 may also lend themselves to centrifugal pumps (Fig. 9.10).

The seals on both reciprocating and centrifugal cryogenic pumps require considerable attention to both design and maintenance. Their objectives are to prevent liquid and pressure loss as well as to prevent extraneous matter such as dust or lubricating oil migrating along the drive shaft. Centrifugal pumps are usually directly coupled to the motor and any process liquid leakage will rapidly reduce the temperature of the motor casing and possibly solidify the lubricants. In reciprocating pumps, the pump drive is usually through a crankcase which converts rotary motion into reciprocating motion. The crank thus moves through an oil bath and if any leakage of cryogenic liquid occurs it could cool the crankshaft excessively and lead to a cooling down of the necessary lubricants. Therefore, careful design ensures that proper heat brakes are in

FIG. 9.9 A reciprocating pump for cryogenic applications (photograph courtesy of Cryomec AG, Allschwil, Switzerland).

FIG. 9.10. A centrifugal pump for cryogenic applications (photograph courtesy of Cryomec AG, Allschwil, Switzerland).

place by using low-conductivity metals and long lengths. Provision of ambient heat inleak into the shaft ensures that the cold does not pass to the oil sump.

It is often advantageous to completely submerge the pump and motor in the pumped liquid. Of course, the motor is encased so that the electrical parts are not submerged in the liquid themselves but the whole motor is subjected to the cryogenic temperatures via conduction through its casing. These pumps are often called 'canned', and are frequently used in LNG pumping service as well as other cryogenic applications, except for oxygen. The advantages of such pumps are

(1) The absence of shaft seals between the cryogenic liquid and the atmosphere;
(2) there are no long drive shafts;
(3) the mechanical bearing design is less complicated;
(4) electrical resistance in motor windings is lower due to the effect of low temperature on the material;
(5) the assembly can be withdrawn through the top of the vessel;
(6) the heat input caused by electrical resistance is more than offset by the reduction in ambient heat inleak.

FIG. 9.11. A canned nitrogen pump (photograph courtesy of Cryomec AG, Allschwil, Switzerland).

Submerged pumps have been used in oxygen service, but these utilize remote motors which tend to add to the complexity and hence cost. A typical canned pump is shown in Fig. 9.11.

9.5 Heat exchangers

9.5.1 *General aspects*

Cryogenic plants generally utilize three types of heat exchangers: shell-and-tube, wound-coil, and plate-fin. The least widely used is probably the shell-and-tube heat exchanger, and the most widely used is the plate-fin. To appreciate why this is so, the three types of heat exchanger are compared in Table 9.2 in terms of area per unit volume, maximum design pressure, practical approach temperatures, number of streams handled in one unit, and materials of construction. The materials of construction have cryogenic service in mind.

9.5.2 *Tubed heat exchangers*

Shell-and-tube and wound-coil heat exchangers have been used in cryogenics since its advent. The shell-and-tube exchanger is still used

TABLE 9.2. Types of heat exchangers used in cryogenic service.

	Shell and tube	Wound coil	Plate-fin
1. Maximum single unit size ($M \times M \times M$)	2 dia. × 25 approx.	4.3 dia. × 65	1.0 × 1.4 × 6.5
2. Maximum design pressure (kPa)	> 20 000	> 20 000	10 100
3. Maximum number of streams	2 (3 in exceptional cases)	5–7	up to 7
4. Temperature limits (°C)	− 269 to + 350 or higher	− 269 to + 350 or higher	− 296 to + 65
5. Construction material	Stainless steel/copper/ brass	Stainless steel/Al/Cu	Al*
6. Typical approach temperature (°C)	6–10	3–5	2–4
7. Area for heat transfer per unit volume (m^2/m^3)	40–80	20–40	800–1200

*Recently, stainless steel designs have emerged for around 2000 kPa design pressure.

today in some instances which are perhaps on the fringes of cryogenics. Such applications include expander plants and reboilers for ethylene plants where a kettle design or a vertical thermosyphon is appropriate because of advantages in configuration, layout, and materials of construction. Such shell-and-tube exchangers for cryogenic use are generally made from stainless steel. Although expensive they compete favourably with aluminium exchangers because of the need to 'box in' aluminium exchangers for fire protection. Where a large number of exchangers are required it may be advantageous to use aluminium plate-fin designs. In some cases shell-and-tube units with copper and brass tubes are made for chilling applications using Freons. Another area where shell-and-tube exchangers are often used today is in the turbo-expander process used to recover NGLs from wellhead gases. This is particularly true of the USA where many such exchangers have been used in processes similar to that shown in Fig. 7.1. The reasons for using a shell-and-tube here rather than a plate-fin exchanger include:

(1) strong economic pressures to make a simple, easy to maintain plant;
(2) plants were usually put into remote areas with little supervision and poor maintenance;
(3) 'free' pressure energy was available and so process thermodynamic efficiency was not the prime consideration;
(4) a greater number of suppliers of exchangers to choose from;
(5) a greater number of fabricators able to work in steels (aluminium fabrication required more skilled welding techniques);
(6) process plant designers standardized the designs to speed up plant delivery to site (plate-fin exchangers generally have a longer manufacturing time).

Regenerators are another form of heat exchanger which adopts the use of two units cycling in parallel. These consist of vessels packed with high heat capacity material through which process gases are passed. Cold gas passes through these regenerators to give up its cold into the vessel contents, and the incoming warm stream is then cooled against the now cold contents. Two vessels must be employed to ensure process continuity. These operate in much the same way as the reversing exchangers described in Chapter 3, and also achieve two things at once: heat transfer and deposition of H_2O and CO_2 onto the cold surfaces by freezing out.

Regenerators are sometimes used today on relatively small air separations (mainly in Eastern Europe) as a means of heat exchange and pretreatment. However, modern large plants tend to use molecular sieves or reversing heat exchangers as pretreatment systems.

Wound-coil heat exchangers are still used today in a number of applications, even though plate-fin heat exchangers have the majority of business. Wound-coil heat exchangers have found favour in the following areas:

(1) very small cryogenic plants and refrigeration where plate-fin technology is more expensive;
(2) high operating pressures—typically above 10 000 kPa—where more than two streams are needed to exchange heat with one another;
(3) large base load LNG plant—very large wound-coil modules can be constructed avoiding the multi-block distribution problems which would arise if plate-fins were used;
(4) many East European processes which are constructed indigenously where plate-fin technology is not available.

Some of the large wound-coil heat exchanger units are equivalent to around 40 maximum sized plate-fin exchanger blocks. Several streams can be catered for in such exchangers, as is shown in the typical example in Fig. 9.12. The streams are separated out into appropriate headers by grouping the tube ends into a tube-plate for each particular stream. Cryogenic plants normally have one high-pressure stream cooling down with several products and process streams warming up. For reasons of mechanical design and economics, the high-pressure cooling stream is put in some of the tubes, the returning high-pressure streams in the remaining tubes, and the low pressure return stream is put in the shell.

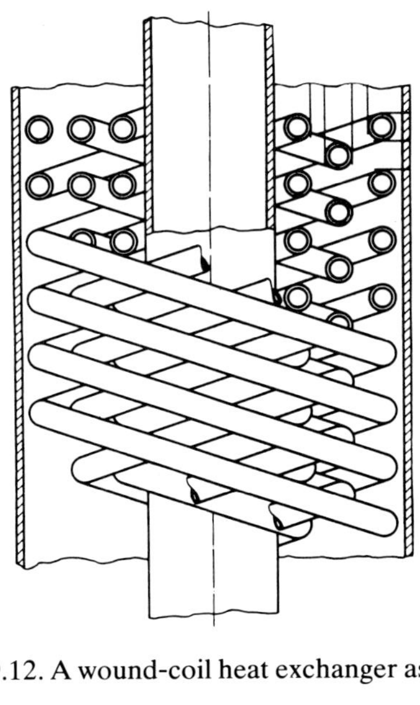

FIG. 9.12. A wound-coil heat exchanger assembly.

9.5.3 *Plate-fin heat exchangers*

Plate-fin heat exchangers are currently the most extensively used form of heat exchange for cryogenic applications. The 1970s and 1980s have seen some major developments in plate-fin technology, mainly in the manufacturing area. Until the early 1980s all plate-fins were made in the following manner from aluminium materials:

(1) parting sheets, side bars, and fins are cut from aluminium;
(2) cut materials are washed;
(3) the exchanger is stacked with brazing foil between layers;
(4) the stack is pre-heated in a chamber;
(5) the stack is dipped in molten salts at about 600°C to braze the layers, side bars, and parting sheets;
(6) the exchanger is cooled;
(7) the block undergoes a thorough washing process to remove salts;
(8) the blocks are then dried;
(9) the headers are then welded on;
(10) the exchanger streams are flow tested and shipped to the customer.

Since the 1980s, the vital steps in plate-fin manufacturing technology have changed dramatically. Steps 1 to 4 are the same, but steps 5, 6, 7, and 8 have been replaced by a vacuum furnace process which heats up the assembly in a large chamber under vacuum. The heat is supplied by an electrical field around the chamber. Because the brazing is carried out in dry, almost surgically clean conditions, there is no need for the washing stage. Some of the older brazed units have not performed adequately because of residues in the core left by incomplete washing after the brazing process. The vacuum-brazed exchangers have proved themselves considerably more reliable in performance than their dip-brazed counterparts. All of the successful vendors of such exchangers used vacuum brazing from the mid-1980s onwards.

Figure 9.13 shows a cutaway sketch of a three-stream plate-fin heat exchanger. Figure 9.14 shows how such blocks are interconnected and Fig. 9.15 shows types of corrugated fins that make up the heat transfer surface.

9.5.4 *Some design aspects of plate-fin heat exchangers*

The sizing of such blocks requires careful consideration of four major areas:

(1) knowledge of maximum design pressure—see Table 9.2;
(2) accurate calculation of heat transfer consideration;
(3) arrangement of layers for best secondary heat transfer;
(4) correct use of distribution aids for single and two-phase streams.

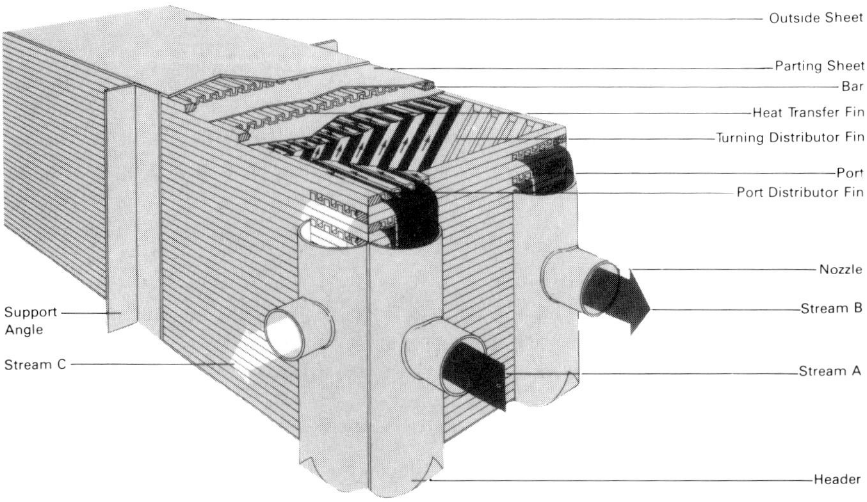

FIG. 9.13. A cutaway drawing of a plate-fin heat exchanger (courtesy of ALTEC International Inc.).

FIG. 9.14. Three plate-fin heat exchangers shown connected (courtesy of ALTEC International Inc.).

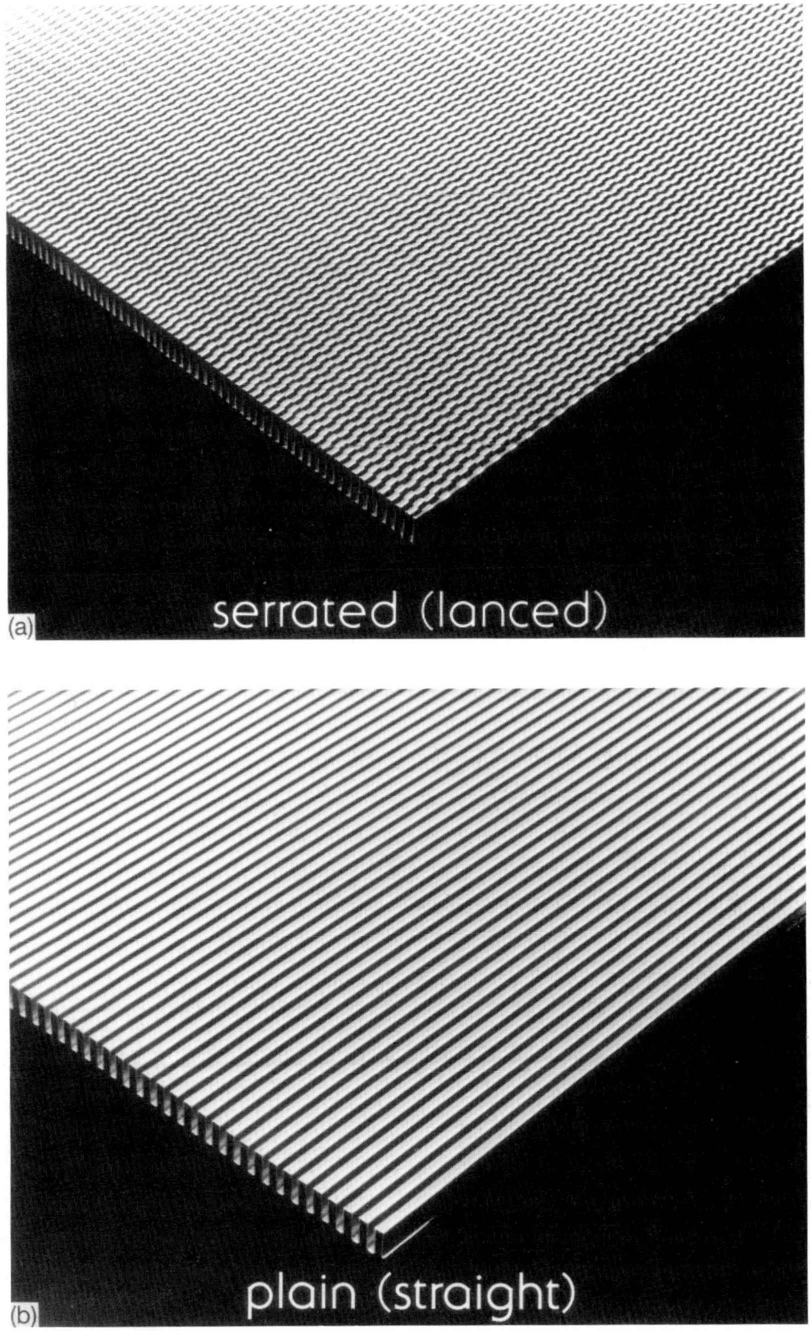

FIG. 9.15. Examples of corrugated heat transfer fins: (a) serrated; (b) plain (courtesy of ALTEC International Inc.).

Design pressures are generally fixed by manufacturers and are usually in line with Table 9.2, but often one stream in the exchanger will be limited to 3000 or 4000 kPa, especially when large header tanks are used. In these circumstances one is usually limited by the thickness of plate that can successfully be welded in aluminium.

Accurate calculation of the heat transfer coefficient and pressure drop is usually based on an experimentally correlated Colburn coefficient, *j*, and friction factor, *f*, for different types of corrugations in plate-fin exchangers. These are generally correlated against the Reynolds number as shown in Fig. 9.16. Each of the manufacturers has his own range of fin types for heat transfer and distribution, and therefore each has his own curves for *j* and *f* factors. Although the values of *j* and *f* for the same range of Reynolds number are similar between manufacturers, they are

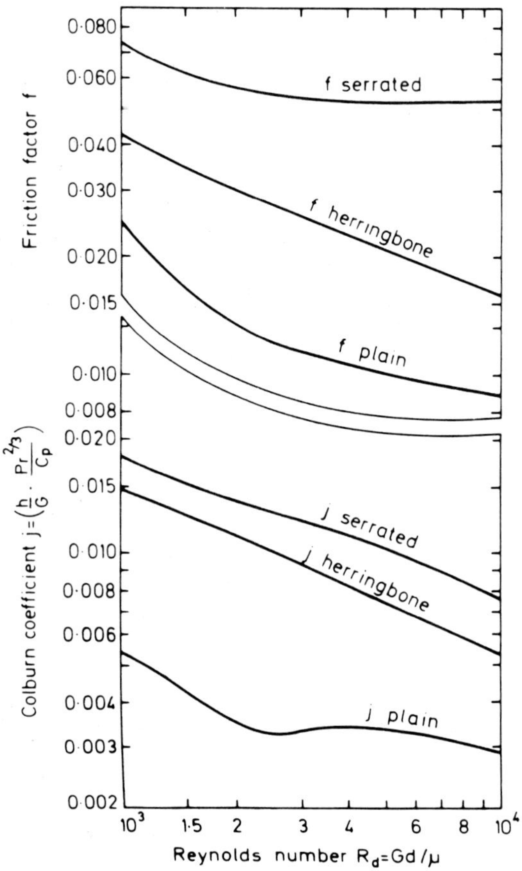

FIG. 9.16. Heat transfer and pressure drop characteristics for typical fins in a plate-fin heat exchanger (courtesy of Marston Palmer Ltd).

not interchangeable. The designer uses these to arrive at the optimum design using a complex computer program and considerable past experience.

The third area which requires serious consideration is closely linked with the heat transfer coefficient, and that is layer arrangement. To illustrate this we can consider a three-stream exchanger where stream A is cooled by return of streams B and C. Although the system is in overall heat balance, let us assume that stream B provides three times as much cold in C. Also, for simplicity we assume that the exchanger is a gas–gas type and flow B is about 3 times flow C. An easy way to configure the layers would be in the order A/B/C/A/B/C/A, etc., if we assume 15 layers in total. This is, in fact, far from ideal, because B must give up three times more cold than C into stream A. Therefore, it must pass cold to A on one side and to A via C on the other side throughout the whole cross-section of the block. This 'secondary' heat transfer is inevitable in plate-fin heat exchangers but should be reduced as far as possible by optimizing the layer arrangement. A better arrangement may be B/A/B/C/B/A/B/C/B/A/B/C/B/A/B. In this situation the stream that gives up most cold to A is on either side of A, while the 'minor' stream C provides secondary heat transfer to A via B in both directions. Suessman describes an effective method of calculating the heat fluxes for each layer and plotting a 'zig-zag' diagram to show the balance of effective secondary heat transfer. Layer arrangement is often underrated in plate-fin heat exchanger design and can lead to operating problems.

Distribution of fluid across the width of each layer also requires very careful attention, especially with two-phase flow. A typical plate-fin layer for single-phase heat transfer is shown in Fig. 9.17.

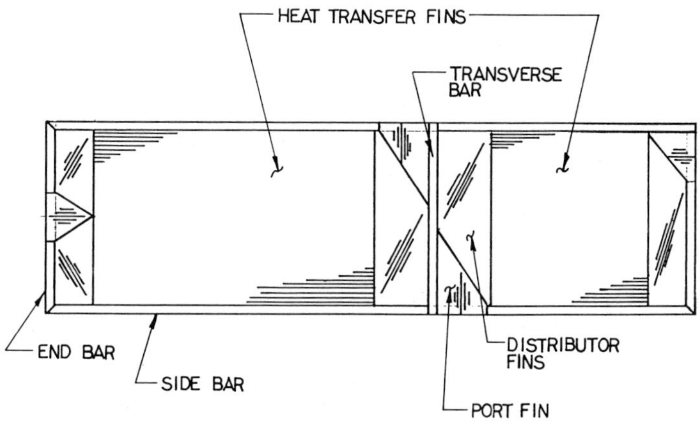

FIG. 9.17. The arrangement of a single layer in plate-fin exchangers (courtesy of ALTEC International, Inc.).

For single-phase flow one still needs effective distribution in each layer to avoid 'dead' zones in the matrix at points X and Y. Therefore, the matrix should have a high pressure drop compared with the end distributor fin to ensure that the 'end' effects are negligible. Selection of the correct type of fin for distribution and the matrix are therefore crucial.

In two-phase distribution, the situation is somewhat different—what is needed here is even distribution of the two phases *into* each layer as well as *across* each layer. Uniform distribution of two phases into each layer can be achieved very simply, but with only partial success, by using perforated impingement plates in the inlet header which adds pressure drop and promotes distribution into the various layers. The exchanger is usually mounted vertically with the two-phase stream entry at the bottom end. This is only effective when turndown requirements are not great (say 60–100 per cent operation) and the distribution or inlet header is not very wide, i.e. the block stack height is not great. In practice, design requirements are seldom like this and alternative methods are sought. Several methods are available and each manufacturer of plate-fins has his preferences. However, virtually all of these rely on separation of the two-phase mixture prior to its admission into the heat exchange matrix. This ensures that the liquid can be admitted evenly across each layer and into all layers for a much wider range of operating conditions. The vapour is allowed to re-mix with the liquid only once both phases have entered the heat exchanger matrix. Typical distribution methods are shown in Fig. 9.18. Four methods are shown, which are mostly self-explanatory. Each has advantages and disadvantages of cost and operability. Method (a) allows the two-phase mixture to enter the exchanger header into the plate and chimney area, whilst (b), (c), and (d) all require a separator outside the exchanger assembly. It is clear from the diagrams that all methods require considerable manufacturing precision when one considers that the layers are usually 6 to 9 mm high.

9.6 Distillation equipment

9.6.1 *General distillation*

In this section, distillation equipment shall be assumed to mean vapour–liquid contacting equipment which carries out separation by absorption or fractionation. Cryogenic technology incorporates both of these techniques in many applications; in particular distillation is used in a great many instances. The instances where absorption and/or distillation are used at subambient temperatures are numerous and include the areas listed in Table 9.3. The table is by no means a complete list, but merely

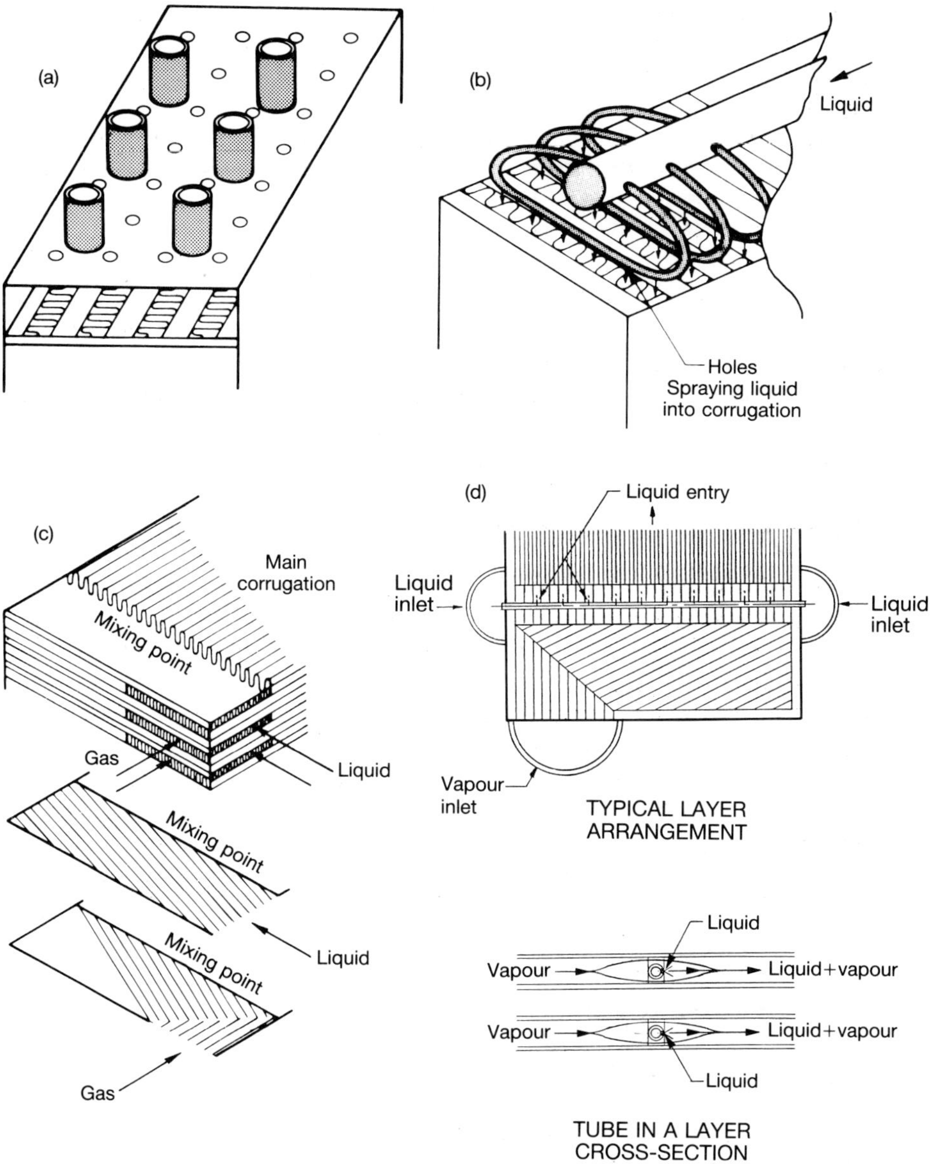

FIG. 9.18. Methods of two-phase distribution in plate-fin exchangers: (a) plate and chimney; (b) sparge pipe; (c) split matrix (courtesy of Marston Palmer Ltd); (d) injection tubes (courtesy of ALTEC International Inc.).

TABLE 9.3. Absorption and distillation applications in cryogenics.

Application	Absorption	Distillation
Air separation	—	inevitable
De-nitrogenation of natural gas	—	inevitable
LPG recovery from natural gas	sometimes	often
D_2/H_2 separation	—	inevitable
H_2/CO_2 separation (methane wash process)	yes	yes
Argon recovery from air separation plants	yes	yes
Argon recovery from ammonia purge gas	often inevitable	—
Rare gas recovery	—	inevitable
Ethylene/ethane separation	—	inevitable
Propylene/propane separation	—	inevitable

an indication of the major areas. Of course, permeation and adsorption are very gradually increasing their range of applications in many areas, but are currently still limited to bulk separation rather than pure-product manufacture.

Although permation and adsorption technologies are playing an increasing role in the above areas, distillation is likely to continue to play a significant part in years to come. By definition, distillation relies on creating a large surface area so that vapours and liquids which are not in equilibrium with each other can come into contact in order to approach equilibrium as closely as possible, in the most cost-effective equipment. By doing this we can purify or separate desired components from mixtures. This can take place at above and below ambient conditions and essentially the equipment is similar for both. In this section we shall emphasize the cryogenic end whilst monitoring the others. Table 9.4 summarizes the different mass transfer equipment that is available.

Figure 9.19 gives more examples of typical loose and structured packing with a sketch of the assembly of a typical tower. Figure 9.20 gives an example of a tray which tends to be used more in low temperature applications. The most common types of trays used in cryogenics are sieve trays, both annular and crossflow. Figure 9.21 shows a section of plate-fin heat exchanger used for distillation purposes. This is a relatively novel technique which offers very high thermodynamic efficiency. In this situation a plate-fin heat exchanger is used to perform a mass- and heat-transfer operation in one unit. This is particularly suitable for a situation where a lower molecular weight gas contains a few percent of heavier components, e.g. 6–10 per cent propanes and heavier. The partly cooled vapour may enter at point A, any condensate is knocked out in V1, the vapour then enters the reflux exchanger, E1, at B, and is cooled by refrigerant DE and product XY. The exchanger is always mounted vertically so that the feed stream cools in an upward

TABLE 9.4. Mass transfer equipment summary.

Method	Comments
Loose packing in towers	Various types. Not used frequently in cryogenics. Uses mainly limited to argon recovery and rare gas recovery where high turndown ratios are not required.
Structured packing	Various types. Not used frequently in cryogenics. Uses mainly limited to argon recovery and rare gas recovery where high turndown ratios are not required.
Valve trays	Various types. Used in hydrocarbons processing where good turndown is required.
Bubble cap trays	Various types. Used in hydrocarbons processing where good turndown is required.
Sieve trays (cross flow and annular)	Used in a variety of cryogenic applications. Probably the most frequently used of all the contacting devices. Lends itself to design for very low tray spacing.
Spray towers and falling films	Seldom used in cryogenics.
Plate-fin refluxing exchangers	Becoming very popular to facilitate simulatenous heat and mass transfer.
'Hi-gee' enhancement of mass transfer by adding centrifugal force	Used in few petrochemical applications. Considered for cryogenics but not as yet tested.

direction to C. As it cools, condensate forms. Instead of the condensate passing up with the main gas flow it runs back down the heat exchanger to the bottom and B. As it runs back it is warmed by the upcoming vapour stream which tends to boil-off light ends from it. Therefore the liquid that emanates from the base B contains only small amounts of light ends and most of the heavy ends. This liquid is approaching the equilibrium condition of the vapour entering at B and *not* the condition at the cold end of the exchanger C.

This is the important feature of the reflux exchanger which makes it more efficient than straight cooling and partial condensation. Several equilibrium stages can be catered for in a single long core, at the same time providing refrigeration at each theoretical stage. This represents the ultimate in distillation.

It becomes apparent that the design of these exchangers is very complex, because the designer has not only to cater for the distribution problems described in section 9.5 for plate-fin exchangers, but also the mass transfer must be designed properly. For this, an appropriate

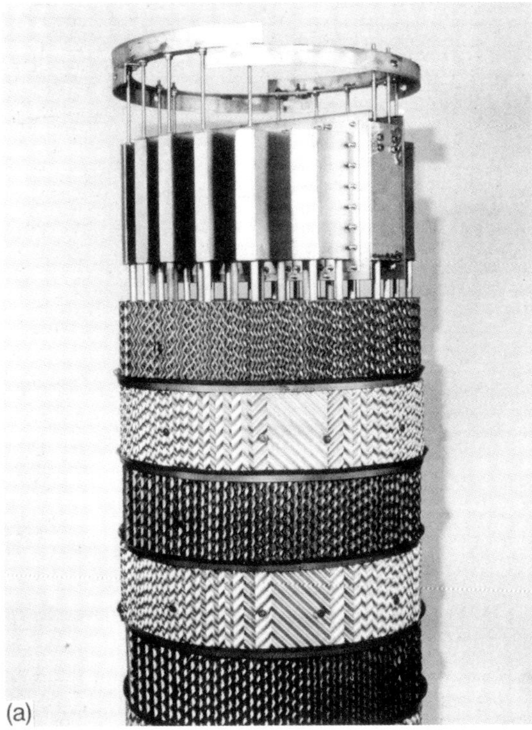

FIG. 9.19. Examples of loose and structured 'Tower Internals': (a) structured; (b) loose (courtesy of ETA Process Plant Ltd).

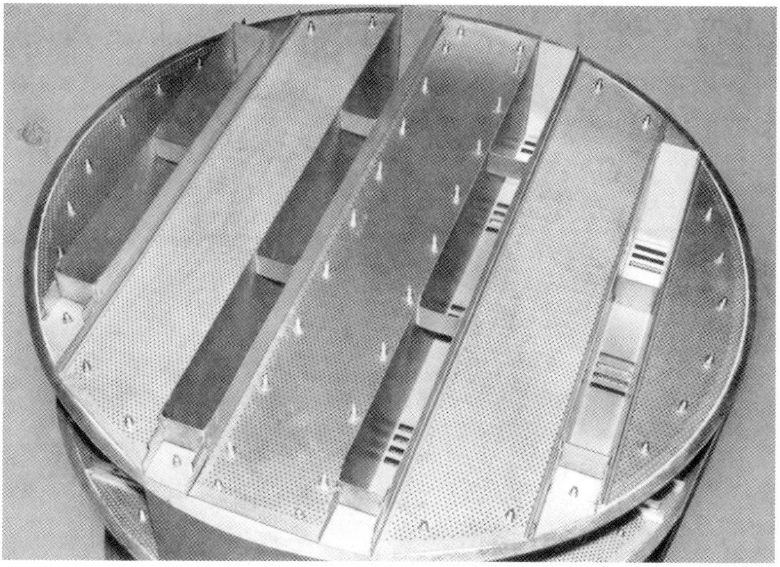

FIG. 9.20. An example of a cross-flow sieve tray (courtesy of Union Carbide Co.).

flooding correlation is required. Many experimentalists have carried out work and arrived at empirical correlations. A typical correlation by Pigford and Hughmark is given below:

$$V_f = C(di)^{0.3} \rho_G^{-0.5} \rho_L^{0.47} \sigma^{0.31} \mu_L^{-0.08} \left(\frac{L}{G}\right)^{-0.2}$$

where V_f = flooding velocity
C = constant
di = tube inside diameter
ρ_G, ρ_L = gas and liquid densities
L, G = liquid and gas superficial velocities
μ_L = liquid viscosity
σ = surface tension

Several other correlations are also available.

9.6.2 *Double columns*

Distillation is a unit operation frequently used in gas separation. A single column with condenser and reboiler as shown in Fig. 9.22(a) is the most common distillation system encountered in chemical engineering. This can potentially achieve total separation of key components. If our components in this example are N_2 and O_2, then we could achieve

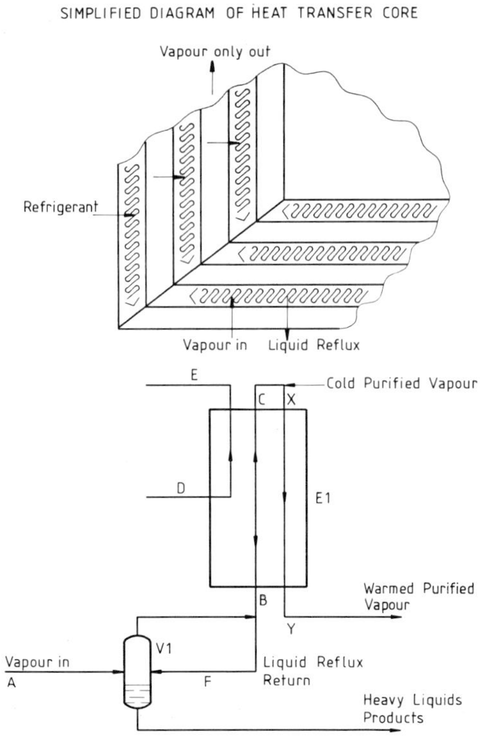

FIG. 9.21. A plate-fin heat exchanger used as a mass transfer device.

complete separation of N_2 and O_2 into N_2 overheads and O_2 bottoms. Although this is theoretically possible, it needs an external heat-pump cycle to allow it to occur, i.e. to provide heat for reboil and cold for condensation. This may be both capital and operating cost intensive, and other methods can be used to achieve the same end.

If nitrogen is the desired product (the more volatile component), then feed air can be admitted to the base of the column as in Fig. 9.22(b). This feed enters after prior cooling and arrives at or just below the gas dewpoint. By taking the liquid from the base of this column, flashing it to low pressure, and feeding it into the overheads condenser, reflux can be provided to the column. This enables one to purify the overheads to pure N_2 product and retain enough driving force in the condenser. This system has two limitations:

(1) only one product, the N_2, can be made pure;
(2) the recovery of this N_2 is dependent on the equilibrium conditions in the base of the column where N_2 is lost in the liquid.

The second point relates to the column operating pressure which in turn

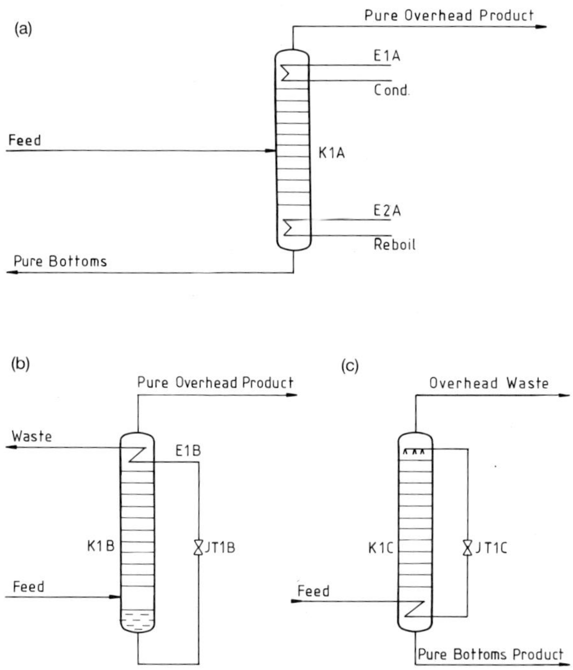

FIG. 9.22. Examples of distillation: (a) a single column with condenser and reboiler; (b) a single column with no reboiler; (c) a single column with no condenser.

relates to the flash pressure downstream of JT1B. For a waste gas pressure of about 130 kPa, the column pressure would be about 700 kPa and thus the approach to equilibrium in the base of K1B is fixed. This means that the recovery of N_2 is fixed.

Taking Fig. 9.22(c) we see that in this case feed gas is used for reboil which at the same time condenses it. This liquid when flashed across JT1C produces reflux for column K1C. This system allows the production of pure oxygen but not pure N_2, and this is a mirror image of Fig. 9.22(b), but instead the bottoms product can be made pure. Again O_2 recovery is limited by equilibrium conditions in the top of K1C.

One important point to notice in the above systems is that only system (a) needs an external cycle to drive condenser E1A and reboiler E2A. Systems (b) and (c) use the process fluids as reboil and condensing media—a more integrated and more efficient thermodynamic solution. When developing the double-column design, the fundamental question was how to make two pure products and not to resort to external refrigeration and reboil cycles. To appreciate the development of the double column it may be easiest to take a look at Fig. 9.22(c). It already has the capability of making one pure product—O_2 at its base. It currently used feed gas to reboil in E1C. If this duty could be supplied by a condensing fluid which may provide reflux in a second column we might 'kill two birds with one stone'. This is exactly what happened in the early development of the double column with the lower column being placed physically underneath the upper column, as shown in Fig. 9.23.

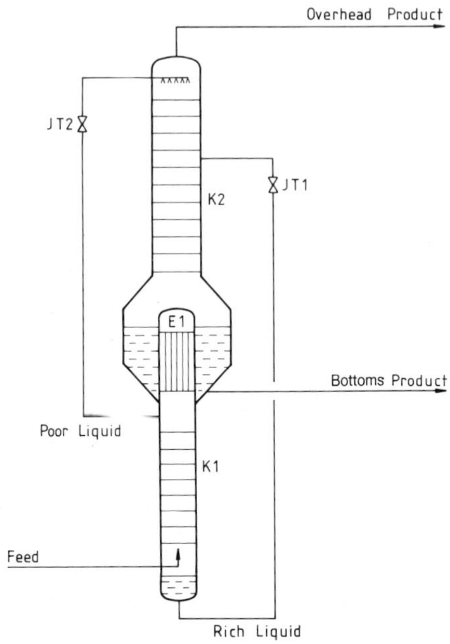

FIG. 9.23. The double-column system.

In the double-column system feed enters the base of K1 at an elevated pressure, say 600 kPa for an air-separation system. K1 behaves like K1B in the single-column N_2 production case, producing pure N_2. The liquids from the base of K1 are flashed in JT1 and passed into K2 from fractionation at a much lower pressure than K1, i.e. about 130 kPa (in K1B this liquid provided only reflux via E1B). As the liquid in K2 runs down the column it is depleted of N_2 and pure oxygen remains in the sump. Reboil to the upper column is provided by E1, which uses heat from N_2 condensing at high pressure in the same exchanger and running down K1, providing it with reflux. Because the upper column is at lower pressure, relative volatilities for N_2/O_2 are better and less reflux needs to be provided to make pure N_2 than in Fig. 9.22(b). Therefore, some pure liquid N_2 can be drawn off from the top of K1, flashed in JT2, and used as reflux to make pure N_2 in K2. Hence we have an integrated system providing two pure products. This double-column system is applied to other gas separation problems, e.g. de-nitrogenation.

9.7 Storage

At some stage product gases and liquids will require storing for distribution to the customers or as a buffer prior to a downstream process. This section will cover the types and methods of storage. These can conveniently be subdivided into small, medium, and large for both gaseous and liquid cryogenic products.

9.7.1 *Small storage equipment*

Both liquids and gases are stored in small containers varying in size from small gas cylinders and dewars for liquids to pallets housing several cylinders. Small amounts of industrial gases such as N_2, O_2, Ar, etc. are normally stored in the gaseous phase in a range of cylinders as shown in Fig. 9.24. These range in size from 0.4 litres (H_2O) to 50 litres in the high-pressure range, to 108.4 litres for low pressure cylinders. For convenience, some of these are put on pallets for larger customer demands, e.g. 15 cylinders on each pallet. For even larger deliveries, tube trailers can be used with a *gas* capacity of 4500 m^3 (at NTP).

Relatively low volatility liquids are also stored and distributed in cylinders such as those in Fig. 9.24. However, these cylinders have to be adapted for liquid withdrawal with goose necks. Usually this kind of storage and distribution method will be limited to fluids such as LPGs, halocarbons, ethylene oxide, etc.

Dewars are used for relatively small amounts of liquid that customers sometimes required. The simplest form of this is the Thermos flask invented by Sir James Dewar for storing cryogenic liquids (simple versions are often used domestically for storing hot liquids). Today's dewars are made of metal and are designed for the higher-boiling-point cryogenic liquids such as N_2, O_2, etc. These are often made from aluminium with a glass-reinforced polymer neck, or completely of stainless steel. Figure 9.25 shows a dewar.

For helium storage, superinsulation is often used. This kind of storage unit is typically used for liquids in quantities around 20–100 litres. They have a boil-off of 1–4 per cent of total contents per day due to heat inleak from the atmosphere. To minimize boil-off rates, the space between the two walls is evacuated and 'getters' are used to remove the remaining gas in the space to provide a near perfect vacuum. Getters can be metals which form compounds with the residual gas. For example hydrides are formed with some metal alloys which absorb hydrogen. Hence if the space was initially filled with hydrogen, these could be used to create a good vacuum after mechanical evacuation.

Metal hydrides can also be used as a means of storing hydrogen in particular. As described in Chapter 8, they have the potential capacity to

FIG. 9.24. Typical cylinders for storing industrial gases (courtesy of T. I. Chesterfield Ltd).

store more hydrogen (volume for volume) than liquid hydrogen storage tanks. For small capacities these compete favourably with cryogenic storage tanks for liquid hydrogen.

9.7.2 *Medium-sized storage tanks*

Medium-sized storage tanks encompass the range of 5 tonnes of cryogenic liquid to about 100 tonnes. They are often transportable tanks on trailers or rail wagons. Larger road tankers typically have a capacity of 17 tonnes of product LPG or about 30–35 mm^3 of carrying volume. Smaller units can range from half of the above amount upwards. The rail tankers can be some 5–6 times the single road tanker capacity. Customers who receive shipments of liquids will have fixed tanks installed

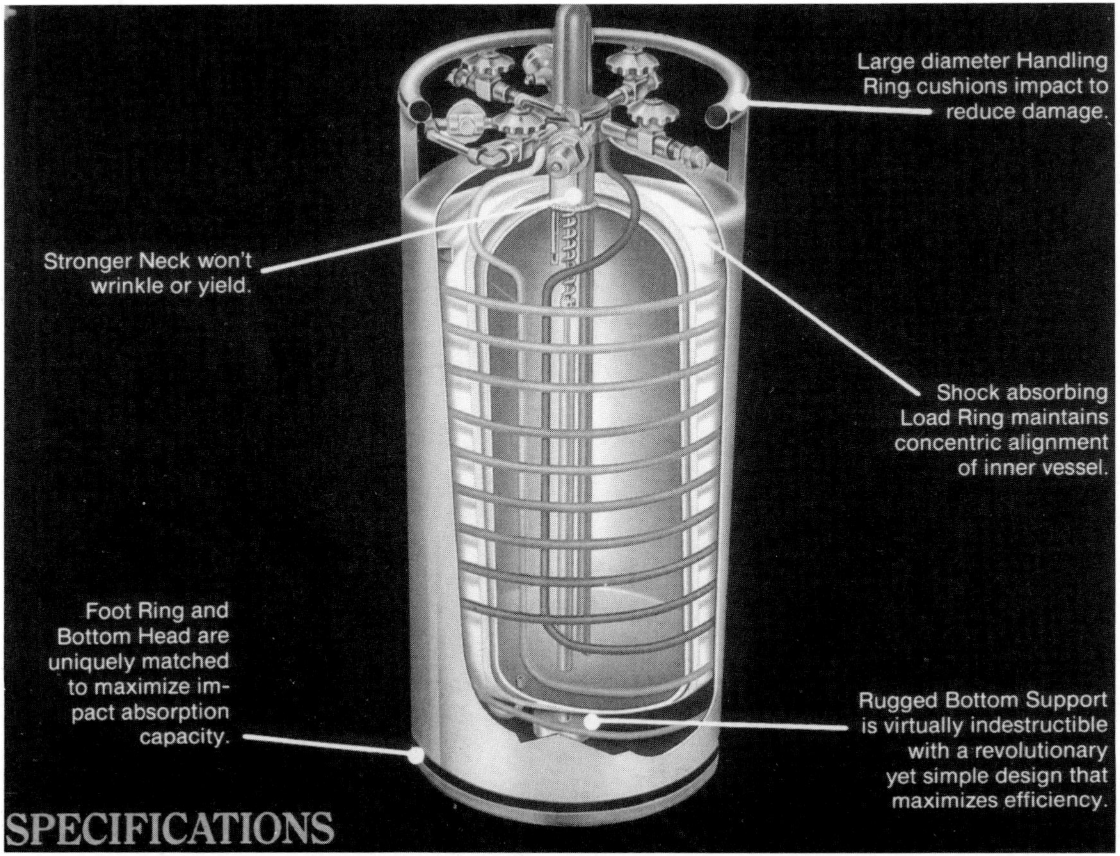

FIG. 9.25. An example of a dewar (courtesy of MVE Cryogenics).

on site with a similar range of capacities, depending on their size and pattern of consumption, which are single-, double, or triple-walled constructions. Single-walled storage tanks are generally limited to LPGs, i.e. propane/butane and mixtures and perhaps other C_3 and heavier compounds. In all cases safety plays a very important role in the design of such tanks. Figures 9.26, 9.27, and 9.28 distinguish between the equipment content and describe each type of tank. These flowsheets apply in essence to both transportable units and fixed units. In this size of storage there is usually no reliquefaction of boil-off and thus there is always a loss of 0.2 to 2 per cent contents per day, depending on the size of tank. This assumes N_2 or O_2 as the stored cryogen in a double-walled tank.

FIG. 9.26. A single-walled storage tank (photograph courtesy of WEFCO (Gainsborough) Ltd).

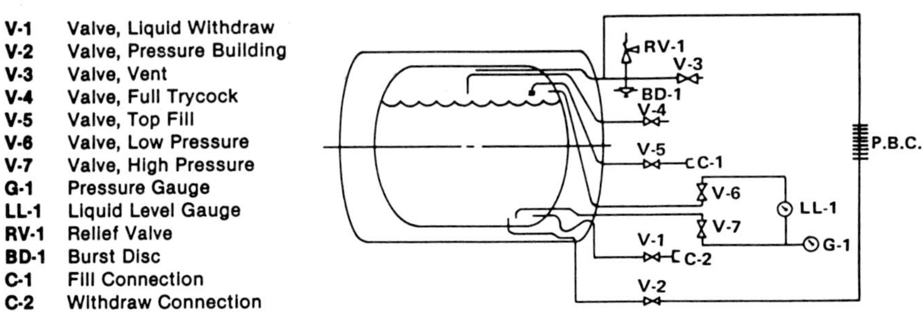

FIG. 9.27. A double-walled storage tank (courtesy of MVE Cryogenics).

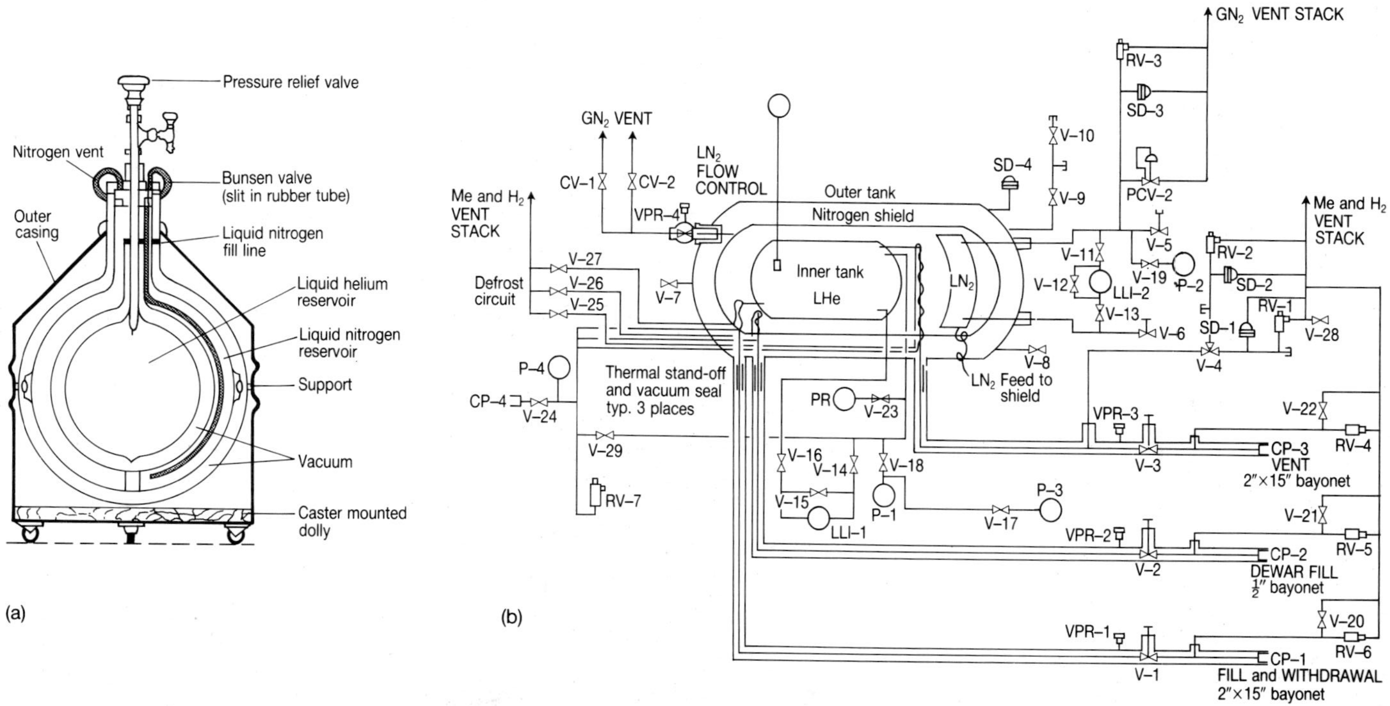

FIG. 9.28. A triple-walled storage tank: (a) small containers; (b) larger rail tankers (courtesy of MVE Cryogenics).

9.7.3 *Large storage facilities*

Usually it is expensive to build high-pressure storage facilities of a very large capacity for cryogenic liquids because of the metallurgy and safety systems required. Therefore, very large storage facilities for cryogenic liquids including N_2, O_2, LPGs, and LNGs are usually designed for near atmospheric operating conditions. Any boil-off is recompressed and reliquefied in the main process to avoid losses of product. These tanks are generally twin-walled; the inner wall is of stainless steel or aluminium and the outer one of carbon steel. The space between is filled with insulation, usually Perlite, or Perlite together with fibre-wool. These tanks also fall into three categories:

(1) spherical, above ground,
(2) cylindrical tanks with domed tops, above ground;
(3) below ground storage: cylindrical pits with stainless steel liner or caverns (natural or man-made).

Spherical tanks range in size from about ~ 5–$4000\ m^3$. Figure 9.29 shows a typical spherical storage tank for LPG. Cylindrical storage tanks

FIG. 9.29. A spherical storage tank for LPGs (photograph courtesy of Chicago Bridge & Iron Co.).

range in size from about 6000–200 000 m^3 storage capacity. Figure 9.30 shows a 21 000 tonnes LNG cylindrical storage tank designed for above-ground construction. There are many underground storage tanks for LNG and LPGs. Examples of this are the facilities at Canvey Island in the UK and Das Island in the Middle East. The objective of below-ground storage is not simply aesthetic but also one of containment of large volumes of liquid in the event of a rupture of the inner and outer walls of the main tank. The above-ground storage units are usually bunded to contain the spillage. In some cases a third, concrete wall is built around the tanks to contain the liquid—this is illustrated in Fig. 9.30.

In the late 1970s and early 1980s considerable interest has grown in underground storage in either used mines or natural or man-made caverns. This has led to commercial installations but has so far been limited to LPG storage. Studies have been carried out on how to store colder liquids but this was still a thing of the future at the time of writing. Figure 9.31 shows an example of a man-made cavern for LPG storage. Several such installations exist in a variety of geological circumstances, including

- salt caverns;
- lined soft rock;
- hard rock;
- used mines (hard rock).

FIG. 9.30. A cylindrical storage tank for LNG, with a prestressed concrete structured dyke system for impoundment (photograph courtesy of CBI).

FIG. 9.31. Underground cavern storage of LPGs. (photograph courtesy of Geostock, Paris, France).

TABLE 9.5. Types and sizes of cryogenic storage facilities.

Description	Application	Typical size range (actual m^3)
(a) Single-walled bottles	LPGs, higher boiling point liquids	0.01–approx. 1
(b) Dewars	Smaller volumes of low boiling point liquids e.g. He, H_2, N_2, Ar, O_2	0.01–0.1
(c) Road trailers with tank	N_2, He, Ar, LPG, etc.	10–35
(d) Rail car with tank	as for (b) or (c)	15–100
(e) Pressurized cylindrical storage, single bullets, double- and triple-walled	as for (b) or (c)	3–50 normally vertical; 50–300 normally horizontal
(f) Spherical storage tanks	as for (b) or (c)	200–2000
(g) Atmospheric storage in cylindrical units	H_2, CO, CH_4, C_2H_4	6000–120 000
(f) Cavern storage	LPGs and possibly N_2	30 000–7 million, though generally economical at above 100 000

Typical capacities considered or in place cover the large scale described above and range from 100 000 m^3 to 5 000 000 m^3 of 'storage space'. Operating units store propane, ethylene, butane, and crude oil. Liquid nitrogen storage has been experimented with but not implemented commercially. Table 9.5 summarizes the various storage methods employed for cryogenic liquids today.

9.8 Sendout and bottling

Most cryogenic liquids stay cryogenic only from the point of view of convenient storage. When needed they are usually vaporized to ambient temperatures. This is done at high or low pressure depending on the pressure required by the user—fuel, process, or bottling plant. Table 9.6 describes the method of operation and typical range of application for each of the vaporizers.

TABLE 9.6. Cryogenic liquid vaporizers.

Type	Description	Application and sizes
Direct fired	Cryogenic liquid boils in coils heated by combustion products from burning fuel gas.	LNG vaporizers on peak shaving facilities C_2H_4 typically.
Semi-direct fired (submerged combustion)	Combustion products from fuel gas bubbles through water bath in which there are coils which contain the boiling cryogen (Fig. 9.32).	LNG vaporizers on base load and peak shaving facilities. C_2H_4 vaporizers typically are steam fired. Sizes typically around 50 000–100 000 $Nm^3\ h^{-1}$
Salt bath	Combustion products heat a salt solution in a fired heater. Salt solution circulated to heat up LNG or other cryogen in separate exchanger.	LNG vaporizers on peak shaving facilities, C_2H_4 typically. Sizes typically smaller than for the direct fired type.
Steam bath	Cryogen vaporized in coils immersed in hot water. Heat added via direct steam injection.	Vaporizing all cryogens including He, H_2, N_2, Ar, CH_4, O_2, etc. Sizes typically 5–3500 $Nm^3\ h^{-1}$.
Ambient air	Cryogen vaporized in finned tubes which gain heat from air (Fig. 9.33).	Vaporizing all cryogens including He, H_2, N_2, Ar, CH_4, O_2, etc. Sizes typically 5–3500 $Nm^3\ h^{-1}$.

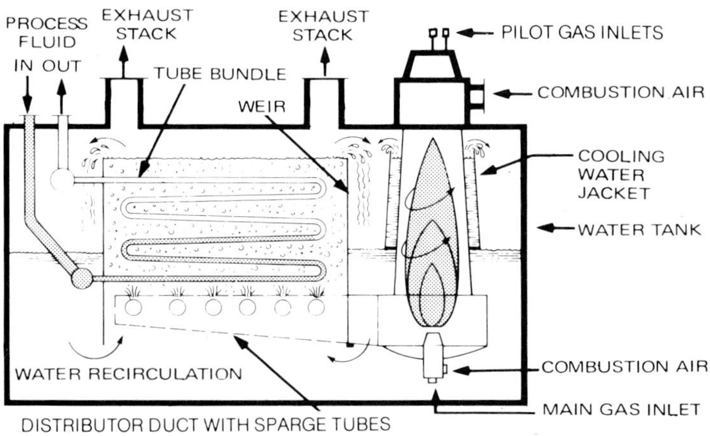

FIG. 9.32. A submerged combustion vaporizer (courtesy of Kaldair Ltd).

FIG. 9.33. An ambient air vaporizer (photograph courtesy of Cryomec AG, Allschwil, Switzerland).

Directly fired heaters are simple but require all electrical and fired equipment to comply with rigorous safety standards. However, intermediate fluid systems (salt bath) have advantages because the fired equipment can be sited in the least hazardous area. The vaporizer part is then put near the product pumps and storage tank. A typical sendout system for LNG would look like that shown in Fig. 9.34.

A liquid nitrogen storage and vaporization facility would look similar except that for N_2 bottling—one can either use the system as shown in Fig. 9.34 or compress vaporized N_2 in a high-pressure bottling compressor. Often N_2, O_2, and other industrial gas vaporizers are ambient air types because of their simplicity. They are however limited by their capacity due to the relatively low ambient air heat transfer coefficient. This can be enhanced a little by using fans, but for larger duties direct or indirect firing methods are generally sought.

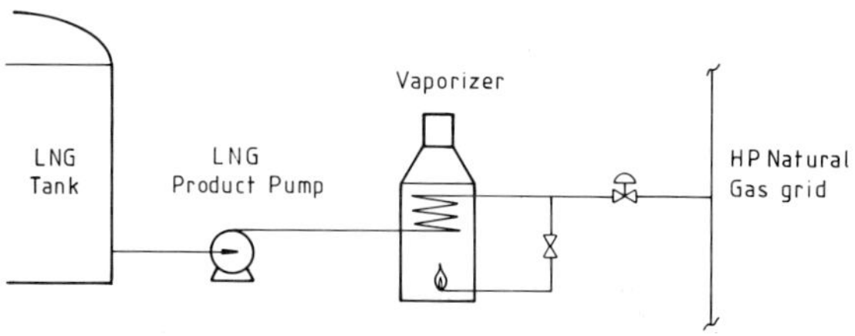

FIG. 9.34. LNG vaporizing and send-out.

Bibliography

BAUERMEISTER, K. J. (1975). 'Turbocompressors for air and oxygen service'. *Hydrocarbon Processing,* (July) 63.

BLAKEY, P. (1981). 'Cryogenic gases'. *Chemical Engineering,* (Oct. 5) 113–24.

CLARKE, M. E. and GARDENER, J. B. (1976). 'Refrigeration with expansion turbines'. *Contemp. Phys.,* **17**(6), 507–28.

DIEHL, J. E. and KOPPANY, C. R. (1969). 'Flooding velocity correlation for gas liquid counterflow in vertical tubes'. *Chem. Eng. Prog. Symp. Series,* **65**(92), 77.

DIERY, W. (1984). 'The manufacture of plate-fin heat exchangers at Linde'. *Linde Reports Sci. Tech.* **37**, 3.

HASELDEN, G. G. (1971). *Cryogenic fundamentals.* Academic Press, London.

HOLM, J. (1981). *The application of turboexpanders for energy conservation.* I. Mech. E. publication no C149/81, 95–105.

HOLM, J. (1985). 'Energy recovery with turboexpander processes'. *C.E.P.,* (July), 63.

KEENS, D. (1980). 'Technology of LNG plants still undergoing a steady evolution'. *Oil and Gas Journal,* (Apr. 7) 84.

PORTER, K. E. and JENKINS, J. D. (1987). 'Distillation 1979–1987'. *Gas Separation and Purification,* **1**, (Sept.), 11.

SCHOLZ, W. H. (1973). 'Coiled tube heat exchangers'. *Linde Reports Sci. Tech.,* **18**, 35.

STRUB, R. A. and MATILE, C. (1975). 'Centrifugal hypercompressors for ethylene service'. *Hydrocarbon Processing,* (June), 67.

SUESSMANN, W. and MANSOUR, A. (1979). 'Passage arrangement in plate-fin heat exchangers'. XV International Congress of Refrigeration, Venice Sept 23, IIF, Paris, France.

TANTAM, D. (1981). 'Application of cryogenic containers'. *Cryogenics,* (Dec.), 691.

THWAITES, A. C. (1978). *The current state of the LNG storage art in British Gas.* United Nations Seminar, SEM.5/R.

VOGEL, E. R., *et al.* (1983). 'TX LNG vaporizer design and operation'. *American Gas Association Transmission Conference,* Seattle, paper 83-T-42.

WOLENTARSKI, W. (1980). 'Turbomachinery limitations for large air separation plants'. *Cryogenic Processes and Equipment in Energy Systems,* p. 25, ASME Publications, USA.

10

DESIGN METHODS

10.1 Introduction

The majority of the preceding chapters concern themselves with the cryogenic and non-cryogenic processes used to recover industrial gases. Something is also said about the types of equipment required for such plant. As there is an emphasis on cryogenic technology in this monograph, it is worthwhile mentioning the prediction of physical and thermodynamic properties of cryogenic fluids and how it affects design techniques. The generation of low temperature requires considerable energy, as can be seen in Chapter 3, Figs 3.9, 3.10, and 3.11, and Table 3.6. The high energy cost of low-temperature refrigeration requires processes with high efficiencies. This means that heat-exchanger temperature differences must be small to reduce the irreversible losses to the minimum economic level. When one is working with small (2–3°) temperature differences, even relatively small absolute errors can become large relative errors. This emphasizes the importance of good thermo-physical data prediction techniques for cryogenic fluids.

Some of the basic foundations for predictive techniques for industrial gases were laid down as early as the seventeenth century. In 1662, Boyle found that a simple relationship between pressure (p) and volume (V) existed. This led to the famous relationship of pV = a constant at constant temperature, which is used today with a reasonable degree of accuracy for gases at low pressures.

Subsequently, in 1802, Gay-Lussac postulated a law stating that at constant pressure, the volume (V) of a fixed mass of any gas increases by the same relative amount (i.e. by 1/273.15 for every degree rise in temperature (T). This gives rise to the equation:

$$\frac{V}{T} = \text{constant}$$

Combining Boyle's and Gay Lussac's law leads to the Boyle–Gay Lussac's Law: pV/T is constant *provided that* the temperature T is expressed as 'absolute temperature' = °C + 273.15. Evidently, this definition of the Kelvin scale of absolute temperature finds its origin in the universal value for the isobaric coefficient of expansion of gases.

10.2 Equations of state

Avogadro's law states that at the same p and T the volume of one mole is the same for all gases. Combination with the Boyle–Gay Lussac law gives, *for 1 mole of any gas,*

$$\frac{pV}{T} = \text{the } \textit{same} \text{ constant} = R$$

$R = 83.1448$ bar cm^3 mol^{-1} K^{-1}

So the equation of state for one mole of gas becomes

$$p = \frac{RT}{V}$$

Gases do *not* normally behave according to this simple equation. Therefore the equation is said to characterize an *ideal gas.* Real gases approximate ideal gas behaviour at relatively low pressures and relatively high temperatures.

Basic data prediction for cryogenic processes depends almost completely on the manipulation of equations of state. Therefore a brief discussion of equation of state for real gases is in order.

Van der Waals (1873) introduced two modifications in the ideal gas equation which at once made it representative, at least qualitatively, of real-gas behaviour. He diminished the molal volume V by a fixed amount b. Physically this means that a test molecule under observation does not find the entire molal volume V at its disposal, since part of it (b) is occupied or shielded from it by other molecules. Secondly, the pressure which the gas exerts on the walls of the vessel is diminished by a certain amount as a result of the attractive forces between the gas molecules, which tend to pull the peripheral molecules back into the body of the gas mass. That contribution was found to be inversely proportional to V^2. Thus van der Waals arrived at the equation of state for real gases:

$$p = \frac{RT}{(V-b)} - \frac{a}{V^2}$$

The equation contains only two substance-dependent parameters, a and b. They can be derived from the critical constants. It should be noted that the equation is of the third order in V which, at least qualitatively, is confirmed by the behaviour of real fluids as shown by isotherms in a p, V diagram. The monumental achievement of the van der Waals equation of state is that with only two constants it can describe the fluid behaviour in both the gas and the liquid phase.

A vast number of essentially empirical variants of the van der Waals equation have been published in later years, such as the Redlich–Kwong equation (1949).

$$p = \frac{RT}{V-b} - \frac{a}{V(V+b)\sqrt{T}}$$

and the Peng-Robinson equation (1976)

$$p = \frac{RT}{V-b} - \frac{a}{V^2 + 2Vb - b^2}$$

Both these equations, and all the other variants, can be further refined by making the parameter a, or both a and b, empirical functions of temperature.

The beauty of these cubic equations is that they have a built-in physical realism: the calculated isotherms in the p,V diagram have qualitatively the right shape, and with properly chosen a and b values the equations reproduce the critical pressure and the critical temperature correctly. Since there are only two parameters they can be applied to substances for which empirical data are scarce. On the other hand, they cannot be expected to be equally accurate everywhere in the p,V,T space. As a rule the predicted critical volume is too high. One can make the equations predict correct vapour pressures, but then the liquid volumes suffer. Nevertheless the cubic equations find wide application in the practice of predicting phase equilibria and thermodynamic properties because they are so easily applicable to a wide range of substances.

A second type of equation of state is the virial equation, introduced by Kamerlingh Onnes in 1901:

$$p = \frac{RT}{V}\left[1 + \frac{B}{V} + \frac{C}{V^2} + \frac{D}{V^3} + \dots\right]$$

By adding more terms to the power series, highly accurate reproduction of empirical data over large areas of the p,V,T space can be achieved. It should be appreciated that the 'virial coefficients' B, C, D, etc., are all functions of temperatures, each of them requiring several constants. So the virial equation contains many empirical constants which have to be fitted statistically to a sufficiently large number of empirical data.

To improve the quality of the virial equation in the liquid phase, Benedict, Webb, and Rubin (1940) added an exponential term. Their equation reads:

$$p = RT\rho + (B_0RT - A_0 - C_0)\rho^2 + (bRT - a)\rho^3$$
$$+ a d\rho^6 + \frac{C\rho^3}{T^2}(1 + \gamma\rho^2)\exp(-\gamma\rho^2)$$

in which $\rho = V^{-1}$, the molal density. In all, the equation contains eight empirical constants. It has been widely used in natural gas and light hydrocarbon technology. Many variants have been published, notably the one by Bender (1973), with 20 constants, and the one by Starling (1972) with 11 constants.

These so-called *extended virial equations* (extended with an exponential term) in a sense are complementary to the cubic equations. They are highly accurate over large areas of the p, V, T space, but they can be used only for a limited number of substances. That is because the constants must be fitted statistically to a large number of experimental data, and for most substances the available data are not sufficiently plentiful.

In one practical aspect the cubic equations are much to be preferred to the extended virial equations. By their nature all equations of state are explicit in p. The usual practical problem, however, is to calculate V for given p and T, so the equation must be solved back to front. For the cubic equations we possess the straightforward algebraic Cardano solution. It cannot fail. The extended virial equation, however, must be solved by iteration. Not only is such a solution much more cumbersome, it also calls for the provision of starting values for the iteration. Particularly in the neighbourhood of the critical point, the iterative solution of a complex virial equation is less robust than the Cardano solution of a cubic equation.

In more recent years the utility of equations of state has been much enhanced by the application of the principle of corresponding states. This principle, laid down by van der Waals in 1880, already found expression in the fact that the two parameters a and b of his equation could be described in terms of the critical temperature and critical pressure. The principle was refined by Pitzer and co-workers (1955), who added a third characteristic constant, the so-called acentric factor ω. Most of the substances dealt with in cryogenics happen to fall in the class of physically well-behaved, so-called 'normal' fluids. According to Pitzer's three parameter corresponding states principle, the behaviour of all substances in that class is exactly the same when expressed in terms of T_c, P_c, and ω. Vice versa, this means that the parameters of equations of state can be expressed in those three characteristic constants. Equations of state with that facility are said to be 'generalized'. In 1972 Soave generalized the Redlich–Kwong equation of state by giving formulae for a and b in terms of T_c, P_c, and ω. The Peng–Robinson equation is also generalized. As regards the virial equations, Starling's BWR equation was generalized, though not in every respect successfully. Very successfull, on the other hand, was a somewhat different generalization procedure introduced by Lee and Kesler.

In practice, an equation of state usually must be applied to a gas or liquid mixture. This adds to the variables p, V, and T the mole fractions $x_1, x_2, x_3 \ldots x_n$ of the n components of the mixture. The values of the constants in the chosen equation of state must now be related to the mixture composition. The common procedure is that for each mixture constant one takes some average of the values of that particular constant for each of the components of the mixture. For each of the mixture constants a specific type of average is precisely prescribed. These are the so-called mixing rules, which are essentially empirical. Clearly a cubic equation of state requires only two mixing rules, whereas the original BWR equation requires eight. This has an interesting consequence. The virial equations, because of their many adapted parameters, are more accurate than the simple cubic equations, *for a single substance.* But since they also need an equally large number of empirical mixing rules, as against only two for the cubic equation, they lose much of their advantage when they are applied to mixtures. As regards mixture properties, the cubic and the virial equations of state in general give results of about the same quality.

10.3 Evaluation of thermodynamic properties

To derive fugacity (for phase equilibrium prediction), enthalpy, and entropy as required by the process engineer, the standard formulae from thermodynamics are applied to the chosen equation of state. These formulae are as follows.

Fugacity f_i of component i in a mixture

$$\ln\left(\frac{f_i}{x_i}\right) = \frac{1}{RT}\int_V^{\infty}\left[\left(\frac{\partial p}{\partial n_i}\right)_{T,\,V,\,n_i} - \frac{RT}{V}\right]\mathrm{d}V - \ln\left(\frac{pV}{RT}\right)$$

with n_i the number of moles of component i in a volume V of mixture.

Mixture enthalpy

$$H = \sum x_i H_i^0 + \int_0^p\left[V - T\left(\frac{\partial V}{\partial T}\right)_{p,\,x}\right]\mathrm{d}p$$

Mixture entropy

$$S = \sum x_i S_i^0 + \int_0^p\left[\frac{R}{p} - \left(\frac{\partial V}{\partial T}\right)_{p,\,x}\right]\mathrm{d}p - R\sum x_i \ln x_i$$

H_i^0 and S_i^0 are the component molal enthalpy and entropy in the ideal gas state. They are functions of temperature only. Constants for Wilhoit's formulae for these functions were given by Harmens for a wide range of normal fluids.

10.4 Problems peculiar to cryogenics

10.4.1 *Hydrates*

When natural gases arrive at the wellhead, they are often mixed with associated hydrocarbon liquids and usually water. Refinery off-gases are generally made up of hydrogen, hydrocarbons, trace non-hydrocarbon impurities, and, often, water. In many cases both these sources of hydrocarbon-rich gases are subsequently subjected to cryogenic processes and thus water must be removed. A fundamental part of the natural gas industry is the control of dewpoints (of both water and hydrocarbons) to avoid condensation in transmission pipelines. This means that it is necessary to reduce not only the water content, but also the hydrocarbon content. For both refinery and natural gas this is often done by chilling using some sort of refrigerant. Given a typical system, one might assume that chilling to just below 0°C followed by adsorption to remove the remaining water would be the most economic solution. This arises from the experience that molecular sieve technology is less economic than chilling for bulk water removal. However, this *cannot* safely be done with hydrocarbon-containing systems because of the formation of *hydrates*.

Hydrates are crystalline solids formed by an association of several water molecules with a hydrocarbon molecule such as CH_4, C_2H_8, i-C_4H_{10}, etc. This phenomenon is not limited to saturated hydrocarbons—H_2S is known to form hydrates, as does ethylene under certain conditions. Smaller molecules such as CH_4 and C_2H_6 can form a body-centred cubic structure. Larger molecules such as C_3H_8 and i-C_4H_{10} form a diamond lattice structure, as do mixtures of hydrocarbons. The number of molecules of water associated with a hydrocarbon molecule varies from just over 5 to 17. In practice we find that some of these form at temperatures well above 0°C. The actual temperature varies depending on hydrocarbon composition. It is therefore important to know at what temperatures hydrates form, in order to avoid this condition in process plant operation. Conditions that promote the formation of hydrates are the presence of free water, high pressures, and low temperatures. Secondary considerations which promote the formation of hydrate crystals are high velocities, pressure pulses, agitation, and introduction of small hydrate crystals.

The range of temperatures over which hydrates can form varies from just above 0°C to around 25°C depending on pressure. A typical natural gas of 0.6 s.g. (or 17.4 molecular weight) forms hydrates with free water at around 1°C at a pressure of 1000 kPa, but if the pressure is raised to 2200 kPa, hydrates can form at about 22°C. If the s.g. of the gas were raised to about 1, then at 2200 kPa hydrates form at about 25°C. This shows that pressure has a more significant effect on hydrate formation than does specific gravity.

To avoid hydrate formation in natural gas transmission lines, operators dry the gas using glycol to depress the water dewpoint below any temperatures the line is likely to experience. This prevents hydrates causing problems in their control systems. In refinery off-gas processes hydrates are avoided by cooling to a temperature just above the hydrate point, thereby taking maximum possible advantage of chilling before resorting to adsorption to remove the remaining water.

10.4.2 *Freezing of less volatile impurities*

Many gas mixtures contain small concentrations of impurities which may freeze in downstream cryogenic processing. Chapter 8 has already discussed the various ways of removing these by absorption, adsorption, permeation, and chilling. The impurities are listed in Table 8.1. Chapter 8 assumes that it is necessary to remove the impurities and in the majority of cases this is true; water and NH_3 are relatively insoluble in the downstream process fluids and would therefore cause blockages in any cryogenic equipment. However, the same is not true for all of these impurities, for example CO_2 and H_2S are relatively soluble in many hydrocarbons. Figures 10.1 and 10.2 give a rough idea of their solubility

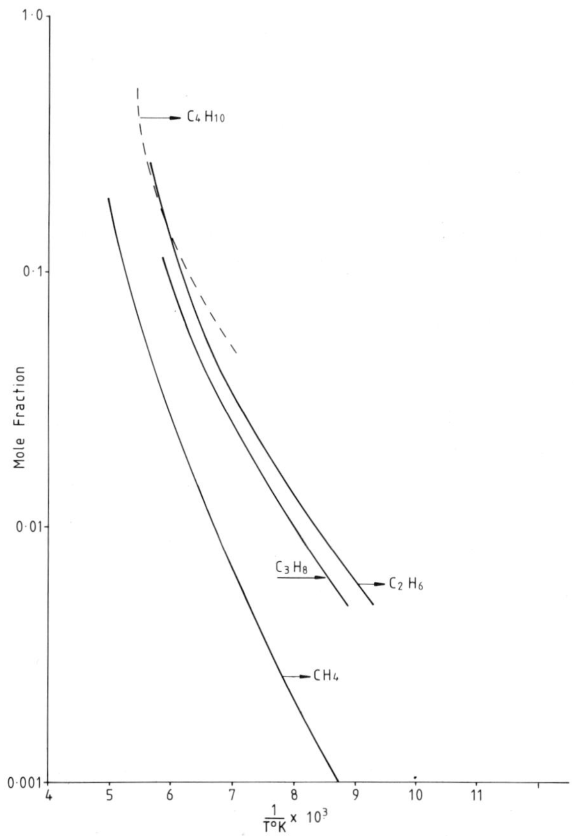

FIG. 10.1. The solubility limits of H_2S in various hydrocarbons.

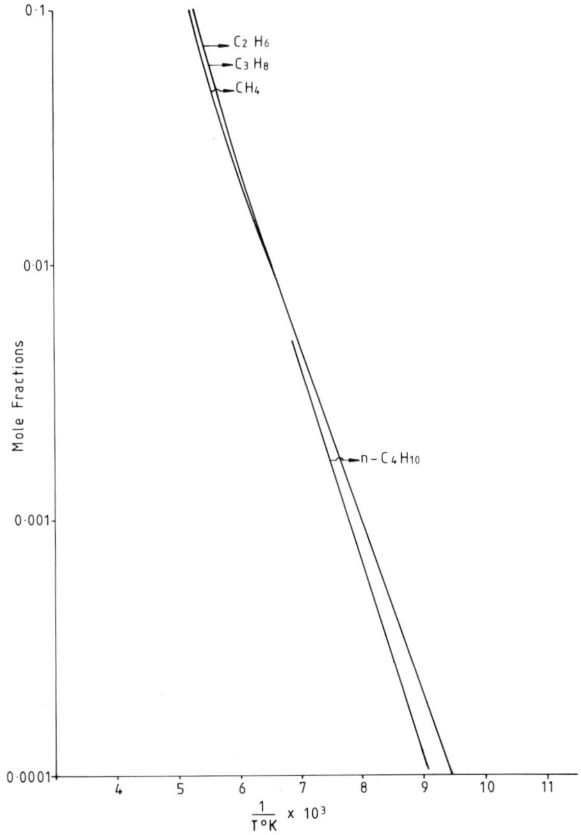

FIG. 10.2. Solubility limits of CO_2 in various hydrocarbons.

in light hydrocarbons from CH_4 to n-C_4H_{10}. Therefore, one may investigate the process conditions in a cryogenic unit and find that substantial amounts of H_2S or/and CO_2 can be admitted without their removal at the pretreatment stage. This can be extended to many other compounds such as benzene, xylenes, methanol, ethanol, etc. The prediction of freezing point data was done experimentally in earlier years, but more recently computer prediction techniques have been developed by cryogenic design establishments.

The experimental work on freezing point prediction led to the discovery of 'eutectics'. One often finds that a mixture can exist as a liquid over a range of compositions at temperatures lower than the freezing point of each of the pure components. The composition which allows it to be at the lowest temperature without solidification is called the 'eutectic'. Knowledge of these compositions and the freezing point

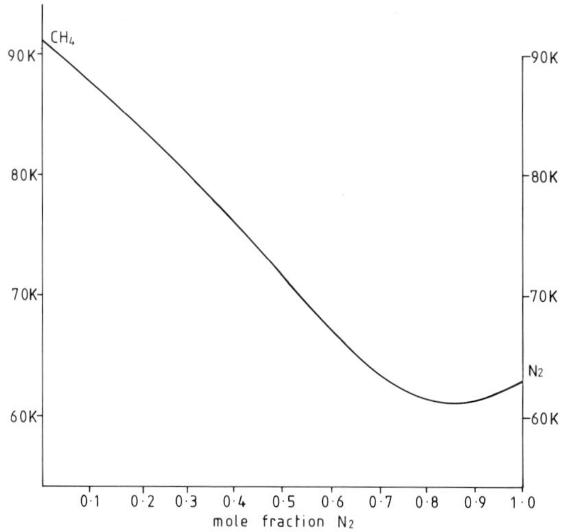

FIG. 10.3. The existence of eutectics in the N_2–CH_4 liquid–solid curve.

curves is used to great advantage in cryogenic plant design. Figure 10.3 gives an example of N_2–CH_4, mixture liquid–solid curves.

10.4.3 *The inverse Joule–Thomson effect of hydrogen and helium*

Many cryogenic processes rely on the fact that when one expands a gas or a liquid across a valve from a high pressure to a low pressure, it cools. This is commonly called the Joule–Thomson effect. However, the expansion of gases does not always reduce the temperature. Cryogenic designers known that under some conditions, hydrocarbon gases and sometimes liquids do not cool when expanded. This applies to light hydrocarbons at pressure below 7000 kPa and at temperatures between ambient and −100°C. It must be noted that this is not a general condition but occurs in only a few select cases.

The cryogenic designer also knows of the inverse Joule–Thomson effect exhibited by hydrogen and helium under certain conditions. Hydrogen and helium have, over the decades of cryogenic plant design, caused considerable consternation in breaking many 'rules'—not cooling when expanded is just one example. It is only in the last decade or so that the thermodynamicists have truly got to grips with hydrogen, giving the designer an accurate tool for data prediction. At very low temperatures (e.g. −100 to −200°C) hydrogen behaves like other gases and cools when expanded, but this behaviour changes as one approaches higher temperatures. Figure 10.4 shows the Joule–Thomson effect variation with temperature for various components. The figure shows the temperature fall or rise when expanding a gas from 5000 kPa to 100 kPa

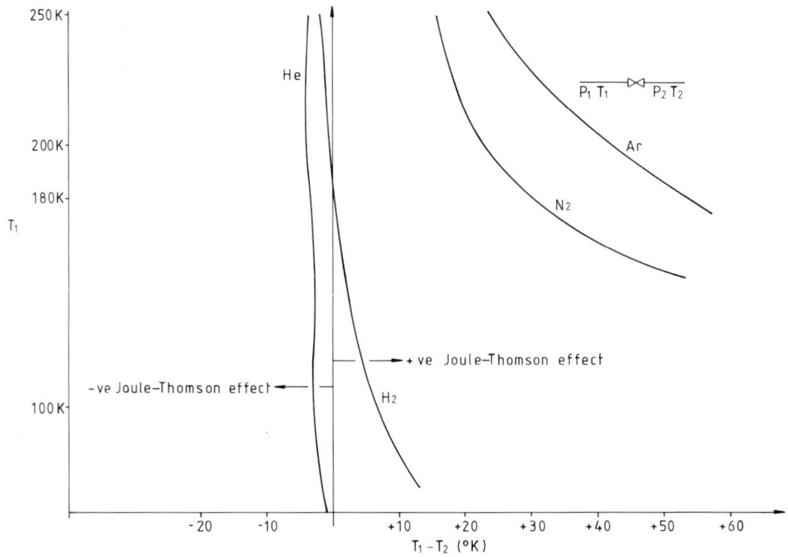

FIG. 10.4. The Joule–Thomson effect in various gases, for $P_1 = 5000$ kPa and $P_2 = 100$ kPa.

as an example pressure ratio for the purposes of comparison. Helium shows a negative Joule–Thomson effect over an even wider range than hydrogen.

Hydrogen and helium purifications often rely on Joule–Thomson expansion to provide refrigeration. This refrigeration is usually provided by the impurities that need to be removed—often a mixture of hydrocarbons or nitrogen and argon, which give a positive Joule–Thomson effect, and counteract the presence of hydrogen or helium. Hence with moderate concentrations of hydrogen these processes can still operate effectively, but with high concentrations of hydrogen they will need external refrigeration. Knowledge of these characteristics is essential for efficient process design of cryogenic plant.

10.5 Computer-aided process design

10.5.1 *Historic background*

Gas separation has always been a technology that depended on equations of state more than any other technology. As a result modern cryogenic and gas separtion technology is usually the result of computer-aided process design. To appreciate the value of computer-aided process design (CAPD) one needs to look back a decade or two, when one can see what was done manually, and how hamstrung the process design engineer was. As a good example let us take a gas mixture at pressure where it is being cooled and partially condensed, and the vapour and liquid phases need separating out. The engineer needs to know several things here if he is to design the cooler and the separator:

(1) the inlet enthalpy of the gas mixture;
(2) the cooling load for the mixture;
(3) the condensing load for the liquid being knocked out;
(4) the shape of the heat load curve for the gas being cooled;
(5) the composition of the vapour and liquid phases at the outlet;
(6) the point at which condensation first occurs—the dewpoint.

Each one of the above points presents the designer with a problem that he had to solve manually in the past. As our example we shall take point number (6). To arrive at a dewpoint the designer needs to know the equilibrium constant for each of the components at conditions of temperature (T) and pressure (P) which he does not yet know. Hence the calculation of dewpoint becomes a process of trial and error.

If the mixture is made up of n components then the composition of the vapour shall be assumed as $y_1, y_2, y_3, \ldots y_n$. The T and P at the inlet of the cooler are known and so is the pressure drop in the exchanger; hence the real unknown is the value of T to be assumed for dewpoint calculation. The first onset of condensation shall occur when

$$\sum_1^n \frac{y_n}{K_n} = 1$$

where n = number of components
y_n = vapour phase content of component n
K_n = equilibrium constant of component n

Equilibrium constants can be arrived by computation, as outlined earlier (the simplest and least accurate method being the ratio of component vapour pressure at the desired conditions and the total pressure). Having arrived at the equilibrium constants for the selected temperature one can test the equation. Several iterations are necessary using this technique to arrive at a reasonably accurate dewpoint. In a similar way, the bubble point of a mixture can be calculated by testing the following relationship:

$$\sum_1^n x_n K_n = 1$$

where x is the liquid phase content of component n, and the other parameters are the same as for the dewpoint.

Flash calculations are done with a similar relationship trial and error basis to arrive at the proper varpour and liquid compositions, as well as the mole fraction condensed. This proved acceptable for single flash calculations, dewpoints, and bubble points. However, one quickly sees how much mundane calculation is involved when it is necessary to carry out a rigorous plate-to-plate distillation column calculation for a 15 component system with the column having 40 trays and 4 feeds on different plates! Decades ago, designers found short cuts through these calculations which gave them satisfactory answers, and having added suitable design margins the operating plants usually worked to everyone's satisfaction. Today's highly competitive environment demands more rigorously optimized plant designs and the assistance of computers is taken very much for granted.

10.5.2 *Process simulation by computer*

Early computer-aided design began with flash, dew, and bubble-point calculations on 'stand-alone' programs. From there, cooling curves for fluids, compressor horsepower, column sizing, tray to tray calculation, etc. could be done. Then as more and more unit operations were calculated by computer programs, people began to think in terms of combining this into one large program which could be used to simulate a string of unit operations in one run. Disadvantages in this system began to appear and interactive design was built into several such 'flow-sheeting packages'. These designs were 'static' designs which assumed conditions at on point in time.

The next stages of development, occurring in the 1980s, was the introduction of time as a variable in flow-sheeting programmes. The simplest of these is probably a batch distillation column, but the real benefit of the time variable is 'dynamic simulation'. This technique is intended to show, the designer the effect of time on his process, and the upsets it may encounter. A good dynamic simulation package is an excellent tool for a designer to review his control system and train operators.

Computer-aided design (CAD) in the late 1980s aimed at providing a link with other engineering functions in a design establishment. For example the process designer may produce his flow-sheet on a CAPD tool which generates a heat and mass balance. Packages are available which can link this information to specialist design computer software. One can see that an even larger computer-aided design tool is formed which links many disciplines electronically. The late 1980s heralded the approach of a 'paper-free' design office (or at least partially paper-free!). Although interesting, this topic is outside the scope of this monograph.

10.5.3 *Structure of a process simulation package*

It is worth describing the structure, nature, and method of use of a typical CAPD package to illustrate the benefits of computer aided design in process selection and optimization of gas separation plants. Figure 10.5 shows a simple block diagram of a typical process flow-sheeting simulation package. This kind of package can be interactive or not, but the former shows tremendous benefit to designers. With interactive packages one can make changes to a particular scheme as it is running. This saves computer costs and lends itself to operation on a micro-computer. The example used to illustrate interactive simulation takes results from HYSIM, a simulation package marketed by Hyprotech, Inc. of Canada.

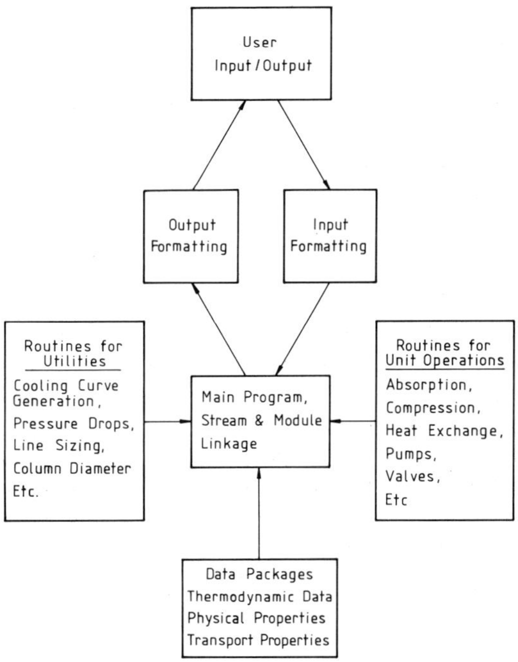

FIG. 10.5. A block diagram of a computer simulation package.

10.5.4 *An example of interactive process simulation*

For our problem let us take a mixture of nine components in a natural gas which we wish to compress from 1000 kPa in a two-stage compressor, cool in a shell-and-tube heat exchanger against cooling water, chill in a gas/gas interchanger, and chill with refrigerant. The resultant condensate is separated and the vapour reheated in the gas/gas interchanger. Figure 10.6 shows the process flow-scheme for this simple system and Fig. 10.7 shows the computer simulation model with unique stream numbers and unit operation or module numbers.

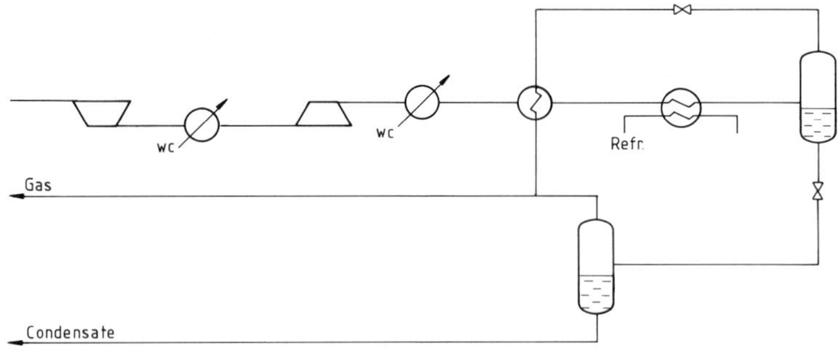

FIG. 10.6. The process flow-scheme of part of a gas-separation plant.

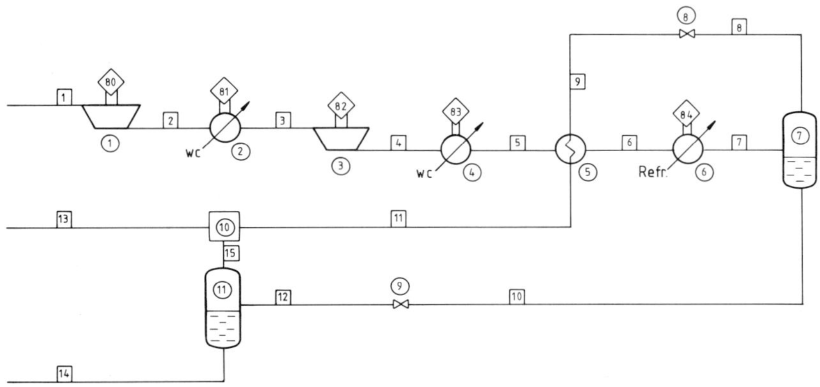

FIG. 10.7. A modular linkage diagram for a simulator.

Stream numbers are marked in the squares on each flow line, module numbers are encircled, and energy (cooling, power) numbers are in 'diamonds'. Initial conditions can be specified by the user interactively on a microcomputer loaded with HYSIM. Initially, stream 1 is defined and desired compressor discharge pressures, cooler outlet temperatures, heat exchanger approach temperatures, pressure drops, and valve outlet pressures are all specified. The job runs as these parameters are fixed. Then one can evaluate the results, change one of the parameters, and allow the program to re-run the parts that have changed. This procedure can be repeated until the desired solution is achieved. Table 10.1 shows the type of output one can produce from an optimization session on the microcomputer. The values marked with an asterisk (*) are those which have been set by the engineer using the program. These set values can be put in and amended interactively.

In addition to the production of straightforward heat and mass balances, many such software packages have other features. For example HYSIM has several thermodynamic data bases including Peng–Robinson, Chao–Seader, Grayson–Streed, and Redlich–Kwong–Soave.

TABLE 10.1. Heat and mass balance output data from a HYSIM run. Values set by the engineer are asterisked.

	Stream				
	1	2	3	4	5
Vapour fraction	1.0000	1.0000	1.0000	1.0000	1.0000
Temperature (°C)	25.0000*	93.5469*	40.0000*	104.9819	40.0000*
Pressure (kPa)	1000.000*	2100.0000*	2070.0000	4090.0000*	4060.0000
Molar flow (kg mol h^{-1})	1000.0000*	1000.0000	1000.0000	1000.0000	1000.0000
Mass flow (kg h^{-1})	18182.3105	18182.3105	18182.3105	18182.3105	18182.3105
Liquid volume flow (m^3 h^{-1})	54.8826	54.8826	54.8826	54.8826	54.8826
Enthalpy (kJ h^{-1})	1.00168E+07	1.26104E+07	1.03997E+07	1.28388E+07	1.00245E+07
Density (kg m^{-3})	7.5219	12.8311	15.1010	24.5791	30.8239
Molecular weight	18.1823	18.1823	18.1823	18.1823	18.1823
Specific heat (kJ kg^{-1} mol^{-1} $°C^{-1}$)	38.7255	42.4497	40.4349	44.3060	42.8019
Thermal conductivity (W m^{-1} K^{-1})	0.0349	0.0454	0.0381	0.0487	0.0401
Viscosity (Cp)	0.0116	0.0140	0.0123	0.0147	0.0128
Z factor	0.9751	0.9760	0.9573	0.9616	0.9198
Surface tension (dyne cm^{-1})	—	—	—	—	—
Standard density (kg m^{-3})	—	—	—	—	—
Nitrogen	0.0510*	0.0510	0.0510	0.0510	0.0510
Methane	0.8810*	0.8810	0.8810	0.8810	0.8810
Ethane	0.0440*	0.0440	0.0440	0.0440	0.0440
Propane	0.0150*	0.0150	0.0150	0.0150	0.0150
n-Butane	0.0040*	0.0040	0.0040	0.0040	0.0040
n-Pentane	0.0010*	0.0010	0.0010	0.0010	0.0010
n-Hexane	0.0010*	0.0010	0.0010	0.0010	0.0010
n-Heptane	0.0020*	0.0020	0.0020	0.0020	0.0020
CO_2	0.0010*	0.0010	0.0010	0.0010	0.0010

TABLE 10.1. (*cont.*)

	Stream				
	6	7	8	9	10
Vapour fraction	0.9877	0.9878	1.0000	0.9985	0.0000
Temperature (°C)	−36.9793	−37.000*	−37.0000	−62.3778	−37.0000
Pressure (kPa)	4030.0000	4000.0000	4000.000	1020.0000*	4000.0000
Molar flow (kg mol h^{-1})	1000.0000	1000.0000	987.7814	987.7814	12.2186
Mass flow (kg h^{-1})	18182.3105	18182.3105	17578.6719	17578.6719	603.6339
Liquid volume flow (m^3 h^{-1})	54.8826	54.8826	53.7592	53.7592	1.1234
Enthalpy (kJ h^{-1})	6.47994E+06	6.48990E+06	6.56958E+06	6.56958E+06	−79682.0391
Density (kg m^{-3})	47.4067	46.9672	45.5260	11.1618	601.4753
Molecular weight	18.1823	18.1823	17.7961	17.7961	49.4028
Specific heat (kJ kg^{-1} mol^{-1} $°C^{-1}$)	47.3548	47.2166	46.4927	36.8552	105.7396
Thermal conductivity (W m^{-1} K^{-1})	—	—	0.0324	—	0.1380
Viscosity (Cp)	—	—	0.0106	—	0.2151
Z factor	—	—	0.7964	—	0.1673
Surface tension (dyne cm^{-1})	—	—	—	—	14.3287
Standard density (kg m^{-3})	—	—	—	—	531.5250
Nitrogen	0.0510	0.0510	0.0516	0.0516	0.0046
Methane	0.8810	0.8810	0.8881	0.8881	0.3031
Ethane	0.0440	0.0440	0.0431	0.0431	0.1149
Propane	0.0150	0.0150	0.0132	0.0132	0.1567
n-Butane	0.0040	0.0040	0.0025	0.0025	0.1273
n-Pentane	0.0010	0.0010	0.0003	0.0003	0.0589
n-Hexane	0.0010	0.0010	0.0001	0.0001	0.0742
n-Heptane	0.0020	0.0020	0.0001	0.0001	0.1592
CO_2	0.0010	0.0010	0.0010	0.0010	0.0011

TABLE 10.1. (*cont.*)

	Stream				
	11	12	13	14	15
Vapour fraction	1.0000	0.2574	1.0000	0.0000	1.0000
Temperature (°C)	33.0000*	−51.2335	32.7816	−51.2335	−51.2335
Pressure (kPa)	990.000	1020.0000*	1000.0000	1020.0000	1020.0000
Molar flow (kg mol h^{-1})	987.7814	12.2186	990.9262	9.0738	3.1448
Mass flow (kg h^{-1})	17578.6719	603.6339	17635.2793	547.0295	56.6045
Liquid volume flow (m^3 h^{-1})	53.7592	1.1234	53.9362	0.9464	0.1770
Enthalpy (kJ h^{-1})	1.01141E+07	−79682.0391	1.01365E+07	−102095.3047	22425.0195
Density (kg m^{-3})	7.0737	98.4959	7.1526	664.0100	10.6707
Molecular weight	17.7961	49.4028	17.7968	60.2868	17.9991
Specific heat (kJ kg^{-1} mol^{-1} $°C^{-1}$)	38.3600	95.7728	38.3663	115.9278	37.6200
Thermal conductivity (W m^{-1} K^{-1})	0.0362	—	0.0362	0.1554	0.0257
Viscosity (Cp)	0.0119	—	0.0119	0.4101	0.0088
Z factor	0.9785	—	0.9782	0.0502	0.9325
Surface tension (dyne cm^{-1})	—	—	—	20.5295	—
Standard density (kg m^{-3})	—	—	—	592.1293	—
Nitrogen	0.0516	0.0046	0.0515	0.0004	0.0167
Methane	0.8881	0.3031	0.8881	0.1022	0.8830
Ethane	0.0431	0.1149	0.0432	0.1276	0.0781
Propane	0.0132	0.1567	0.0133	0.2047	0.0181
n-Butane	0.0025	0.1273	0.0025	0.1707	0.0022
n-Pentane	0.0003	0.0589	0.0003	0.0793	0.0002
n-Hexane	0.0001	0.0742	0.0001	0.1000	0.0000
n-Heptane	0.0001	0.1592	0.0001	0.2143	0.0000
CO_2	0.0010	0.0011	0.0010	0.0009	0.0017

TABLE 10.1. (*cont.*)

	Stream				
	80	81	82	83	84
Vapour fraction	2.0000*	2.0000*	2.0000*	2.0000*	2.0000*
Temperature (°C)	0.0000*	0.0000*	0.0000*	0.0000*	0.0000
Pressure (kPa)	0.0000*	0.0000*	0.0000*	0.0000*	0.0000*
Molar flow (kg mol h^{-1})	0.0000*	0.0000*	0.0000*	0.0000*	0.0000*
Mass flow (kg h^{-1})	0.0000	0.0000	0.0000	0.0000	0.0000
Liquid volume flow (m^3 h^{-1})	0.0000	0.0000	0.0000	0.0000	0.0000
Enthalpy (kJ h^{-1})	2.59360E+06	2.21062E+06	2.43907E+06	2.81433E+06	−9964.7617

Unit operations

1 comp	1	−2	80		6 cool	6	−7	−84	Eff—	75.0000		
2 cool	2	−3	−81		7 sepa	7	−10	−8	DP—	30.0000		
3 comp	3	−4	82		8 valv	8	−9		Eff—	75.0000		
4 cool	4	−5	−83		9 valv	10	−12		DP—	30.0000		
5 heat	5	−6	9	−11	10 bala	11	15	−13	DP1—	30.0000	DP2—	30.0000
					11 sepa	12	−14	−15	UA—	248291.5000		
									DP—	30.0000		

In terms of simple modules for the building of flowsheets it includes the usual ones like exchangers, coolers, separators, reactors, mixers, splitters, pumps, compressors, expanders, valves, and columns. In addition the HYSIM package has a number of 'utility' subroutines which give the designer rapid access to detailed simulation tools such as rigorous distillation columns, heat-exchanger cooling curve production, phase diagram generation, etc.

As a further aid to help the designer get an idea of plant size there are a number of other 'utility' subroutines covering preliminary plant size, including column diameters and line sizing. With these facilities built in the designer has a very powerful tool for carrying out process optimization.

Table 10.2 and Figs. 10.8 and 10.9 give examples of selected outputs including overall heat and mass balance, critical envelope, and heat-exchanger cooling curves.

One might therefore think that such tools eliminate the need for the highly trained cryogenic engineer because the computer can provide all the know-how and thermodynamic data. This is in fact far from the truth. Most of these tools have a preferred range of conditions over which they give accurate data. Many cryogenic processes operate at high pressure and very low temperatures. They often process mixtures which are highly non-ideal, i.e. they do not closely follow the equations of state. The specialist's knowledge of a cryogenic plant designer is essential in order to appreciate the scope and limitations of such software packages. It is only with this knowledge and experience that such packages can be applied to the mutual satisfaction of the designer and the operator of the plant.

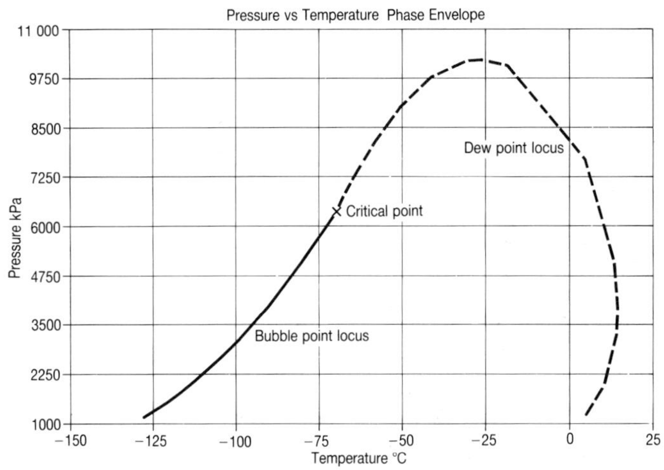

FIG. 10.8. The critical envelope generated by HYSIM.

TABLE 10.2. Modular flowsheet and selected output from a simulator (courtesy of Hyprotech Ltd).

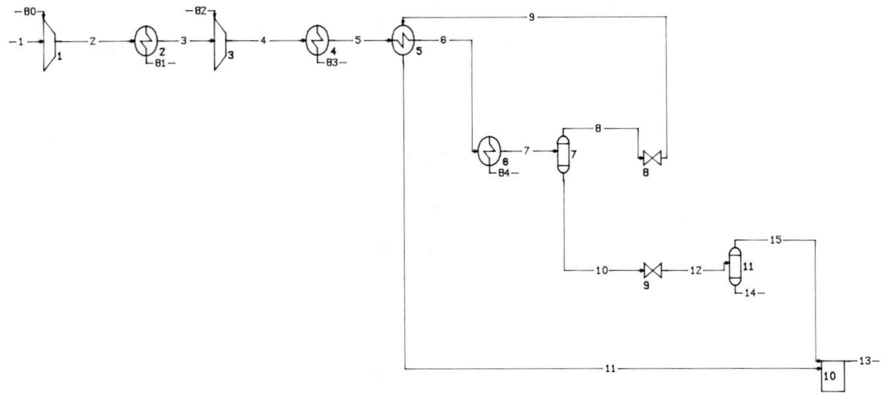

	Stream							
	1	2	3	4	9	10	11	12
Vapour fraction	1.0000	1.0000	1.0000	1.0000	0.9985	0.0000	1.0000	0.2574
Temperature (°C)	25.0000	93.5469	40.0000	104.9819	−62.3778	−37.0000	33.0000	−51.2277
Pressure (kPa)	1000.000	2100.0000	2070.0000	4090.0000	1020.0000	4000.0000	990.0000	1020.0000
Molar flow (kg mol h^{-1})	1000.0000	1000.0000	1000.0000	1000.0000	987.7814	12.2188	987.7814	12.2188
Mass flow (kg h^{-1})	18182.3105	18182.3105	18182.3105	18182.3105	17578.6719	603.6339	17578.6719	603.6339
Liquid volume flow (m^3 h^{-1})	54.9932	54.9932	54.9932	54.9932	53.8676	1.1257	53.8678	1.1257
Enthalpy (kJ h^{-1})	1.00168E 07	1.26104E 07	1.03997E 07	1.28388E 07	6.56958E 06	−79682.0391	1.01141E 07	−79682.0391
Nitrogen	0.0510	0.0510	0.0510	0.0510	0.0516	0.0046	0.0516	0.0046
Methane	0.8100	0.8810	0.8810	0.8810	0.8881	0.3031	0.8881	0.3031
Ethane	0.0440	0.0440	0.0440	0.0440	0.0431	0.1149	0.0431	0.1149
Propane	0.0150	0.0150	0.0150	0.0150	0.0132	0.1567	0.0132	0.1567
n-Butane	0.0040	0.0040	0.0040	0.0040	0.0025	0.1273	0.0025	0.1273
n-Pentane	0.0010	0.0010	0.0010	0.0010	0.0003	0.0589	0.0003	0.0589
n-Hexane	0.0010	0.0010	0.0010	0.0010	0.0001	0.0742	0.0001	0.0742
n-Heptane	0.0020	0.0020	0.0020	0.0020	0.0001	0.1592	0.0001	0.1592
CO_2	0.0010	0.0010	0.0010	0.0010	0.0010	0.0011	0.0010	0.0011

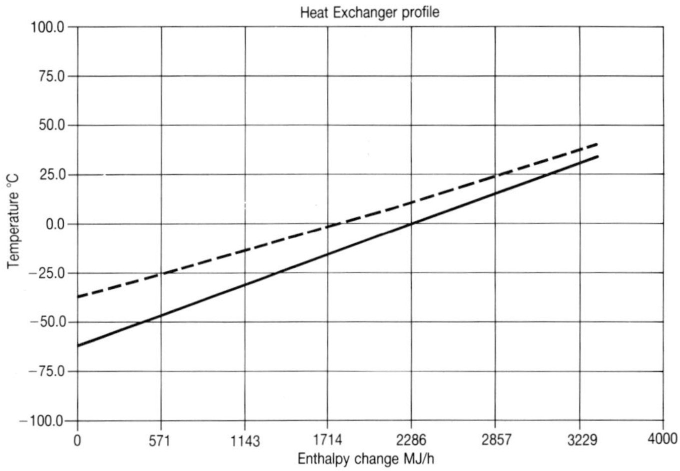

FIG. 10.9. Heat exchanger curves for HYSIM output.

Bibliography

AMERICAN CHEMICAL SOCIETY (1968). *Applied Thermodynamics Symposium,* Ind. Eng. Chem.

BARNES, F. J. and King, C. J. (1974). 'Synthesis of cascade refrigeration and liquefaction systems'. *Ind. Eng. Chem. Proc. Des. & Dev.* **13**(4), 421.

BENDER, E. (1973). 'An equation of state for predicting vapour-liquid equilibria of the system N_2–Ar–O_2'. *Cryogenics,* **13**, 11.

BENDICT, M., WEBB, G. B., and RUBIN, L. C. (1949). 'An empirical equation for thermodynamic properties of light hydrocarbons and their mixtures'. *J. Chem. Phys.,* **8**, 334.

CHIU, C. (1978). 'Evaluate separation for LNG plants'. *Hydrocarbon Processing,* (Sept.), 266.

DENBIGH, K. (1968). *The principles of chemical equilibrium* (2nd edn). Cambridge University Press.

GLASSTONE, S. and LEWIS, D. (1966). *Elements of physical chemistry* (2nd edn). Macmillan & Co., London.

GPSA (1987). *Engineering data book* (10th edn). Vol. II. GPSA, Tulsa, Oklahoma.

HARMENS, A. (1979). 'Wilhoit's formulae for ideal gas storage thermodynamic properties'. *Proceedings Conf. on Chem. Thermodyn. Data on Fluids and Fluid Mixtures,* p. 112. IPC Science and Technology Press, Guildford.

HYSIM (1987). *Users guide and reference manual,* Version 3.3. Hyprotech Ltd, Calgary, Alberta, Canada.

KAISER, V., *et al.* (1978). 'Analyse mixed refrigerant cycles'. *Hydrocarbon Processing,* (July) 163.

KISTENMACHER, H. (1987). 'Overview of the methods used for the

calculation of separation processes during the last 6 decades'. *Linde Reports Sci. Tech.* **43**.
KLOTZ, H. C., *et al.* (1983). 'What's the impact of bad data on NRU designs?'. *Hydrocarbon Processing,* (Apr.), 84.
LEE, B. I. and KESLER, M. G. (1975). 'A generalized thermodynamic correlation based on three-parameter corresponding states'. *A.I.Ch.E.J.* **21**, 510.
LIMB, D. I. and HEALY, M. J. (1984). 'Loss analysis aids process selection'. *Hydrocarbon Processing,* (Aug.), 87.
LUKS, K. D. and KOHN, J. P. (1981). 'Avoid freeze-up in LNG/LPG work'. *Hydrocarbon Processing,* (Apr.), 135.
PARADOWSKI, H. and DUFRESNE, J-P. (1983). 'Process analysis shows how to save energy'. *Hydrocarbon Processing,* (July), 103.
PENG, D. Y. and ROBINSON, D. B. (1976). 'A new two-constant equation of state'. *Ind. Eng. Chem. Fundam.,* **15**, 59.
PITZER, K. S., LIPPMANN, D. Z., CURL, R. F., HUGGINS, C. M., and PETERSEN, D. (1955). 'The volumetric and thermodynamic properties of fluids, II'. *J. Am. Chem. Soc.,* **77**, 3433.
REDLICH, O. and KWONG, J. N. S. (1949). 'On the thermodynamics of solutions, V'. *Chem. Revs.* **44**, 233.
SOAVE, G. (1972). 'Equilibrium constants from a modified Redlich–Kwong equation of state'. *Chem. Eng. Sci.* **27**, 1197.
STARLING, K. E. and HAN, M. S. (1972). 'Thermo data refined for LPG'. *Hydrocarbon Processing,* **51**, 129.
VICTORY, D. J. and VALENCIA, J. A. (1987). *The CFZ process: direct methane–CO_2 fractionation.* 66th Annual GPA Conv., Denver.

INDEX